Foundations of
Administrative Law

Interdisciplinary Readers In Law
ROBERTA ROMANO, *General Editor*

Foundations of Corporate Law
ROBERTA ROMANO

Foundations of Tort Law
SAUL LEVMORE

Foundations of Administrative Law
PETER SCHUCK

Foundations of Contract Law
RICHARD CRASWELL AND **ALAN SCHWARTZ**

Foundations of
Administrative Law

PETER H. SCHUCK

New York Oxford
Oxford University Press
1994

Oxford University Press

Oxford   New York   Toronto
Delhi   Bombay   Calcutta   Madras   Karachi
Kuala Lumpur   Singapore   Hong Kong   Tokyo
Nairobi   Dar es Salaam   Cape Town
Melbourne   Auckland   Madrid

and associated companies in
Berlin   Ibadan

Library of Congress Cataloging-in-Publication Data
Foundations of administrative law / Peter H. Schuck, editor.
p. cm. — (Interdisciplinary readers in law)
ISBN 0-19-508525-6 (cloth). —
ISBN 0-19-507813-6 (pbk.)
1. Administrative law—United States.
I. Schuck, Peter H.
II. Series.
KF5402.A5F68   1994   342.73′06—dc20
[347.3026]   93-7243

1 2 3 4 5 6 7 8 9

Printed in the United States of America
on acid-free paper

Contents

Introduction, 3

I The Theoretical Foundations of the Administrative State, 9

James Landis, *The Administrative Process, 13*

Stephen Breyer, *Regulation and Its Reform, 17*

Charles Wolf, *A Theory of Nonmarket Failures, 22*

Mark Seidenfeld, *A Civic Republican Justification for the Bureaucratic State, 25*

Notes and Questions, 27

II The Historical Foundations of Administrative Law, 31

Stephen Skowronek, *Building a New American State: The Expansion of National Administrative Capacities, 1877–1920, 33*

Robert Rabin, *Federal Regulation in Historical Perspective, 39*

Notes and Questions, 49

III The Administrative Procedure Act, 53

Cass Sunstein, *Factions, Self-Interest, and the APA: Four Lessons Since 1946, 55*

Martin Shapiro, *APA: Past, Present, Future, 60*

Arthur Bonfield, *State Law in the Teaching of Administrative Law: A Critical Analysis of the Status Quo, 69*

Notes and Questions, 72

IV The Determinants of Agency Behavior, 75

Jerry Mashaw, *Explaining Administrative Process: Normative, Positive, and Critical Stories of Legal Development, 77*

Terry Moe, *"The Politics of Bureaucratic Structure", 89*

James Q. Wilson, *"Beliefs", 95*

Charles Lindblom, *The Science of "Muddling Through", 104*

Notes and Questions, 108

V Models of Procedural Justice and Effective Governance, 111

Jerry Mashaw, *Due Process in the Administrative State, 113*

John Thibaut and Laurens Walker, *A Theory of Procedure, 115*

Charles Reich, *The New Property, 119*

Robert Rabin, *Reflections on "The New Property", 123*

Peter Schuck, *When the Exception Becomes the Rule: Regulatory Equity and the Formulation of Energy Policy Through an Exceptions Process, 125*

Robert Kagan, *Adversarial Legalism and American Government, 135*

Notes and Questions, 151

VI Controlling Administrative Discretion, 155

1. The Problem of Administrative Discretion, 163

    Kenneth Davis, *Discretionary Justice, 163*

    Theodore Lowi, *The End of Liberalism, 166*

    Jerry Mashaw, *Prodelegation: Why Administrators Should Make Political Decisions, 177*

    Notes and Questions, 183

2. Review by Courts and Specialized Tribunals, 186

    Richard Stewart and Cass Sunstein, *Public Programs and Private Rights, 186*

Cynthia Farina, *Statutory Interpretation and the Balance of Power in the Administrative State, 193*

R. Shep Melnick, *Administrative Law and Bureaucratic Reality, 200*

Peter Schuck and E. Donald Elliott, *To the* Chevron *Station: An Empirical Study of Federal Administrative Law, 212*

Harold Bruff, *Specialized Courts in Administrative Law, 218*

Notes and Questions, 227

3. Presidential and Congressional Review, 231

Peter Strauss, *The Place of Agencies in Government: Separation of Powers and the Fourth Branch, 231*

Alan Morrison, *OMB Interference with Agency Rulemaking: The Wrong Way to Write a Regulation, 241*

Christopher DeMuth and Douglas Ginsburg, *White House Review of Agency Rulemaking, 245*

Louis Fisher, *"Micromanagement by Congress: Reality and Mythology", 248*

Notes and Questions, 257

4. Internal Controls: Management, Culture, and Professional Norms, 260

Jerry Mashaw, *The Management Side of Due Process, 260*

James Q. Wilson, *"Culture" and "Compliance", 269*

William Simon, *Legality, Bureaucracy, and Class in the Welfare System, 274*

Notes and Questions, 288

5. Public Participation, 290

Richard Stewart, *The Reformation of American Administrative Law, 290*

Jerry Mashaw, *Bureaucratic Justice: Managing Social Security Disability Claims, 297*

Philip Harter, *The Role of Courts in Regulatory Negotiation—A Response to Judge Wald, 301*

Notes and Questions, 310

6. Market-Oriented Controls, 312

    Ronald Cass, *Privatization: Politics, Law, and Theory, 312*

    Notes and Questions, 318

VII Comparative Administrative Process, 321

James Q. Wilson, *"National Differences", 323*

Rudolph Schlesinger, Hans Baade, Mirjan Damaska,
    and Peter Herzog, *Comparative Law, 338*

Notes, 361

VIII The Future of Administrative Law, 363

Martin Shapiro, *Administrative Discretion: The Next Stage, 365*

E. Donald Elliott, *The Dis-Integration of Administrative Law:
    A Comment on Shapiro, 380*

Notes and Questions, 384

Foundations of
Administrative Law

Introduction

Administrative law, it has rightly been said, is "a field that seems forever in search of itself." Ronald A. Cass and Colin S. Diver, *Administrative Law: Cases and Materials* (Boston: Little, Brown, 1987), at xxi. This restless, self-conscious quest for identity is not a sign of delayed adolesence; administrative law is no longer young. The first texts on the subject were published more than sixty years ago, and it has been an important part of the law school curriculum for at least half that time. The perpetual identity crisis is not neurotic but intellectual; it is driven by new scholarship, new substantive challenges, and new legal doctrines and structural reforms.

Public law scholarship, of which administrative law is a main branch, has exploded in recent years. As a result, we now know more than our predecessors did about many dimensions of the administrative state: its historical origins and political functions, the behavior of its agencies, its policy impacts, and the relations among the various institutions and processes that constitute it. Much of this scholarship is a product of disciplines other than law relying on methodologies other than case analysis. The research embraces the positive and normative, the empirical and theoretical, the doctrinal and behavioral, the critical and apologetic. This work reveals a remarkably rich variety of structural, legal, political, and policy settings in which agencies operate, a diversity that belies the monolithic notions of "the agency" and "the administrative process" that most commentators (including me) use as shorthands from time to time.

Unprecedented challenges confront the agencies. Convulsive technologi-

3

cal and market changes necessitate new policy agendas that often render the agencies' procedures, and sometimes their missions, problematic. Rapid advances in pharmaceuticals and medical devices, for example, raise new ethical dilemmas and strain the Food and Drug Administration's already inadequate approval and enforcement processes. Abundant new telecommunication technologies are making the Federal Communications Commission less a regulator than a referee of competitive struggles. Changing energy markets and environmental values make policy coordination within the executive establishment more difficult. Constantly shifting political relationships between the agencies and the president, Congress, courts, states, and private groups threaten traditional institutional understandings and make formal administrative law principles less relevant. All of this has led to a new crisis of legitimacy for the administrative state.

These pressures inevitably generate calls for changes in the ground rules. A recurrent struggle in the legislative, judicial, and bureaucratic arenas concerns the role that the market should play in advancing the state's programmatic goals. Various forms of incentive-based regulation, deregulation, and privatization of governmental activities have been adopted. In response, the courts have had to devise doctrinal accommodations while attempting to safeguard traditional process values. A second struggle concerns the administrative agency's legitimate role in a constitutional regime of separated powers. Most commentators had assumed that the New Deal decisively settled this issue. Indeed, in his introduction to an earlier administrative law reader published in 1979, Robert Rabin decried the field's traditional preoccupation with questions of legitimacy, viewing them as "increasingly anachronistic with the passage of time." Rabin, *Perspectives on the Administrative Process* 11 (1979).

Yet after almost a quarter century of divided government, the issue returned to center stage in the 1980s. Far from being an academic debate, it has precipitated a series of important Supreme Court rulings on separation of powers; indeed, four dissenting justices in a 1991 decision seemed willing to question the constitutionality of the independent regulatory agencies. Freytag v. Commissioner of Internal Revenue, 111 S. Ct. 2631, 2660–61. Widespread criticisms of agencies have also prompted Congress to adopt far-reaching statutory reforms of agency structure, authority, and practice. These debates, moreover, are likely to persist. Among other reasons, they are so normatively charged that empirical evidence can never fully resolve them.

Precisely because administrative law is constantly searching for its intellectual bearings, those who map the terrain labor under a special burden to chart an intelligible course. Like administrative law scholarship generally, the leading casebooks grew more sophisticated, innovative, and ambitious during the 1980s. (In contrast, the treatises remain much the same). These casebooks contain inventive, eclectic mixes of traditional case law, regulatory theory, substantive policy analysis, case studies, and research findings from a variety of disciplines. Their editors have demonstrated considerable pedagogical imagination in compiling them.

Yet these innovations are still insufficient to capture the institutional complexity and human drama of the administrative process and to initiate students into its subtle mysteries. Even the most resourceful casebook writers candidly acknowledge as much in their prefaces. Their frustration partly reflects the casebook's inherent limitations as a pedagogical form, but it also seems endemic to the subject. As already noted, agencies govern an extraordinarily diverse set of activities, institutions, and legal regimes that often have surprisingly little in common. Whatever one's inquiry, it matters a great deal whether the agency specializes in public utility pricing, child welfare administration, occupational licensure, disability claims, public assistance, immigration control, police work, hospital reimbursement, higher education, or myriad other functions. Each agency's behavior reflects its distinctive history, statutory scheme, policy problematic, legislative politics, bureaucratic culture, judicial review pattern, private interests, and specialized bar. Indeed, an important question about administrative law is whether it can encompass this diversity without becoming so abstract that it is normatively meaningless and practically irrelevant.

This concern helps to account for what E. Donald Elliott in Chapter VIII calls the "dis-integration of administrative law." By this he refers to Congress's growing tendency to create distinct bodies of procedural law to govern different agencies, rather than subjecting them to the more universal principles of the Administrative Procedure Act. Together with the generality of much administrative law doctrine and the strong norm of judicial deference to agency expertise, this fragmentation also helps to explain why much of the doctrine often seems peripheral to what courts and agencies do, serving to rationalize decisions based on other considerations. But if the doctrine has little "bite" and imposes few constraints on agency behavior, how are administrative lawyers to understand the process and predict the decisions it will generate?

Several premises inform the structure and content of this book. First, although a student's mastery of administrative law doctrine is a necessary condition for answering this question, it is not sufficient. For the most part, this book leaves to the many administrative law casebooks the important task of analyzing doctrine, which is their main concern. My distinctive mission here is to help the student explore the larger themes and issues that frame the diverse contexts in which doctrine is embedded and the deep conflicts that doctrine is meant to resolve.

The readings that illuminate those themes and issues have been organized under the following eight questions, each of which is the subject of one of the remaining chapters: (1) What are the administrative state's normative foundations? (2) How did it evolve historically into its present form? (3) How have changing interpretations of the Administrative Procedure Act affected that evolution? (4) What factors determine how agencies in general and in particular behave in defining and implementing their missions? (5) Which visions of justice and effective governance do and should guide the organization and performance of the administrative state? (6) What techniques are available for controlling administrative power and how effective are they? (7) What can we learn from how other liberal democracies have organized their administrative states? (8) What is likely to be the future course of administrative law?

A second premise is that the law's traditional boundaries, preoccupations, and methodologies are too parochial to enable the student to answer these questions in a satisfying way. At a minimum, she must also be informed about how social scientists view the administrative state. As many of the following readings show, their researches constitute invaluable sources of data and insight on the subject. Thus although this book is primarily meant to provide supplementary readings for law school courses in administrative process, it should also be of interest to students of politics and public administration, organization theory, sociology, and other fields to which the theory, history, structure, and performance of the administrative state are pertinent.

Third and related, the law schools' obsession with the courts, which persists despite many decades of criticism, constitutes a serious impediment to an understanding of administrative process and policy. While judicial doctrine is important, administrative law course and casebooks are already so preoccupied with caselaw that there is little danger, and much opportunity, in a book of this kind largely ignoring doctrine in favor of a more institutional and contextual orientation. In this way, the student (especially the law student) can learn that agencies' behavior is shaped far less by the episodic decisions of reviewing courts than by institutional politics and other nonlegal phenomena such as agency culture, market forces, and professional norms.

Finally, the focus should be on *federal* agencies. This is a reluctant concession to the same practical considerations, not all of them intellectually defensible, which account for the almost universal emphasis on federal agencies in administrative law courses and casebooks. The neglect of state agencies is regrettable. Here, as in many other public policy domains, the states are often the innovators, testing grounds, and harbingers. Indeed, as state governments continue to reform and strengthen their executive establishments, as state supreme courts become more independent centers of judicial innovation, and as federal authority (if not revenue) devolves to the state and local levels in important policy areas, the states are likely to generate even more important administrative law in the future. Administrative law teachers and their students would do well to learn more about what is happening "out there." In recognition of this fact, Chapter III contains an article describing some of the distinctive patterns found in state administrative procedure.

The editor has selected the readings that follow with these intellectual and pedagogical imperatives very much in mind and wishes to acknowledge the generous assistance of many other administrative law scholars in culling a regrettably limited number of readings from several large and rich literatures. Except for a few instances in which the footnotes and references in the original selection provided useful bibliographical references to literatures with which the reader may be unfamiliar (political science and comparative law), they have been deleted in the interests of enhancing accessibility. The interested reader is urged to read the original works, including references, in their entirety.

For those who will be using this reader without having taken a course on administrative law or administrative process, a very summary outline of the subject may be useful.

Administrative law refers to the legal doctrines—a complex mixture of constitutional, statutory, regulatory, and "common law" principles—that govern the structure, decision processes, and behavior of administrative agencies. A course in administrative law, which usually focuses on the federal government, typically begins with a consideration of (1) how and why agencies were created; (2) how they differ from legislatures and courts; (3) how—in light of these differences and the Constitution's failure to define the agency's legal status—the delegation of legislative, judicial, and administrative powers to agencies can be legitimated; and (4) the formal and informal mechanisms through which the legislature and executive exercise control over agencies. An important theme that runs through all of these issues is the continuing tension between the constitutional principles of separation of powers and checks and balances on the one hand, and the values of expert decisionmaking, policy coordination, and due process on the other.

Next, such a course would consider the different ways in which agencies exercise their delegated powers. The federal Administrative Procedure Act (APA) distinguishes between two types of decision processes—rulemaking and adjudication—and the courts have generally permitted agencies, subject to legislative direction, to choose which type to use for a particular decision. Each type is further divided into a formal and an informal mode. *Formal adjudication* employs a quasi-judicial process in which an agency order is issued after a court-type hearing and is based on a trial-type record. *Formal rulemaking* employs a similar process in order to produce a rule of future effect designed to implement, interpret, or prescribe law or policy. *Informal rulemaking* employs a quasi-legislative process in which the agency publishes notice of a proposed rule (again, of future effect) and invites public comment on the rule by all interested parties, after which the agency publishes a revised (usually final) rule that becomes effective shortly thereafter. Procedures for *informal adjudication,* a residual category that includes most agency decisions and actions, are not specified by the APA. In general, judicial review is more stringent for formal than for informal proceedings, and for questions of law than for questions of fact or exercises of agency discretion.

Much administrative law doctrine is concerned with the nature of the hearing rights, broadly defined, that parties enjoy in different proceedings and for different issues. These rights are protected not only by the APA's procedural requirements, but also by constitutional principles of due process, which restrict the manner in which an agency may act to limit one's property or liberty interests. An agency is restricted in the ways in which it may compile decision records; obtain information from, and release information to, parties outside the agency; allocate decisional authority, channel information, and otherwise organize decision processes within the agency; respond to legislative and executive pressures; impose sanctions; find facts; and render decisions.

Finally, administrative law doctrine is concerned with the availability, timing, form, standards, and procedural incidents of judicial review of agency actions (and inactions). It also prescribes the types of remedies against agency illegality or error that the reviewing courts may establish and enforce.

The Theoretical Foundations of the Administrative State

Why is there an administrative state? From the perspective of a post-Great Society observer, this question might seem odd, even impertinent. Administrative agencies, after all, are so ubiquitous in American society that we take them utterly for granted—rather like VCRs, suburbs, and advertising. New ones are always being created, old ones seldom expire, and their permanence seems secure despite the Constitution's relative silence about them.

The question, however, is perhaps even more urgent today that it was in the New Deal era. Today, the agencies' legal legitimacy seems firmly settled, but their *political* legitimacy is not. As their power to command social resources has grown, public concern about their role in our scheme of government has intensified. In contrast to most of the rest of the world (including most democracies), Americans have never been comfortable with the administrative state and have therefore always demanded that it be justified afresh. "Why is there an administrative state?" has never been a merely academic question; it reflects the deepest anxieties of our political culture.

In principle, socially valued goods and activities can be produced in three ways—through private action, state action, or some mixed form. All societies treat some goods as public, others as private; each of them classifies particular goods differently. For most goods, the United States privileges private provision in an important sense: a strong public–private distinction is deeply embedded in both law and political tradition, and there is a strong presumption favoring private provision such that proposals for state provision or for state regulation of private provision bear a burden of justification, a burden that is often politically difficult to sustain. In contrast, some other democratic societies with more statist traditions reverse this presumption; there the public–private distinction is less ideologically charged, public provision is the rule, and private provision the exception.

The American public sector comprises a smaller share of the national economy than most other postindustrial countries. Many goods that are publicly provided elsewhere are partly or largely private in the United States; health care, voter mobilization, higher education, airlines, and religion are some examples. Moreover, even when the U.S. government decides to treat a good as public, it is more likely than other governments to mandate or subsidize private provision as distinct from providing the good itself. (The precise configuration of these patterns is in flux in many countries, as evidenced by the United State's expansion of publicly provided health care and higher education and by other countries' privatization of national airlines and other government-owned enterprises).

Why has the United States defined the domains of "public" and "private" as it has? Why are those lines drawn so differently in the United States than elsewhere? These are extremely difficult and important questions, and many of the selections in this book are intended to help the reader frame some answers of her own. Chapters II, VI (section 6), and VIII, which are concerned with the administrative state's historical foundations, its market-oriented alternative forms, and comparative administrative law, are especially pertinent. Suffice it to say at this point that the distinctive American conception of the administrative state and its purposes reflects a political culture grounded in a combination of unique historical conditions, constitutional structures, institutional growths, and ongoing struggles among competing interests and values.

But this political culture reflects more than the outcomes of political competition constrained and colored by historical and structural factors. Rational arguments about the appropriate domains of private action and of the administrative state have also played their part. Drawn both from normative theories of political philosophy and economy and from positive theories of administrative capacity and implementation, these arguments are essential rhetorical weapons in the perennial democratic discourse over the nature of the good society, the proper conception of the public interest, and the appropriate domains of law, politics, and administration. Political actors deploy these arguments explicitly or implicitly, in one form or another, at every opportunity. The readings in this chapter present some of the leading formulations. The notes and questions at the end discuss some others.

Perhaps the most influential rationale for the agency was articulated during the New Deal by James Landis, who was a leading regulatory theorist and practitioner as well as a law professor and dean. Landis advanced a techno-cratic justification for agencies, emphasizing their ability to solve the Great Depression's industrial problems and market failures in ways that common law courts no longer could.

Four decades later, Landis's heady optimism about regulation seemed one-sided and hopelessly naive. By the late 1970s, sharp dissatisfaction with the New Deal and Great Society agencies was widespread. Analysts like Stephen Breyer and Charles Wolf were criticizing the administrative state from sophisticated versions of the economistic perspective derived from clas-sic liberal theory. In this view, society's main goals are to protect individual liberty from governmental intrusion and to pursue economic efficiency (usu-ally defined as maximization of net social benefits or of wealth). Since com-petitive markets conduce to both liberty and efficiency, agencies are justified to the extent that they overcome market failures. In the excerpts printed here, Breyer, who helped design airline deregulation and is now a federal judge, reviews the kinds of market failures that can qualify, while Wolf, an economist with RAND, reminds us that regulatory agencies fail, not just markets. Breyer's analysis is directed primarily at "economic" regulation, broadly de-fined to include not only price and entry controls but also health, safety, and environmental controls—that is, much of what is sometimes called "social" regulation. It is not directed at regulation with "noneconomic" goals such as antidiscrimination law. In contrast, Wolf's analysis purports to have universal application.

The recent revival of interest in civic republicanism has led law professor Mark Seidenfeld to offer another justification for the administrative state. Society, in his view, has more substantive (and egalitarian) goals than the economists suppose. Agencies are the best institutional locus for achieving these goals because of their capacity to foster deliberation about the common good and to enrich citizenship by enlarging public participation in policy decisions.

The Administrative Process

JAMES LANDIS

If in private life we were to organize a unit for the operation of an industry, it would scarcely follow Montesquieu's lines. As yet no organization in private industry either has been conceived along those triadic contours, nor would its normal development, if so conceived, have tended to conform to them. Yet the problems of operating a private industry resemble to a great degree those entailed by its regulation. The direction of any large corporation presents difficulties comparable in character to those faced by an administrative commission. Rates are a concern, likewise wages, hours of employment, safe conditions for labor, and schemes for pensions and gratuities. There must follow the enforcement of pertinent regulations as well as the adjudication of claims of every nature made not only by employees but also by the public. This is in fact governance. It is true that the sanctions available to the governing bodies of industry to enforce decisions differ from those traditionally employed by government; but, partly because of the rapidity and directness of their execution, the penalties that private management can impose possess a coercive force and effect that goverment even with its threat of incarceration cannot equal. The management of a business like the United States Steel Corporation has wide powers to affect the economic security, stability, and subsistence level of its two hundred thousand employees. It has power, too, to influence the lives of its numberless customers. But more than this, such a corporation either by itself or in combination with its contemporaries can virtually determine what policies with reference to the production and sale of steel we shall pursue as a nation.

The significance of this comparison is not that it may point to a need for an expanding concept of the province of governmental regulation, but rather that it points to the form which governmental action tends to take. As the governance of industry, bent upon the shaping of adequate policies and the development of means of their execution, vests powers to this end without regard to the creation of agencies theoretically independent of each other, so when government concerns itself with the stability of an industry it is only intelligent realism for it to follow the industrial rather than the political analogue. It vests the necessary powers with the administrative authority it creates, not too greatly concerned with the extent to which such action does violence to the traditional tripartite theory of governmental organization. The dominant theme in the administrative structure is thus determined not primarily by political conceptualism but rather by concern for an industry whose economic health has become a responsibility of government. . . .

Viewed from this standpoint, it is obvious that the resort to the administrative process is not, as some suppose, simply an extension of executive power. Confused observers have sought to liken this development to a pervasive use of executive power. But the administrative differs not only with regard to the scope of its powers; it differs most radically in regard to the responsibility it possesses for their exercise. In the grant to it of that full ambit of authority necessary for it in order to plan, to promote, and to police, it presents an assemblage of rights normally exercisable by government as a whole. Moreover, its characteristic is this concept of governance, limited, of course, within those boundaries derived from its constituent statutory authority. But administrative power, though it may begin as an effort to adapt and make efficient police protection within a particular field, moves soon to think in terms of the economic well-being of an industry. The creation of that power is, in essence, the response made in the light of a tripartite political theory to the demand that government assume responsibility not merely to maintain ethical levels in the economic relations of the members of society, but to provide for the efficient functioning of the economic processes of the state.

A survey of existing administrative agencies reveals how they were called into being when the political power of our democratic institutions found it necessary to exercise some control over the varying phases of our economic life. The many characteristic illustrations that the field affords demonstrate this primary administrative thesis; namely, the existence of a growing need for vesting in a public authority supervision over the economic integrity of industries and their normal development. It is significant to note this trend in the development of the administrative process; it is equally significant to note departures from it. For, indeed, administrative agencies have been created whose jurisdiction related less to a particular type of industrial activity than to a general social and economic problem which cut across a vast number of businesses and occupations. . . . They possess more nearly the character of tribunals, of business and labor courts, where the function is one more closely akin to policing as distinguished from promoting.

The distinction is significant. From it may flow some basis for differentiation in the nature and composition of administrative agencies and in their relationship to the other branches of government. . . .

The advantages of specialization in the field of regulatory activity seem obvious enough. But our governmental organization of the nineteenth century proceeded upon a different theory. Indeed, theorists have lifted the inexpertness that characterized our nineteenth-century governmental mechanisms to the level of a political principle. Such a practical politician as Andrew Jackson took occasion to urge the Congress to take measures against permitting the civil servants of the government a "long continuance" in office. But expertness cannot derive otherwise. It springs only from that continuity of interest, that ability and desire to devote fifty-two weeks a year, year after year, to a particular problem. With the rise of regulation, the need for expertness became dominant; for the art of regulating an industry requires knowledge of the details of its operation, ability to shift requirements as the condition of the industry may dictate, the pursuit of energetic measures upon the appearance of an emergency, and the power through enforcement to realize conclusions as to policy. . . .

I have mentioned a broad distinction which underlies types of administrative agencies now in existence. That distinction relates to the difference between those administrative bodies whose essential concern is the economic functioning of the particular industry and those which have an extended police function of a particular nature. Although it is dangerous to deal in motives, yet the reasons which prompted a resort to the administrative process in the latter area would seem to be reasonably clear. In large measure these reasons sprang from a distrust of the ability of the judicial process to make the necessary adjustments in the development of both law and regulatory methods as they related to particular industrial problems.

Admittedly, the judicial process suffers from several basic and more or less unchangeable characteristics. One of these is its inability to maintain a long-time, uninterrupted interest in a relatively narrow and carefully defined area of economic and social activity. As Ulpian remarked, the science of law embraces the knowledge of things human and divine. A general jurisdiction leaves the resolution of an infinite variety of matters within the hands of courts. In the disposition of these claims judges are uninhibited in their discretion except for legislative rules of guidance or such other rules as they themselves may distill out of that vast reserve of materials that we call the common law. This breadth of jurisdiction and freedom of disposition tends somewhat to make judges jacks-of-all-trades and masters of none.

Modern jurisprudence with its pragmatic approach is only too conscious of this problem. To its solution it brings little more than a method of analysis, a method that calls upon the other sciences to provide the norms. It thus expands rather than contracts areas of inquiry. If the issues for decision are sociological in nature, the answers must be on that plane. If the problem is a business problem, the answer must be derived from that source. But incredible areas of fact may be involved in the disposition of a business problem that calls not only for legal intelligence but also for wisdom in the ways of industrial operation. This difficulty is intrinsic to the judicial process. It is enhanced under our constitutional system which permits judges to disregard those solutions reached by other governmental agencies, such as the legislative and the administrative, when the solution appears to them unfair, unreasonable, or unwise. . . .

To these considerations must be added two others. The first is the recognition that there are certain fields where the making of law springs less from generalizations and principles drawn from the majestic authority of textbooks and cases, than from a "practical" judgment which is based upon all the available considerations and which has in mind the most desirable and pragmatic method of solving that particular problem. . . .

The second consideration is, perhaps, even more important. It is the fact that the common law system left too much in the way of the enforcement of claims and interests to private initiative. Jhering's analysis of the "struggle for law"—the famous essay in which he indicated that the process of carving out new rights had resulted from the willingness of individuals as litigants or as criminal defendants to become martyrs to their convictions—pointed only to a slow and costly method of making law. To hope for an adequate handling of the problem of allowable trade practices by the sudden emergence of a host of Pyms and Hampdens was too delightfully visionary to be of much practical value. The retaliatory powers of business associates or competitors is today such an immensely powerful force that few persons care to run the risk of its offensive vengeance in the effort to secure what they might deem to be their legal rights. . . .

The agency's power to initiate action exists because it fulfils a long-felt need in our law. To restrict governmental intervention, in the determination of claims, to the position of an umpire deciding the merits upon the basis of the record as established by the parties, presumes the existence of an equality in the way of the respective power of the litigants to get at the facts. Some recognition of the absence of such equality is to be found in rules shifting the burden of proof and establishing prima facie and even conclusive presumptions of fact or of law. In some spheres the absence of equal economic power generally is so prevalent that the umpire theory of administering law is almost certain to fail. . . .

One other significant distinction between the administrative and the judicial processes is the power of "independent" investigation possessed by the former. The test of the judicial process, traditionally, is not the fair disposition of the controversy; it is the fair disposition of the controversy *upon the record as made by the parties*. . . . But, in strictness, the judge must not know of the events of the controversy except as these may have been presented to him, in due form, by the parties. . . .

On the other hand, these characteristics, conspicuously absent from the judicial process, do attend the administrative process. For that process to be successful in a particular field, it is imperative that controversies be decided as "rightly" as possible, independently of the formal record the parties themselves produce. The ultimate test of the administrative is the policy that it formulates; not the fairness as between the parties of the disposition of a controversy on a record of their own making. . . .

Equally deserving of notice is the power that the administrative possesses to conduct independent explorations as a prelude either to the fashioning of policies, by way of case decision and specific regulations, or to the obtaining of additional powers from the legislature in order to achieve more effective control over an industry. . . .

It is in the light of these broad considerations that the place of the administrative tribunal must be found. The administrative process is, in essence, our generation's answer to the inadequacy of the judicial and the legislative processes. It represents our effort to find an answer to those inadequacies by some other method than merely increasing executive power. If the doctrine of the separation of power implies division, it also implies balance, and balance calls for equality. The creation of administrative power may be the means for the preservation of that balance, so that paradoxically enough, though it may seem in theoretic violation of the doctrine of the separation of power, it may in matter of fact be the means for the preservation of the content of that doctrine.

There is no doubt but that our age must tolerate much more lightly ineffi-ciencies in the art of government. The interdependence of individuals in our civilization magnifies too greatly its cost. The pressure for efficiency has led elsewhere to concentrations of power on a scale that beggars the ambitions of the Stuarts. Our democratic processes have been evolving a different answer and in the form and extent of that answer may lie much of relevance to the matter of their endurance.

Regulation and Its Reform

STEPHEN BREYER

The major objectives of most economic regulatory efforts fall within one of the categories discussed in this selection. The *justification* for intervention arises out of an alleged inability of the marketplace to deal with particular structural problems. Of course, other rationales are mentioned in political debate, and the details of any program often reflect political force, not rea-soned argument. Yet thoughtful justification is still needed when programs arc evaluated, whether in a political forum or elsewhere. Usually this justifica-tion is one (or more) of the following.

The Control of Monopoly Power

The most traditional and persistent rationale for governmental regulation of a firm's prices and profits is the existence of a "natural monopoly." Some industries, it is claimed, cannot efficiently support more than one firm. Electricity producers or local telephone companies find it progressively cheaper (up to a point) to supply extra units of electricity or telephone service. These "economies of scale" are sufficiently great so that unit costs of service would rise significantly if more than one firm supplied service in a particular area. . . .

Rent Control or "Excess Profits"

What Is a Rent?

Since "economic rent" is often confused with "monopoly profit," the problem of rent control is often confused with that of controlling monopoly power. Yet rents and monopoly profits are very different.

A firm will earn an economic rent if it controls a source of supply that is cheaper than the current market price. It is a rent and not a monopoly profit if the cheap source could not supply the entire market. . . .

Unlike a monopoly profit, the existence of a rent does not mean that there is "inefficiency" or "allocative waste." . . . Rents are common throughout the economy, in competitive and noncompetitive industries alike. Any firm that finds a more efficient production process, that finds an unusually cheap supply source, that luckily buys a machine at a time when they are cheap, that has unusually effective managers—but that cannot expand to the point of satisfying a significant share of industry demand at prices that reflect its lower costs—earns a rent. To discourage the earning of rents is highly undesirable, for it would impede the search for efficiency. In many instances it seems perfectly fair that rents should accrue to producers who, through talents or skill, produced them.

Compensating for Spillovers (Externalities)

A considerable amount of regulation is justified on the ground that the unregulated price of a good does not reflect the true cost to society of producing that good. The differences between true social costs and unregulated price are "spillover" costs (or benefits)—usually referred to by economists as "externalities." If a train emits sparks that occasionally burn the crops of nearby farmers, the cost of destroyed crops is a spillover cost imposed upon the farmers by those who ship by train—so long as the shipper need not pay the farmer for the crop lost. Similarly, if honeybees fertilize nearby apple orchards, the beekeepers provide a spillover benefit to the orchard owners—so long as the latter do not pay the former for their service. Spillover benefits have sometimes been thought to justify government subsidy, as when free public education is argued to have societal benefits far exceeding the amount which students would willingly pay for its provision. Yet when one considers regulatory systems, spillover costs—not benefits—are ordinarily encountered. . . . The more readily monetary values of the commodities or objectives can be estimated, the more directly useful "spillover" characterization is likely to prove. In other cases, one is better off speaking directly of noneconomic reasons for and against taking a particular action rather than explicitly invoking the notion of "spillover."

Moreover, since there is always some possible beneficial effect in reversing a market-made decision, one can always find some (broadly defined) spillover cost rationale for regulating anything. Thus, the rationale, if it is to be intellectually useful, should be confined to instances where the spillover is large, fairly concrete, and roughly monetizable.

Inadequate Information

For a competitive market to function well, buyers must have sufficient information to evaluate competing products. They must identify the range of buying alternatives and understand the characteristics of the buying choices they confront. At the same time, information is a commodity that society must spend resources to produce. The buyer, looking for alternative suppliers, spends time, effort, and money in his search. The seller spends money on research, labeling, and advertising to make his identity and his product's qualities known. In well-functioning markets, one would expect to find as much information available as consumers are willing to pay for in order to lower the cost or to improve the quality of their choices. . . .

Excessive Competition: The Empty Box

A commonly advanced rationale for regulation of airlines, trucks, and ships is that competition in those industries would otherwise prove "excessive." This rationale has been much criticized as incoherent or at least inapplicable to the transportation regulation it is meant to justify. In fact, the difficulty with the term is that it has been used to describe several different types of rationale—some of which are no longer acceptable justifications for regulation. The notion common to all those rationales is that prices, set at unprofitably low levels, will force firms out of business and result in products that are too costly. . . .

Other Justifications

The reader should be aware of several other possible justifications for regulatory systems. While important, they have been used less often in the United States than elsewhere to justify governmental regulation of individual firms. *Unequal bargaining power.* The assumption that the "best" or most efficient allocation of resources is achieved by free-market forces rests in part upon an assumption that there is a "proper" allocation of bargaining power among the parties affected. Where the existing division of such bargaining power is "unequal," it may be thought that regulation is justified in order to achieve a better balance. It is sometimes argued, for instance, that the "unequal bargaining power" of small sellers requires special legislative protection. While in principle one might regulate the "monopoly buyer" in order to protect these sellers, the more usual congressional response is to grant an exemption from the antitrust laws, thus allowing the sellers to organize in order to deal more effectively with the buyer. This rationale underlies the exemption granted not only to labor, but also to agricultural and fishing cooperatives.

Rationalization. Occasionally governmental intervention is justified on the ground that, without it, firms in an industry would remain too small or would lack sufficient organization to produce their product efficiently. One would ordinarily expect such firms to grow or to cooperate through agreement, and to lower unit costs. But social or political factors may counteract this tendency. In such circumstances, agencies have sought to engage in industrywide "planning." In the 1960s, for example, the Federal Power Commission argued that increased coordination in the planning and operation of electric power generation and transmission facilities would significantly lower unit costs. The commission felt that environmental, political, regulatory, and managerial problems make it difficult for firms to plan jointly. The result was a relatively unsuccessful federal agency effort to encourage industrywide rationalization.

Moral hazard. The term "moral hazard" is used to describe a situation in which someone other than a buyer pays for the buyer's purchase. The buyer feels no pocketbook constraint, and will purchase a good oblivious to the resource costs he imposes upon the economy. When ethical or other institutional constraints or direct supervision by the payer fail to control purchases, government regulation may be demanded.

The most obvious current example is escalating medical costs. As medical care is purchased to an ever greater extent by the government or by large private insurers (with virtually no constraint on the amount demanded by the individual users), medical costs have accounted for an ever greater proportion of the national product. The fact that purchases are paid for by others frees the individual from the need to consider that using more medical care means less production of other goods; thus, he may "unnecessarily" or "excessively" use medical resources. If one believed that too much of the gross national product is accounted for by medical treatment, and also believed that the problem of moral hazard prevents higher prices from acting as a check on individual demand for those resources (which in turn reduces incentive to hold down prices), one might advocate regulation to keep prices down, improve efficiency, or limit the supply of medical treatment.

Paternalism. Although in some cases full and adequate information is available to decisionmakers in the marketplace, some argue that they nevertheless make irrational decisions and that therefore governmental regulation is needed. This justification is pure paternalism: the government supposedly knows better than individuals what they want or what is good for them. Such distrust of the ability of the purchaser to choose may be based on the alleged inability of the lay person to evaluate the information, as in the case of purchasing professional services, or the belief that, although the information could be accurately evaluated by the lay person, irrational human tendencies prevent this. The latter may be the case where small probabilities are involved, such as small risks of injury, or where matters of life and death are implicated, such as when those suffering from cancer will purchase a drug even though all reasonably reliable information indicates that it is worthless or even harmful. Whether the brand of paternalism based on mistrust of consumer rationality is consistent with the notions of freedom of choice that underlie the free market is questionable. However, it plays an important role in many governmental decisions.

Scarcity. Regulation is sometimes justified in terms of scarcity. Regulation on the basis of this justification reflects a deliberate decision to abandon the market, because shortages or scarcity normally can be alleviated without regulation by allowing prices to rise. Nonetheless, one might decide to abandon price as an allocator in favor of using regulatory allocation to achieve a set of (often unspecified) "public interest" objectives, such as in the case of licensing television stations. Sometimes regulatory allocation is undertaken because of sudden supply failures: to rely on price might work too serious a hardship on many users who could not afford to pay the resulting dramatic price increases, as in the case of the Arab oil boycott. "Scarcity" or "shortage" calling for regulation may also be the result of the workings of an ongoing regulatory program, as when natural gas must be allocated because of rent control or when an agency awards licenses to enter an industry.

The Mixture of Rationales

Many existing regulatory programs rest upon not one but several different rationales. Thus, for example, one might favor regulation of workplace safety for several reasons. . . .

The importance of distinguishing rationales lies in the extent to which different rationales may suggest different remedies. Thus, one who believes that the primary problem is informational will tend to favor not classical regulation, but governmental efforts to provide more information. Although one who accepts a paternalistic rationale may disagree with one who believes the problem is informational, the clear statement of their points of difference can form the basis of empirical work that will lead them toward agreement upon the basic rationale and thus help choose the regulatory weapon best suited to the problem at hand.

A Theory of Nonmarket Failures

CHARLES WOLF

Arguments between defenders of the market and advocates of government intervention to correct the market's shortcomings are characterized by a curious asymmetry. . . .

The existing theory of market "failures" provides a powerful, accessible, and convenient instrument for attacking the market. . . .

There is no such guide for those who wish to consider the shortcomings of nonmarket mechanisms and activities. I define *nonmarket* activities as those undertaken by governments and other institutions whose sources of revenue come principally from taxes, donations, or other nonprice sources, rather than from charging prices in markets—where buyers can choose what to buy, as well as whether, when, and how much. Governments are the prinicipal nonmarket organizations. . . .

I want to outline such a theory to facilitate comparisons between the inevitable shortcomings of the market and the no-less-inevitable shortcomings of nonmarket efforts to remedy them. A more distant aim is to improve the choices that exist among leaving admittedly imperfect markets alone, supplanting them with nonmarket mechanisms (perhaps made less imperfect by anticipating their own likely shortcomings), and designing combinations of market and nonmarket mechanisms that will prove to be less imperfect than either alone. . . .

Why Nonmarkets Fail

. . . The answer lies in the distinctive supply and demand characteristics that differentiate nonmarket outputs from market outputs.

On the supply side, there are several such characteristics. *First, the "products" of nonmarket activities are usually hard to define in principle, ill defined in practice, and extremely difficult to measure independently of the inputs which produce them.* . . .

Another characteristic of nonmarket "supply" is that evidence of output quality is elusive, in part because the information that in the market would be transmitted by consumer behavior is missing. . . .

The absence of sustained competition is a third factor contributing to the difficulty of evaluating the performance of nonmarket production. . . .

Finally, nonmarket activities generally have no "bottom line" for evaluating performance comparable to the profit-and-loss statements of market activities.

Reprinted with permission of the author from *The Public Interest,* No. 55 (Spring 1979), pp. 114–33. © 1979 by National Affairs, Inc.

Nor, in the case of nonmarket output, is there a reliable mechanism for terminating nonmarket activities if they are unsuccessful. . . .

There are also distinctive characteristics applying to the *demand* for nonmarket activities, and to the process by which these demands become effective. *One such characteristic is the enormous expansion over the past few decades of public awareness of the shortcomings of market outcomes*—an awareness that has resulted from the activity, perhaps hyperactivity, of information media, environmental groups, and consumer organizations. . . .

. . . The political effectiveness of public demands can lead to nonmarket activities with infeasible objectives and redundant costs.

The supply and demand characteristics of the nonmarket sector are fundamental to the theory of nonmarket failure. They provide an explanation, clues about where to look for specific types and sources of such failure, and a basis for formulating a typology of nonmarket failure analogous to that which already exists for market failure. In both cases, the "failures"—whether market or nonmarket—are evaluated against the same criteria of success. Nonmarket remedies "fail" to the extent that they, too, result in outcomes that depart from the efficiency or distributional goals by which market outcomes are judged to fail. But the ways in which nonmarket solutions "fail" differ from those in which market outcomes fail. There are four types of nonmarket failure. They result from the distinctive demand and supply characteristics of nonmarket output.

. . . Lacking the direct performance indicators available to market organizations (from consumer behavior and the profit-and-loss bottom line), public agencies must develop their own standards. *These standards are what I will call "internalities": the goals that apply within nonmarket organizations to guide, regulate, and evaluate agency performance and the performance of agency personnel.* I refer to these internalities synonomously as "private" goals because they—rather than, or at least in addition to, the agency's "public" purposes—provide the motivations behind individual and collective behavior. This structure of rewards and penalties constitutes what Kenneth Arrow refers as "an internal version of the price system." . . .

Of the specific types of internalities that often accompany nonmarket activities—and bring with them nonmarket failures and distortions—let us examine three:

Budget growth. Lacking profit as a measure of performance, a nonmarket agency may view its budget as the proxy goal to be maximized.

A distinctive variant of the budget internality is the tendency of public agencies, eschewing or precluded from profit maximization as their objective, to try to maximize the size of their employee staffs. . . .

Technological advance. Compatible with the budget internality is one favoring "advanced," "modern," "sophisticated," or "high" technology. . . .

The technological internality can have perverse consequences, not only in excessive zeal for what is complex and novel, but in mindless opposition to what is simple and familiar. . . .

A bias *against* new technology can, of course, just as easily lead to nonmarket failure. Parts of the American educational system, for example, seem to resist even the development and experimental use of such new technology as videotaping for presentations to large classes, computer-aided instruction, and performance contracting—all of which might reduce the demand for teachers. . . .

Information acquisition and control. Another element in the utility functions of some nonmarket organizations is information. Frequently in nonmarket (and market) organizations, information is readily translated into influence and power. Consequently, information becomes valued in its own right—an internality for guiding and evlauating the performance of agency members. . . .

Other Types of Nonmarket Failure

Redundant and rising costs. Whether policy takes the form of regulation, administering transfer payments, or direct production of public goods, *there is a tendency for nonmarket activities to exhibit redundant costs, and for costs to rise over time.* . . .

Derived externalities. Government intervention to correct market failure may generate *unanticipated side effects, often in an area remote from that in which the public policy was intended to operate.* Indeed, there is a high likelihood of such derived externalities because government tends to operate through large organizations using blunt instruments whose consequences are both far-reaching and difficult to forecast. In the Russian proverb, "When elephants run, other animals tremble." . . .

Ditributional Inequity in Nonmarkets

Nonmarket activities, whether intended to overcome the distributional inequities of market outcomes, or to remedy other inadequacies in the market's performance, may themselves generate distributional inequities. . . .

The role of nonmarket activities in producing distributional inequities, whether maldistributions of power or of income, derives from specific demand and supply characteristics associated with nonmarket output. *On the demand side, the principal causal characteristic is heightened public awareness of the inequities generated by the market* and the resulting clamor for redistributive programs, often without prior consideration of the inequities that may be generated by these programs themselves.

On the supply side, the principal characteristics leading to distributional inequities are the typical monopoly of nonmarket output in a particular field and the related absence of a reliable feedback process to monitor agency performance. . . . These parallel categories are suggestive, but should not be misunderstood. *The inadequacies or "failures" of nonmarket activities are not exact analogues of those associated with market activities.* . . .

Two comments apply to both lists. First, for the several types of market and nonmarket failures, it is much easier to estimate signs than magnitudes.

Table I.1 Market and Nonmarket Failures

Market	*Nonmarket*
1. Externalities and public goods	1. "Internalities" and "private" goals
2. Increasing returns	2. Redundant and rising costs
3. Market imperfections	3. Derived externalities
4. Distributional inequity (income and wealth)	4. Distributional inequity (influence and power)

Estimating magnitudes requires careful consideration of specific cases and contexts. . . .

It should also be noted that the types and sources of market failure indicate the circumstances in which government intervention is worth contemplating, and in which alternative policies are worth analyzing as possible remedies. Similarly, the types and sources of nonmarket failure indicate the circumstances in which government intervention may itself misfire, and in which other potential correctives are worth analyzing as possible remedies for the likely shortcomings of government intervention.

What is the "bottom line" on nonmarket failure? Like that of market failure, it must be entered in red ink, although its numerical sum is unclear. Whether the red-ink entry for the one is greater or less than that for the other can't be answered in general terms. Sometimes one may be greater, sometimes the other. The answer will depend on the specifics of particular cases.

A Civic Republican Justification for the Bureaucratic State

MARK SEIDENFELD

Recently, commentators have proffered an alternative understanding of the constitutional constraints on government regulation—the "civic republican" theory of constitutional democracy. Modern civic republicans view the Constitution as an attempt to ensure that government decisions are a product of deliberation that respects and reflects the values of all members of society. Civic republicanism promises democratic government that does not exclude or coerce citizens whose backgrounds and values differ from those of mainstream society. The civic republican model rejects the pluralistic assertion that government can, at best, implement deals that divide political spoils according to the prepolitical preferences of interest groups. Instead, government's primary responsibility is to enable the citizenry to deliberate about altering preferences and to reach consensus on the common good.

One prominent proponent of civic republicanism, Cass Sunstein, has noted that it tolerates regulation by an extensive bureaucracy more than does pluralist theory. Yet Sunstein adheres to the traditional principle that "basic value judgments should be made by Congress." Sunstein thus seems to accept the bureaucracy as an unavoidable aspect of modern government and has set about the task of constraining administrative policy-setting discretion to fit within civic republicanism to the extent possible.

By contrast, I believe that civic republicanism provides a strong justification for the assignment of broad policymaking discretion to administrative agencies. I agree with Sunstein that "[t]he exact amount of desirable agency discretion . . . will depend on the context." But I contend that, on the whole, civic republicanism is consistent with broad delegations of political decision-making authority to officials with greater expertise and fewer immediate political pressures than directly elected officials or legislators. Moreover, given the current ethic that approves of the private pursuit of self-interest as a means of making social policy, reliance on a more politically isolated administrative state may be necessary to implement something approaching the civic republican ideal.

In sum, I view the civic republican conception as providing an essential justification for the modern bureaucratic state. This article argues that although the Congress, the president, and the courts retain an important reviewing function, having administrative agencies set government policy provides the best hope of implementing civic republicanism's call for deliberative decisionmaking informed by the values of the entire polity.

This thesis has several implications for public policy. First, it suggests that congressional and judicial efforts to limit agency discretion and thereby eliminate perceived problems with the legitimacy of agency policymaking are often misguided. Second, if administrative policy setting is to achieve the civic republican ideal, agency decisionmaking processes must proceed in a manner consistent with civic republican theory. Hence, my thesis suggests the need for numerous changes in administrative law. For example, Congress should amend the Administrative Procedure Act to require public involvement in the early stages of agency policy formulation. Congress should also require that its members and the White House staff reveal all of their interactions with agency personnel. Courts should abandon the rigid dichotomy that they draw between agency decisions of law and decisions of policy, and should review both for persuasiveness in light of pragmatic limitations. . . .

The Administrative State as a Means of Fulfilling the Civic Republican Promise

Administrative agencies—the so-called fourth branch of government—may be the only institutions capable of fulfilling the civic republican ideal of deliberative decisionmaking. Congress adheres primarily to pluralistic norms and responds most directly to factional influence. Although one proponent of civic republicanism, Cass Sunstein, has sought to revitalize Congress's deliberative processes through more active judicial review, the size, structue, and historically rooted decisionmaking procedures of Congress render the prospect of revitalization unlikely. Perhaps for this reason, another proponent of civic republicanism, Frank Michelman, has called upon the judiciary to define directly the values that underlie governmental policy and are embodied in law. Courts, however, are too far removed from the voice of the citizenry, and judges' backgrounds are too homogenous and distinct from those of many Americans to ensure that judicially defined policy will accord with the public values of the polity.

Administrative agencies, however, fall between the extremes of the politically overresponsive Congress and the overinsulated courts. Agencies are therefore prime candidates to institute a civic republican model of policymaking. Some recent administrative resolutions of tough policy choices illustrate the role that agencies can play. For example, although the American public, experts, and government officials all agreed the the United States should close some military bases, Congress was unable to close any, or even set the criteria for deciding which bases should be closed. Too many representatives found the prospect of a base closing in their district politically unacceptable. A special commission, however, was able to order base closings and do so in a fashion that took into account efficiency concerns, the need for national defense, and the economic dislocations in areas where bases will close.

I believe that the success achieved by the Defense Base Closure and Realignment Commission was not an anomaly. The place of administrative agencies in government—subordinate and responsible to Congress, the courts and the president—allows for the checks on agency decisionmaking that ensure politically informed discourse and prevent purely politically driven outcomes. The bureaucratic structure of administrative agencies and the processes by which they frequently decide questions of policy also foster deliberative government. Consequently, as the remainder of this section outlines, the administrative state holds greater promise for a "civic republican resolution" to many questions of policy than either the judicial or legislative alternatives.

NOTES AND QUESTIONS

1. Two seminal articles in the intellectual lineage leading to the works excerpted here are Ronald Coase, "The Nature of the Firm," 4 *Economica* 386 (1937), and George Stigler, "The Theory of Economic Regulation," 2 *Bell Journal of Economics & Management Science* 1 (1971). Coase's article, although chiefly concerned with the economic determinants of organizational form and behavior in private firms, stimulated interest in what came to be known as "principal-agent" or "agency" theory, which explores the "agency cost problem." This theory analyzes the different incentives facing principals and their agents and the cost structure of various strategies for

seeking to maximize the principal's benefits. It has obvious relevance for administrative agencies, which may be (depending upon one's political theory) agents of two principals, Congress and the president. For an introduction to principal-agent theory, see a symposium book edited by John Pratt and Richard Zeckhauser, *Principals and Agents: The Structure of Business* (Boston: Harvard Business School, 1985).

Stigler's article has spawned a vast literature on the theory of public choice, which applies economic concepts to the analysis of political behavior, and a specific implication of that theory known as "agency capture," which hold that agency decisions are likely to be dominated by those narrow interests that can organize most cheaply to exploit officials' private incentives. The public choice approach has flourished among political scientists, economists, and some legal scholars. Other early contributions include Anthony Downs, *An Economic Theory of Democracy* (New York: Harper & Row, 1957), and *Inside Bureaucracy* (Boston: Little, Brown, 1967); Mancur Olson, *The Logic of Collective Action: Public Goods and Theory of Groups* (Cambridge, Mass.: Harvard University Press, 1965); and William Niskanen, Jr., *Bureaucracy and Representative Government* (Chicago: Aldine Publishing Co., 1971). A general introduction to public choice theory, aimed especially at law students, appears in Richard Posner, *Economic Analysis of Law* (Boston: Little, Brown, 4th ed., 1992), Part VI. Posner, adopting and then extending Stigler's approach, argues there that agencies are established to implement deals struck by electorally driven legislators supplying services demanded by organized interest groups and that judicial review of agency action is intended to secure and enforce those legislative deals in the long term. A symposium on public choice theory appears in 74 *Virginia Law Review* 167 (1988). The interdisciplinary trio of Matthew McCubbins, Roger Noll, and Barry Weingast (known collectively and affectionately as "McNollgast") have been especially energetic in developing, and to some degree testing, the implications of this theory for the structure and behavior of the administrative state. The McNollgast analysis of the administrative process is discussed and criticized in Jerry Mashaw's article in Chapter IV; their response to Mashaw is cited in the first note to the chapter.

2. One of the focal points in the debate over public choice theory is the concept of the public interest (or the general good). Is there a public interest that transcends the competing claims and the pluralistic bargaining process for resolving them that shape administrative decisions? If so, what is its content and how do we recognize it when we see it? To what extent are official decisions motivated by it? Although these are phrased as positive questions, they clearly raise normative ones as well. In Michael Levine and Jennifer Forrence, "Regulatory Capture, Public Interest, and the Public Agenda: Toward a Synthesis," 6 *Journal of Law, Economics & Organization* 167 (1990), the authors contrast the public interest and capture theories of political behavior with "postrevisionist theories" that use agency and information theory to demonstrate "slack"—the limits on the political accountability of the administrative state. They describe this approach as follows:

> Thus, voters may have difficulty monitoring and controlling a legislature, a legislature may have similar difficulties monitoring and controlling its committees, and so on through the relationships between oversight committees and regulatory agencies and between agency heads or commissioners and their bureaucracies. Similarly, these organizations need to gather information about the world that they attempt to regulate. The costliness and uncertainty of information about that world and about the effects of policy in it affect relationships among themselves and with lobbyists, voters, and others.

This way of looking at governing organizations does not treat them as a "black box," as do public-interest and capture theories, but rather attempts to describe the control and influence each actor in a hierarchical political system has over another. The theory describes mechanisms of supervision; it does not directly predict substantive outcomes (p. 7).

3. Landis, of course, was writing as a committed apologist for the New Deal and as a member of President Roosevelt's team, not simply as a legal scholar. (He taught at Harvard Law School and would later be its dean.) Yet he was hardly writing on a clean slate; the Interstate Commerce Commission was already a half-century old and had already exhibited much of the regulatory pathology ("marasmus," as one scholar described it) that would spawn calls for deregulation of surface freight transportation in the 1970s and 1980s, calls that Congress to some degree accepted. Other agencies had been established—and severely criticized—for decades, as Skowronek's and Rabin's essays in Chapter II explain. Does Landis adequately address those criticisms? What might his candid answer be?

4. Wolf's article is paradigmatic of a large genre of commentary and analysis critical of the administrative state, most of it by economists and political scientists and much of it cited in Stephen Breyer's book, also excerpted here. In view of the remarkable diversity in the structure, missions, cultures, and politics of administrative agencies, how generally applicable are Wolf's points? In this respect, one might contrast Breyer's more highly differentiated, case study treatment of regulation. What are the limitations of this approach? Which is likely to be more accurate and useful?

5. Wolf maintains that nonmarket failure is ubiquitous, perhaps as much so as is market failure, and suggests that decisionmakers must compare the risks and benefits of these allocation techniques. Is the dichotomy between "market" and "nonmarket" allocations always so clear? If other analytical distinctions should be made, what are they? For example, should a theory of nonmarket failure be sensitive to the different forms that governmental intervention may take? See Cass Sunstein, *After the Rights Revolution: Reconceiving the Regulatory State* (Cambridge, Mass.: Harvard University Press, 1990), especially chapter 3. Does Wolf provide a workable method for determining which type of allocation is on balance more desirable? Can one do so in the absence of far more sophisticated positive and normative theories of politics than he offers here? In this connection, how might Wolf respond to the claim that the public cares at least as much about the distributional and process aspects of the competing allocation methods as it does about the efficiency aspects that he emphasizes? Is such a claim plausible? Is Breyer's list of justifications for regulation an exhaustive set? Can you think of others? These issues are discussed in Chapter V.

6. In the introduction to his book, Breyer notes that his analysis applies to "economic regulation" as broadly conceived to deal not only with prices and entry, but also with health, safety, and the environment" and that it "may have only limited applicability to areas that are less obviously "regulatory" (such as the awarding of welfare grants) or that plainly advance noneconomic social goals (such as enforcement of antidiscrimination laws)" (p. 7). It is useful to consider why, and to what extent, his analysis cannot be extended to such areas.

7. Like public choice theory, but for very different reasons, civic republican models of politics and administration have enjoyed a considerable academic vogue in recent years. See, for example, "Symposium: The Republican Civic Tradition," 97 *Yale Law Journal* (1988). Seidenfeld's article is an example of such work. Consider his central claim that agencies "fall between the extremes of the politically overresponsive

Congress and the overinsulated courts" and that therefore they are "prime candidates to institute a civic republican model of policymaking." What assumptions underlie his use of the terms "overresponsive" and "overinsulated," and how could we know whether they are valid? Does the fact that he must invoke an exotic form of agency—a special, single-issue, short-term commission—to support his argument cast some doubt on the generality of the claim? Is it necessary to expound a theory of civic republicanism in order to justify the prescriptions for changes in administrative law that he mentions?

II

The Historical Foundations of Administrative Law

Today's administrative state did not evolve in a gradual, deliberate fashion, nor is it the product of rational processes of institutional design or of sober evaluation of competing political theories. Instead, it developed spasmodically in a characteristically American way: rough-and-ready, pragmatic, intended as much to resolve the bitter political struggles of the day as to add a coherent element to the constitutional-governmental structure. One can no more understand the behavior of administrative agencies without knowing something about their histories than one can understand an individual without knowing about her upbringing and experiences.

The development of the administrative state can be divided very roughly into four periods: an early stage culminating in the Progressive-era regulatory agencies and administrative reforms; the New Deal watershed; the post-New Deal period, including the consolidation of the administrative state and its vast expansion in the social regulation initiatives beginning in the mid-1960s; and the deregulation movement beginning in the mid-1970s and accelerating in the Reagan years.

This is not to say, of course, that administratvie power did not raise important issues of constitutional legitimacy, judicial control, and effective governance long before the establishment of the first federal regulatory agencies in the 1880s. In the sixteenth century, the Tudors created an extensive, notoriously oppressive system of regulation in England, one that led to the first great landmark decision legitimating judicial review of agencies, *Dr. Bon-*

ham's Case, 8 Coke Rep. 107 (1610). And in the United States, the decision establishing the principle of judicial review for constitutionality, Marbury v. Madison, 5 U.S. (1 Cranch) 137 (1803), also involved a suit against an agency official, here the secretary of state. Between 1816 and the Civil War, as James Q. Wilson points out, the executive branch grew nearly eightfold, although almost all of this growth was confined to the Post Office where bureaucratic discretion, a major complaint about agencies today, was not a great problem.

The excerpt from Stephen Skowronek sounds the dominant themes of administrative development in the United States from the close of Reconstruction through 1920. His detailed studies of the long process of reforming the civil service, the army, and business regulation show how the embryonic national state of "courts and parties" gave way after protracted political struggles—first to "patchwork" reforms that proved wholly inadequate by 1900, and then to a genuine reconstitution of the administrative apparatus under presidents Roosevelt, Taft, and Wilson. This apparatus, however, was one that neither the president nor Congress could effectively control. The Rabin excerpt picks up the story in 1920, carrying it to the mid-1980s and the deregulation movement. He traces the gradual legitimation of the administrative state by Congress and (after 1937) the courts, the different visions of regulatory power that agencies instantiated at various times, and the complex interactions between the courts, agencies, and Congress beginning in the 1960s. Rabin concludes by noting the remarkable continuity of controversy, now spanning more than a century, over the appropriate scope and mechanisms of federal regulatory authority.

Building a New American State: The Expansion of National Administrative Capacities, 1877–1920

STEPHEN SKOWRONEK

The United States was born in a war that rejected the organizational qualities of the state as they had been evolving in Europe over the eighteenth century. Indeed, it has been argued that an underlying cause of the Revolution was the gradual development of a more concentrated, specialized, and penetrating state apparatus in Britain. One may read in the Declaration of Independence a rejection of the substantive results of this organization. Standing armies, centralized taxing authorities, the denial of local prerogatives, burgeoning castes of administrators—these were the forces against which a new political culture found definition. . . .

Federalist and Whig party programs notwithstanding, the most striking operational characteristics of early American state organization were a radical devolution of power accompanied by a serviceable but unassuming national government. The national government throughout the nineteenth century routinely provided promotional and support services for the state governments and left the substantive tasks of governing to these regional units. This broad diffusion of power among the localities was the organizational feature of early American government most clearly responsible for the distinctive sense of statelessness in our political culture.

The national institutions that penetrated the territory are one indication of the rather innocuous role played by central authority in the government of the nation. Land offices, post offices, and customhouses were the most important of these national institutions, and they illustrate the orientation toward basic services that routinely dominated federal concerns. Except in war, the activities of these field offices could bring in more than enough revenue to keep the national government solvent. Another example of the radically decentralized organizational scheme is provided by the army. The early American state routinely relegated to its tiny regular army the tasks of securing the frontier and aiding economic development with the talents of its corps of engineers. American national defense relied upon a militia system of citizen-soldiers organized and controlled by the several states. The role of the national government in the militia system consisted of providing a modest subsidy and offering perfunctory administrative guidelines. Finally, the primaary concern of the national government in providing support for regional centers of decisionmaking is confirmed in a general overview of the corpus of nineteenth-century law. One finds the most fundamental social choices—from the organization of capitalism to the regulation of family life—firmly lodged in state legal codes.

The success of the early American state came to depend on the working rules of behavior provided by courts and parties. These two nationally integrated institutional systems tied together this state's peculiar organizational determinants and established its effective mode of operations. They coordinated action from the bottom to the top of this radically deconcentrated governmental scheme. Under a Constitution designed to produce institutional conflicts and riddled with jurisdiction confusions, they came to lend order, predictability, and continuity to governmental activity. . . .

The judiciary not only helped define the terms of internal governmental activity, it also helped define relations between state and society. It fell to the courts at each level of government to nurture, protect, interpret, and invoke the state's prerogatives over economy and society as expressed in law. Most notable in this regard was the way nineteenth-century courts filled a governmental vacuum left by abortive experiments in the administrative promotion of economic development. The generally inept supervision of public works projects and joint public–private ventures combined with a depression in 1837 to make state legislatures increasingly reluctant to invest in large and costly promotional activities. By the late 1830s, an alternative means of fostering economic development had come to the fore—the widespread distribution of special corporate charters. These charters were designed to promote and channel private economic ventures, yet they left to the courts a fairly loose reign over the state's police powers. By interpreting the charters' reserve clauses for the protection of the public interest, the judiciary became the chief source of economic surveillance in the nineteenth century. Over time, courts molded the prerogatives of government into predictable but flexible patterns of policy toward capital accumulation.

This system of control was well established by 1850, when the race for railroad access was becoming the centerpiece of national economic development. The courts had become the American surrogate for a more fully developed administrative apparatus. The corporate law they articulated during the nineteenth century stands next to the party machinery as a distinctive and key element in America's effective mode of governmental operations. . . .

It was not until the pressures of industrialism created irresistible demands for a permanent concentration of governmental controls that this became a serious problem in itself. The state of courts and parties had been organized to provide a regional focus for governmental action. Its working structure presumed a highly mobilized democratic polity and the absence of national administrative capacities. Its mode of operations was not geared for an interdependent industrial society. . . . The hold that the party machines had gained over American institutions would have to be broken before new centers of national institutional authority could be built. The expansion of national administrative capacities thus threatened to undermine party government in the nation that first gave it full expression. At the very time that parties were developing in Europe to challenge the hegemony of more traditional state institutions, state-building efforts in America aimed at the disintegration of party hegemony as it had developed over the course of a century.

It should come as no surprise that the power of the courts grew with the power of the party machines. The judiciary's governing capacities were stretched to their limits in the late nineteenth century to fill the "void in governance" left between party hegemony and rapid social change. As Congress tore down the Civil War institutional apparatus with one hand, it vastly expanded the jurisdiction of the federal courts with the other. The sweeping powers granted were an open invitation to the federal judiciary to assume the role of stern policeman for the new national economy. In 1890, the federal judicial structure itself was reinforced with the establishment of a national appellate court system. The expansion of federal judicial power in the late nineteenth century was the natural response of the early American state to demands for national authority in the industrial age. . . .

. . . Ironically, the more thoroughly courts and parties structured relations between state and society, the less appropriate that structure became; and the more thoroughly courts and parties controlled the internal operations of government, the less room remained for alternative kinds of institutional relations between state and society to take over. The hegemony of these highly developed but functionally limited supports for democracy and capitalism gave the illusion of statelessness in America an ironic twist. These once innocuous pillars of governmental order now appeared as imposing obstacles to any new departures in institutional development. . . .

An entrenched state organization that presumed the absence of national administrative capacities could, at best, accomodate only a bastardized version of the cosmopolitan reform program. Between 1877 and 1900, America's state-building vanguard waged a series of losing struggles to restructure American government around a national bureaucratic regimen. Their efforts were frustrated just at those critical junctures where a timely administrative innovation would have altered the foundations of power in the state apparatus and recast the boundaries of national institutional politics. They were employed in government only so far as they supplemented the arrangements they sought to supplant. They found themselves unwittingly implicated in the implementation of stopgap measures to defuse immediate governing problems without penetrating to the problem of American government itself.

America's institutional response to industrialism in the late nineteenth century was delimited at every turn by the consolidation of the state of courts and parties. State building in this period is marked by the appearance of a series of notable institutional adaptations, each of which was caught in the unresolved tension between the governing demands of a new age and the triumph of this old governmental order. Alternative governing arrangements designed to support national administrative capacities intruded into existing institutional relationships, only to be held in an awkward state of suspension. The question of whether or not America was going to build an independent arm of national administrative power was contested but answered in the negative. . . .

The Rise of Administrative Power and the Disintegration of Governmental Order

At the dawn of the twentieth century, opposition to party bosses and imperious judges was being voiced by a number of reform movements that otherwise had little in common. Populists, socialists, and corporate liberals railed against the way capitalism and democracy had developed under the aegis of courts and party machines and called for placing relations between the American economy and the American polity on a new foundation. . . . The central question in institutional development was correspondingly altered. It was no longer a question of whether or not America was going to build a state that could support administrative power but of who was going to control administrative power in the new state that was to be built.

Considered against the alternatives, the special appeal of the bureaucratic remedy is not difficult to discern. It alone offered both a measure of moderation and a measure of continuity. . . .

Yet, the appeal of the bureaucratic remedy and the politics of bureaucratic development were two very different things. . . . The result was the halting development of a new state that institutionalized governmental disarray as it spawned new kinds of services and supports for a burgeoning industrial society. . . .

The Fundamentals of Structural Change: Electoral Realignment and the Changing Shape of Institutional Politics

. . . The universe of institutional opportunities, incentives, risks, and constraints informing the actions of government officials changed dramatically in the wake of the electoral realignment of the 1890s. . . . Even after 1911 there was no return to the level of partisan division and rotation in governing institutions that had characterized the late nineteenth century. Between 1875 and 1896, parties either exchanged control in one of the houses of Congress every two years or maintained divided control. Between 1896 and 1920, the parties divided control of Congress in only one two-year interval (1911–13) and exchanged control only two other times (1913 and 1919). The president's party matched the majority party in both houses of Congress in only three separate two-year intervals between 1875 and 1896, whereas the parties matched in all but two separate two-year intervals between 1896 and 1920. . . . Newfound electoral strength and stability released institutional innovation from the political constraints of the previous era. . . .

Political Strategy and Political Vision: Three Elusive Models for a New Governmental Order

The dimensions of the politics of reconstitution may be elaborated a bit further by focusing on the changing position of one of its principal actors, the president. The president had never risen far above the status of a clerk during the heyday of party competition. . . .

The situation was quite different in the post-1896 period. Theodore Roosevelt, for example, faced a unique set of opportunities, incentives, risks, and constraints in formulating his institutional strategy, especially after a landslide election in 1904 gave him a second term with a solid Republican Congress. Security in office opened new possibilities for asserting executive prerogatives long preempted by Congress, for forging distinctly executive political alliances outside established party channels, and for pursuing institutional innovations that would rebuild the governmental resources of the executive branch. . . .

. . . The creative dimension in state-building politics—carried largely by new cadres of professionals standing outside established centers of institutional power in the late nineteenth century—was taken up in the executive office by three consecutive presidents in the early twentieth century. These three presidents—a reform politician, a scholarly judge, and a political science professor—each stood a bit outside the regular party organizations. When in office, each entered the struggle for power and position on the side of administrative reform. . . .

The institutional initiatives that secure Roosevelt's reputation as the premier state builder of his age were concentrated in the period between his landslide election of 1904 and the congressional elections of 1906. From a position of electoral strength and political security, the executive-professional reform coalition pursued a course of redistributing institutional powers and prerogatives away from Congress and the courts toward the president and the bureaucracy. By 1907, however, Congress had clearly had enough. . . .

. . . Taft was skeptical of Roosevelt's vision of executive stewardship. The idea that a president could do anything not expressly forbidden was, from the new president's point of view, not only inappropriate institutional strategy but also unsound political theory. Taft believed that a strict interpretation of the Constitution provided the only legitimate guide for rearticulating lines of authority and reestablishing governmental order while expanding national administrative capacities. This neo-Madisonian model for reconstitution implied trimming the abrasive edges off the stewardship theory, but it did not imply a return to the governmental order of the late nineteenth century. . . . Administrative expansion was to be carried on with an eye toward separating new bureaucratic powers and reconciling them with the original constitutional design.

Was there and institutional strategy capable of realizing such a vision in the wake of the Roosevelt challenge? Taft surely chose a reasonable course in reestablishing cordial relations with Old Guard Republicans in Congress. With due deference to their interests and prerogatives, he was able by 1910 to gain their explicit endorsement for his own institutional reform initiatives. But the president's respect for the Old Guard tended to leave moderate Republicans and Progressives isolated. Tensions mounted as the interests of Roosevelt purists in the bureaucracy were compromised and moderates in Congress were presented with the choice between subordination or insurgency.

When the Democrats gained control of Congress in 1911, Taft's institutional reform program became tangled in the very constitutional conflicts it had been designed to avoid. What in 1909 had seemed a reasonable strategy for moving institutional reform toward a Madisonian reconstitution was now a serious liability. Insurgent Republicans joined resurgent Democrats in open revolt against the administration. The shifting sands of electoral politics and institutional coalition building pushed the president into a series of watershed constitutional confrontations with Congress, each of which focused on the authority to control new national administrative powers. By 1912, the executive–professional reform coaltion had been placed in jeopardy, the president had been placed on the defensive, and Congress had seized the initiative in controlling the administrative machinery. . . .

. . . Current Republican divisions and a new Democratic Congress suggested that [Wilson] might take control of the legislative process, make a substantive appeal for the allegiance of Progressive voters, and give his party a new claim to majority status in national politics. In the old Jeffersonian style, he used his party to forge a cooperative partnership between Congress and the president. . . .

Under the Wilson strategy, national administrative development became an extension of party development and was worked through the president's cooperative partnership with fellow partisans in Congress. In this way, the Labor Department, a newly fortified Agriculture Department, and the Federal Trade Commission emerged to welcome a number of different constituencies into a new national Democratic coalition. . . .

The outbreak of war in Europe undermined Wilson's efforts to forge a national electoral coalition of Democrats and Progressives and left him an increasingly uneasy partner in his new system of party government. . . . Makeshift arrangements were improvised for the duration of the war. Afterward, the return of Republican control to Congress and the newfound interest of the president in executive prerogatives opened an undirected scramble for the control of administrative power. The postwar consolidation of the administrative advances of the Progressive era forged a new institutional politics in American national government, a politics organized around administrative power and a stalemate of constitutional controls. . . .

. . . Roosevelt, Taft, and Wilson each facilitated the growth of national administrative power, but their neo-Hamiltonian, neo-Madisonian, and neo-Jeffersonian answers to the problem of rearticulating lines of internal governmental authority each fell short of the mark.

Federal Regulation in Historical Perspective
ROBERT RABIN

World War I and the 1920s: The Associational Idea

. . . A strong case can be made that Hoover espoused the single most influential and coherent regulatory philosophy between the Progressive era and the New Deal. In 1917, his views were already clearly formulated. He envisioned the federal administrator's role as that of a vigorous facilitator of cooperative private enterprise. To Hoover, efficiency was an overriding objective, but one which required a middle course between government planning and a laissez faire economy. In his view, production and marketing systems needed a rational order; and that order could best be achieved by voluntary, cooperative efforts by enterprisers in discrete areas of commercial activity. Hoover's approach was distinctive in the important sense that he envisioned government playing a major organizing role—one which blurred the line between public and private—in rationalizing business activity. Federal administrators were to be the energizing force in creating private communities of interest. . . .

The War Industries Board was run by another distinguished outsider to governmental service, Bernard Baruch. Like Hoover, Baruch's instinct was to steer a hybrid course between rule by edict and reliance on the market, a course that meshed well with Wilson's irresolution. . . . In a singular fashion, the line between public and private spheres of activity was blurred.

After a year of deliberation, Congress passed the Transportation Act of 1920, the furthest reaching federal regulatory statute of its day. Earlier railroad legislation had focused almost exclusively on transportation rates, reflecting a guiding assumption that the essence of the railroad problem was price discrimination against shippers and localities. The 1920 act gave the ICC [Interstate Commerce Commission] even greater authority in this area—empowering the agency to set minimum and even exact rates—and more importantly, unveiled a far broader conception of federal regulatory power. The provisions on pooling, consolidation, extension, and termination of service created the framework for restructuring the industry, under regulatory guidance and supervision, along lines designed to make railroad service more efficient. . . . No previous federal regulatory legislation rested on so bold an assertion of market failure.

Yet the 1920 act was seriously defective as a regulatory mechanism for planning a reconstituted transportation network. . . .

The ICC was *required* to do little more than it had done before—police ratemaking practices. And in the political ambience of the 1920s, its authority was exercised in ways that made the 1920 act entirely unremarkable. The agency failed to establish a consolidation plan and refused to develop its own strategies on market-sharing, service curtailment, and extension questions. Instead, it followed the line of least resistance, generally acquiescing in the unsystematic efforts of various railroads to merge, combine resources, or alter patterns of service. . . . Although the associational vision may have departed from a policing model of the regulatory enterprise, it did at bottom reflect a deep-seated aversion to an interventionist style of government regulation—a style characterized by positive public efforts to influence the level or character of the aggregate supply and demand of various commodities. . . .

In historical perspective, the New Deal appears as a distinct break from the past. The regulatory initiatives of the Populist and Progressive eras were largely discrete and limited measures—not dissimilar in kind from common law tort prohibitions against unfair trade practices. They were aimed largely at particularized fields of activity in which vigorous competition led to sharp market practices. Pre-New Deal regulatory initiatives rested on the common law assumption that minor government policing could ensure a smoothly functioning market. But the Depression put to rest this constrained view of national power. Even the more traditional regulatory aspects of the New Deal conceived of government activity as a permanent bulwark against deep-rooted structural shortcomings in the market economy.

And the New Deal ventured considerably beyond the regulatory model developed in the Commerce Act. the NIRA [National Industrial Recovery Act] and AAA [Agricultural Adjustment Act] represented countervailing strategies to the atomistic tendencies of the market—tendencies that appeared invariably to trigger downward price spirals. Along similar lines, the NLRA [National Labor Relations Act] served as a buffer against inequality of bargaining power in the labor market. It openly rejected free market assumptions about the mobility of labor. The TVA [Tennessee Valley Authority] embraced comprehensive governmental planning as a tool for developing a social infrastructure in a regional economy suffering from perpetual depression. Once more, the legislation directly controverted assumptions of a self-correcting mechanism regulating the flow of market transactions.

The New Deal's distributional programs further demonstrated how far the new approach to goverment intervention departed from the old. The public works and social insurance programs undertaken by the New Dealers put the federal government squarely in the position of employer and insurer of last resort. Instead of indirectly creating incentives for changes in private market behavior, the new government programs established a reliance principle: The public came to look upon government as its guarantor against acute economic deprivation. As a result, the spheres of public and private activity were intermingled in ways that would have a pervasive effect on succeeding waves of administrative reform. . . . By the late 1930s, after a half century of uncharted growth, the federal administrative system was viewed in a new light—the ongoing critique of regulatory politics turned from concerns about the appropriate realm of administrative action to a focus on agency decisionmaking processes. . . .

The reaction against agency processes reached its peak in the late 1930s. In a widely publicized report published in 1938, Roscoe Pound, chairman of the special committee of the ABA [American Bar Association] on administrative law, excoriated the regulatory system for "administrative absolutism" and catalogued the suspect "tendencies" of administrative agencies, among them: (1) to decide without a hearing, (2) to decide on the basis of matters not before the tribunal, (3) to decide on the basis of preformed opinions, (4) to disregard jurisdictional limits, (5) to do what will get by, (6) to mix up rulemaking, investigation, and prosecution, as well as the functions of advocate, judge, and enforcement authority.

The ABA's public criticism was coupled with a lobbying effort which led in 1940 to congressional passage of the Walter-Logan bill, an act which was both blatantly political (exempting "favored" agencies) and quite restrictive (significantly enhancing the role of judicial and intra-agency review). But Roosevelt's veto of the bill was upheld, the war intervened, and for the moment federal regulatory processes were forgotten.

Roosevelt had not been deaf to the rising tide of procedural criticism, however. In 1939, he instructed his attorney general to appoint a committee to report on the "need for procedural reform in the field of administrative law." After the war ended, this report served as the foundation for the drafting of the APA [Administrative Procedure Act], which passed both houses unanimously in 1946.

Although the APA, as a generally applicable code of administrative procedure, was indisputably of great practical significance, it was nonetheless—in an important sense—a symbolic gesture. The act was a formal articulation of agency due process in return for the newly recognized powers of wide-ranging administrative intervention in the economy. Symbolic benefits aside, the APA had considerable limitations. It did not purport to reshape the role of administrative government in the federal system; it established no new substantive areas of agency responsibility. Nor did the act purport to alter common law judicial review principles establishing the allocation of authority between court and agency for deciding questions of law and fact. In addition, the act failed to address the vast field of informal agency action—the entire range of interactions between agencies and regulated parties that take place outside the context of a formal hearing or a rulemaking proceeding. Finally, the act spoke in the broad terms of a charter—"substantial evidence," "arbitrary and capricious," "statement of basis and purpose," and so forth—employing language sufficiently vague to allow the greatest leeway in the scope of administrative discretion to fashion regulatory policy in a particularized context.

The structure established by the APA was to be the subject of a barrage of criticism a generation later. At the time, however, and throughout the 1950s, the APA served as a hallmark of a quiescent period. Prosperity reigned after World War II. Although technological development continued to trigger new regulatory schemes, no new flurry of regulatory reform activity clouded the horizon. The impulse further to redefine the relationship between the public and private sectors was temporarily in abeyance.

On the judicial side, the post-*Schechter* [*A.L.A. Schechter Poultry Corp. v. U.S.*] era ushered in a period of unprecedented goodwill toward the regulatory system. With the final legitimation of the New Deal came the acceptance of a central precept of public administration: faith in the ability of experts to develop effective solutions to the economic disruptions created by the market system. Although this premise was not new—it had served as a foundation for early railroad regulation and as a basic tenet of Progressive thought—it came to political fruition with the New Deal agencies. and for the first time, the courts came to grips with the implications of expertise for the allocation of decisionmaking authority between court and agency. . . .

The political meaning of these developments almost lost at the time. The Supreme Court's principal decisions on questions of availability, timing, and scope of judicial review marked off a vast field for unconstrained administrative policymaking. This vast discretionary authority exerted a powerful influence over agency-constituency relations for the succeeding two decades. Similarly, the major holdings on internal agency processes—holdings, for example, on institutional decisionmaking, choice between rulemaking and adjudication, and hearing procedures—provided substantial leeway for agencies to define their own priorities and initiatives, and develop close informal relationships with constituent groups, relatively free from judicial interference. A backlash would occur against this pervasive sense of judicial restraint—but not until a new era dawned in which substantive regulatory reforms again took precedence on the political agenda.

Regulatory Politics and the Great Society

. . . Even more than the New Deal, the Great Society was a manifestation of the twentieth century conception of presidential leadership. Like Roosevelt's program, the War on Poverty was not a governmental response to an indigenous reform movement. Rather, it was the product of a strong president, surrounded by a latter-day "brain trust," who translated the perception of a major social problem into a program of federal administrative initiatives. While Johnson drew upon the New Deal ideology of government responsibility, however, he took it to new heights. . . .

Around 1970, however, a remarkable resurgence of regulatory reform activity occurred that is not so easily traced to a particular source—and which, nonetheless, reshaped the federal regulatory system. Although the major preoccupations of the period are clear enough—an upsurge of interest in health, safety, and conservation issues—it is anything but self-evident why these concerns emerged precisely when they did and why they evoked such fierce passions. In fact, it is no small matter even to identify the coalitions which fashioned the newly conceived agenda for change. . . . As we have seen, the heated debate about the need for greater control over the discretion exercised by New Deal agencies had been temporarily resolved by the passage of the Administrative Procedure Act in 1946. The enactment of the APA, however, created only a brief respite from criticism. By the early 1960s, a consensus existed that something was amiss.

Two groups of critics emerged. One group contended that the agencies were ignoring their mandate to establish clear and consistent policy guidelines—that economic regulation was adrift in a sea of irresolution. In the classic exposition, Henry Friendly challenged the agencies to abandon their practice of deciding major policy issues almost exclusively through case-by-case adjudication, and exhorted regulators to take advantage of their unique capacity to engage in long-term planning through administrative rulemaking.

Another school of critics expressed concern about the oppressive tendencies of the regulatory system. In a widely noted essay, Charles Reich stressed the dramatic fashion in which government largess had come to exercise a pervasive influence over the basic needs of the individual by setting the terms on which one might pursue an education, practice an occupation, or realize the expectation of economic security in later years. While Friendly advocated greater reliance on rulemaking, Reich argued the need for more adequate procedural rights in adjudicatory settings. . . . Friendly, Reich, and like-minded critics converged in their reliance on the traditional procedural mechanisms of rulemaking and adjudication for effecting regulatory reform. Just as the APA resorted to a simple dichotomous model—likening agency adjudication to the judicial process and agency rulemaking to the legislative process—so did the critics in the early 1960s pick up on the unfinished business of 1946.

Viewed in this context, NEPA [National Environmental Policy Act] represented a wholly different strategy for controlling administrative discretion. Most importantly, the act had a powerful substantive impetus that was absent from earlier plenary regulatory reform proposals. In essence, APA-type reform took the agency's "mission" as given, and aimed at creating a formal decisionmaking methodology which would be more conducive to careful deliberation, and to achieving "correct" outcomes. By contrast, NEPA directly challenged the premise that mission-oriented agencies were discharging their responsibilities with a proper regard for "the public interest," as long as they failed to give focused consideration to the impact of their decisions on the environment. NEPA anticipated an altered process of decision rather than simply better procedures for decision; the environmental impact statement was to be "action-forcing" in that every federal regulatory agency was to reassess its mandate in view of the environmental consequences of any major decision it might reach.

In addition to this highly significant substantive requirement, NEPA brought a different procedural perspective to regulatory reform. The environmental impact statement requirement constituted a departure from the classical approaches to controlling discretion just discussed. From the APA through the New Property, procedural reformers had turned to the adversary model of decisionmaking and proposed more extensive trial-type requirements for agencies. By contrast, NEPA was silent about hearings. . . .

The primary thrust of NEPA—which underscores its distinctive character—was its reliance upon an internal management technique that owed a greater debt to organization theory than to administrative law. Under the most optimistic scenario, the routinization of an impact statement requirement would necessitate specialized administrative personnel and the establishment of new channels of communication and information-flow within an agency. If these bureaucratic operating procedures were conscientiously pursued, a traditionally mission-oriented agency could exhibit a new sensitivity in defining its organizational goals. . . .

The second pathbreaking piece of environmental regulatory legislation was the Clean Air Amendments of 1970 (CAA). In many ways, the design of this landmark pollution control scheme is at the polar extreme from NEPA. Where NEPA is a broadly stated charter of environmental rights consisting of barely a page of text, the CAA is an extraordinarily technical document as lengthy as a decent-sized novel (although hardly as readable). Where NEPA was designed to influence the mandate of every federal agency, the CAA is addressed to a single specialized agency headed by a sole administrator. Where NEPA is directed exclusively at the federal agencies, the CAA allocates major implementation responsibilities to the states. And finally, where NEPA is designed to effectuate internal management reforms within the federal bureaucracy, the CAA relies upon traditional rulemaking and adjudicative enforcement procedures—including private rights of action—and establishes various avenues for judicial review.

What unites the Clean Air Act with NEPA as a real innovation in regulatory design is congressional recourse to an action-forcing principle. The CAA, like NEPA, rejects the prevailing New Deal wisdom that agency experts could best bring their technical expertise to bear on problems of public policy if they were pointed in the right direction, whether allocation of air traffic routes, design of river rechannelization projects, or whatever, and told to regulate in "the public interest." . . . NEPA was meant to widen the administrator's horizons.

The CAA posed a very different challenge to the New Deal perspective. In setting stringent deadlines for administrative action, the CAA questioned the very will of the regulatory agencies to act. . . .Whereas NEPA's response was to rewrite the substantive mandate of the agencies, the CAA took the different tack of requiring that precise standards be established and explicit compliance timetables be met. . . .

Putting aside the singular implementation strategies associated with NEPA and the CAA, certain generalizations can be made about the salient characteristics of Public Interest era regulatory reform from a historical perspective. At the time of enactment both NEPA and the CAA stirred up almost no controversy, despite the arguably draconian implications of each regulatory scheme. In both cases the most significant political compromises were negotiated in order to iron out differences between forces friendly to the legislation, rather than to pacify a hostile opposition. . . .

As James Q. Wilson has argued, much of the public interest regulation passed in the 1970s can be characterized politically as involving concentrated costs (on industry) and dispersed benefits (to "the public")—not a scenario that traditionally augured well for the enactment of regulatory legislation. But in the 1970s, big business was truly on the defensive as the public seemed responsive to a wide variety of concerns about the quality of life. An entire series of initiatives resulted—on auto safety, product design, air and water pollution control, scenic conservation, and occupational health and safety, to mention only the most significant—which manifested a distinct bias against economic growth. The political climate made it virtually impossible to oppose such programs in principle—and focused objections can always be pursued in the process of agency implementation.

A second striking feature of the Public Interest era legislation is that it was not the product of a social movement for reform, nor even the outcome of pluralistic, interest-group politics. As Wilson noted, the passage of NEPA and the CAA might well be characterized as instances of "entrepreneurial politics"—situations in which astute politicians adopted anticipatory strategies, setting the agenda for regulatory action prior to clearly articulated interest-group demands for change.

This phenomenon has been much more common than is generally recognized. . . .

The distinctive aspect of Public Interest era regulation, however, was the extent to which exceedingly diffuse concerns about health, safety, and conservation were enacted into law with virtually no presidential initiative. . . .

A third aspect of the Public Interest era that warrants attention is the relatively limited ideological thrust of the reform measures that characterize the period. This assertion may seem at odds with the antigrowth theme evident throughout the era. Consider, however, that no substantial wealth redistribution impulse fueled the Public Interest reform efforts, and no discernible challenge was mounted against the autonomy of a market-based economy. Instead, the key legislation of the period suggested a return to the policing model of Progressivism. In the areas of pollution control, occupational safety, and consumer protection, the prevailing ideology anticipated the internalization of the previously unrecognized costs of industrial growth—a market-corrective strategy that posed no challenge to the premises of an exchange economy. . . .

The New Deal agenda, by contrast, is more problematic to characterize. . . . Under the rubric of a "return to normalcy," the New Deal contained the seeds of a new regulatory order—an order that might have dramatically altered the market economy.

Within its particular range of concerns, the Great Society even more explicitly rejected the concept of limited federal government interference in private affairs. The community action and legal service programs used federal resources to monitor and reorder federal–state aid and assistance programs, blurring the distinction between public and private spheres of activity. By contrast, Public Interest era agencies reflected the classical dichotomy between public regulation and private activity, and employed traditional command-and-control rulemaking and adjudication powers to achieve the substantive health and safety goals established by Congress.

Even more centrally, however, the Great Society programs built upon the nascent redistributional impulses of the New Deal. . . . In this regard, Public Interest era regulation seems more closely akin to the narrow premises of pre-Depression ideology. Once again, the regulatory system—whether dealing with health, safety, or conservation interests—operated primarily in an enforcement or policing mode. . . .

The Public Interest Era in the Courts

[In an omitted position of this section, Rabin describes some of the early landmark "public interest" decisions—*Office of Communication of United Church of Christ v. FCC, Scenic Hudson Preservation Conference v. FPC,* and *Citizens to Preserve Overton Park v. Volpe*—in which the D.C. and Second Circuits and the Supreme Court, respectively, exhibited a striking loss of faith in administrative expertise and a growing concern about how agencies made and justified their discretionary judgments. The court's concern was especially acute with respect to regulatory decisions involving the clash of environmental and other intangible values with well-organized economic interests. In these and other cases, the courts devised a number of techniques to implement a far more searching, demanding, skeptical mode of judicial review.]

If steadfast protection of intangible values was one major underpinning of the new judicial activism, the other principal source was the heightened awareness of the problem of scientific uncertainty about risks to health and safety. The importance of this phenomenon can hardly be overstated. . . .

Risks to health and safety involve uncertainty which is strikingly different from that found in typical economic regulation. . . .

Not only were administrative decisions necessarily based on projections from scattered episodes at much higher exposure levels or on inferences from experimental data based on animal studies, but inaccurate, insufficiently protective, administrative decisions might lead to irreversible long-term risks to society of devastating magnitude. . . . Although the FDA had faced dilemmas such as these for many years in issuing new drug clearances, the new regulatory legislation was so pervasive in scope and addressed such a wide range of previously unacknowledged health and safety risks that it once again posed a distinctive challenge to assumptions about administrative expertise. . . .

Thus, the era of judicial deference that had spanned three decades was abruptly brought to an end. Regulation was once again viewed with a skeptical, if not jaundiced, eye. But it would be a serious mistake to regard the Public Interest era as a return to the early New Deal sentiments of *Schechter* and its forerunners. As is evident, the new social regulation—of which the CAA was an integral part—posed no threat to the existing common law jurisdiction of the courts. Nor, as indicated in the preceding section, did it portend a major redistribution of political or economic power.

The legitimacy of the regulatory enterprise was not in question. Rather, the courts were centrally concerned with the question of how to control effectively the exercise of administrative discretion in the singularly perplexing cases of scientific and technological complexity. Deference to traditional processes of informal rulemaking and adjudication in such cases appeared to be tantamount to surrendering the function of judicial review. . . . In the early 1970s, judicial activism was frequently extended to the point of requiring agencies to provide something approximating a Right Answer to problems of scientific and technological uncertainty. This is not to say that the courts were so naive as to think that the regulatory agencies could absolutely eliminate uncertainty through sufficiently scrupulous factfinding. Rather, the courts were pressing toward a decisionmaking methodology that would closely resemble the adversarial search for truth so congenial to the judicial mind—a methodolgy that demanded joinder of issue, elaborate articulation of all contending values, impartiality of perspective, and a carefully reasoned decision.

In contrast, *Vermont Yankee [Nuclear Power Corporation v. Natural Resources Defense Council]* tilted uneasily toward the unstructured political model of agency policymaking incorporated in the notice-and-comment provisions of the APA. Nonetheless, the Court was understandably reluctant to embrace a decisionmaking methodology which, as a practical matter, would eliminate any checks on administrative discretion (other than constitutional safeguards). Instead, the Court opted for what might be labeled a Best Efforts approach. Taking into account that the agency's statutory mission might appropriately create a political "tilt," and recognizing the quasi-legislative character of agency rulemaking, the Court sought assurance of good faith consideration of the issues by the regulator rather than demanding a pristine search for truth.

As rough approximations, these two categories reflect contrasting perspectives which by the late 1970s, as the Best Efforts approach reached its peak, came to characterize a reassessment of the judicial activism of the preceding few years. There was a retreat along all of the major fronts of judicial review to the more secure ground of requiring only that the regulatory agencies offer some assurance of good faith consideration of the issues. . . .

With due regard for the right-to-hearing cases, the core of Public Interest era activity was firmly embedded in the new social regulation. Perhaps for the first time in a century of steady administrative expansion, the coordinate institutions—legislative and judicial—acted in tandem rather than moving in opposing directions. In the early 1970s, Congress recognized the need to address a variety of low-visibility collateral costs to health, safety, and the environment that were the legacy of relatively unimpeded long-term industrial growth. At the same time, the courts heralded the importance of these newly recognized values, and in doing so conducted a searching reexamination of fundamental principles of judicial review of administrative discretion. . . . Then, as the decade wore on, the approach shifted back to a less intrusive (although still insistent) requirement that agencies demonstrate their competence rather than simply have it taken for granted. . . .

A Postscript on Deregulation—Unresolved Tensions in the
Administrative State

. . . By the late 1970s, however, the expansionist period of the Public Interest
era had, in turn, run its course. For the first time in a century, a discernible
political movement sought to reassess the need for regulatory programs that
administered markets as a means of promoting the health of particular industries. This movement was exceedingly widespread: The regulatory system
came under close scrutiny by policy institutes and journals, academic disciplines, and politically influential public officials who all came to focus on a
clear and dominant emerging theme—deregulation.

. . . The increased focus on deregulation was not principally a reaction to
the latest wave of regulatory reform—the social regulation of the Public
Interest era. No serious effort was mounted to revoke the recent congressional initiatives in the areas of health, safety, and environmental protection—let alone to reassess the need for earlier Progressive era efforts to
establish policing controls on the market. Instead, the new criticism was lev-
eled at administrative activity extending beyond the policing model; it constituted an attack—unparalleled in vigor—on price-and-entry regulation.

The economic regulatory agencies were chastised, as in the days of the
NIRA, on the grounds that widespread government intervention in support of
price-and-entry regulation was suppressing competition and fostering economic waste. In unprecedented fashion, Congress responded with a series of
deregulatory initiatives: the Airline Deregulation Act of 1978 (providing for
total deregulation, in phases, of fare setting and entry restriction in air travel);
the Motor Carrier Reform Act of 1980 and the Staggers Rail Act (relaxing
rate and entry regulation of motor carriers and rail traffic, respectively); and
the Depository Institutions Deregulation and Monetary Control Act of 1980
(eliminating interest ceilings on time and savings deposits).

The congressional embrace of deregulation marked a pause after a century
of steady growth in the federal administrative system. Moreover, the rising
storm of criticism over excessive regulation extended far beyond the field of
legislative activity. Traditional administrative rulemaking—so-called command and control rulemaking—came under general attack for its economic
inefficiency in failing to discriminate between low-cost and high-cost compliance activity. Critics excoriated administrative adjudication for its expense
and delay. They further challenged agency priorities in establishing policy as
failing to give precise consideration to the costs and benefits of various regulatory options—including the possibility of taking no action at all.

Lending a sympathetic ear, the Carter and Reagan administrations succes-
sively sought to sustain further the deregulatory mood through economic
efficiency measures at the agency implementation stage—such as allowing the
offsetting and trading of air pollution emissions credits under the Clean Air
Act, and requiring cost-benefit justification from the agencies for "major"
regulatory initiatives prior to administrative action. Indeed, the Reagan ad-

ministration appeared to pursue a broader-based de facto deregulation policy by appointing unsympathetic agency administrators and proposing drastic budget cuts in regulatory programs untainted by congressional disapproval.

Thus, a half-century after the New Deal—a century after the birth of the federal regulatory system—the world-wise French aphorism seems not entirely out of place, *"plus ça change, plus c'est la même chose."* The system has grown by leaps and bounds, yet it remains devoid of any coherent ideological framework. In the political sphere, while a broad-ranging commitment to government intervention seems a continuing legacy of the New Deal, no consensus exists on the appropriate scope of federal regulatory activities. The long-standing tension between a regulatory system dedicated to effective policing of the market and one which would stimulate cooperation among private interests—a tension as venerable as the federal regulatory presence itself—has never been conclusively resolved. For the present, the minimalist policing model of regulation appears to be in the ascendancy, but only the most self-confident seer would predict the future. Indeed, even the welfare sector of the administrative system—an especially deep-rooted outgrowth of the New Deal—has not been immune from attack in the years since the demise of the War on Poverty.

In the courts, the legitimacy of regulation is no longer seriously questioned. But here, too, widespread recognition of the need for an expansive regulatory system has simply posed critical questions rather than resolved them. The courts have not developed a consistent approach to controlling agency discretion. Such an approach would have to draw on a theory of administrative expertise that dealt coherently with the technical and political dimensions of the regulatory process. Lacking an intelligible theoretical framework, the Supreme Court has oscillated between activism and restraint in reviewing agency decisions. Like Congress, the judicial system gets higher marks for pragmatism and flexibility in dealing with each successive wave of regulatory reform than it does for intellectual coherence and certainty of approach.

NOTES AND QUESTIONS

1. Many scholars have risen to the challenge of tracing the development of the administrative state. Gabriel Kolko, an historian with a Marxist orientation, has advanced a provocative account that contradicts traditional public interest explanations. In several works, most notably *The Triumph of Conservatism* (New York: Free Press of Glencoe, 1963), Kolko argues that the Progressive-era regulatory agencies were established not in a spasm of populist indignation but as a result of concerted efforts by large corporate interests. These interests, Kolko claims, succeeded in creating and controlling a government-enforced cartel (the agency), which would reduce competition by discouraging new entry, mandating minimum prices, and adopting other anticompetitive policies. One is struck by the convergence at many points among Kolko's interpretation, the political model offered by the public choice theorists, who tend to locate

near the other end of the ideological spectrum, and the more middle-of-the-road critiques advanced by the regulatory reformers of the 1970s.

A more traditional interpretation is offered by legal historian William E. Nelson in *The Roots of American Bureaucracy, 1830–1900* (Cambridge, Mass.: Harvard University Press, 1982). Nelson maintains that the bureaucratization of Congress, the executive branch, and even the courts during the post–Civil War period reflected a broad public effort to protect pluralism and individual and minority interests, in contrast to the majoritarian, party-governed norms that dominated the prewar system. According to Nelson, the mugwump and progressive advocates of an administrative state, many of whom had abolitionist and scientist backgrounds, were neither self-interested class advocates nor utopian redistributionists. They sought instead to create a bureaucracy that would protect a pluralism that seemed in jeopardy.

A brief historical survey by James Q. Wilson stresses several other developmental patterns. He notes the absence of any widely perceived "bureaucracy problem" throughout most of American history—that is, until the New Deal. He also emphasizes the "clientelism"—agencies' active promotion of the interests of favored special interest groups, including state and local governments—that has come to characterize much of our administrative state. He views this development as antithetical to the values of Madison and the other Framers and suggests that it is fostered by, if not endemic to, the separation of powers and the particularistic, localistic nature of American politics. Wilson, "The Rise of the Bureaucratic State," 41 *The Public Interest* 77 (1975).

2. Does either Skowronek or Rabin offer evidence that refutes Kolko's thesis? What kinds of evidence could successfully do so? Consider, for example, the kind of economic evidence that Breyer and many other critics of misguided regulation have adduced documenting the often deleterious effects of regulatory programs on the profits and competitiveness of the regulated industries. Interstate Commerce Commission regulation of the railroads is perhaps the classic example, but there are others, including natural gas regulation, which Breyer analyzes in chapter 13 of his book. Consider also evidence about regulatory programs such as rent control, which transfer wealth from a politically well-organized regulated industry to a relatively diffuse, weakly organized group of beneficiaries.

3. Does the proliferation of "social regulation" at the federal level during the 1960s and 1970s, discussed by Rabin, fit Kolko's model? Note, for example, that most social regulation programs adopted in recent years, such as equal employment opportunity law or occupational health and safety controls, tend to apply to all industries, not just to one or a few related industries, as the Securities and Exchange Commission and the Interstate Commerce Commission do. Note also the political effects of new modes of public participation represented by "public interest" group advocacy and litigation, developments discussed below, especially in Chapter VI, section 5.

4. Skowronek finds in the leading Progressive-era reformers three conceptions of the administrative state: neo-Hamiltonian (Theodore Roosevelt), neo-Jeffersonian (Wilson), and neo-Madisonian (Taft). Rabin's discussion of Hoover suggests a fourth model, which Rabin calls "associational" but which one might also term corporativist. What are the characteristic features of each of these models, and how accurately do the first three represent the views of their eponyms? What are the models' essential constitutional presuppositions? (Here, I use "constitutional" to refer not only to the institutional arrangements and values contained or immanent in the document, but also to the basic structures of political and social power, such as parties, communica-

tions, social structure, etc.). How compatible is each with the U.S. Constitution and what constitutional changes would be necessary to instantiate each model and render it effective? What are the distinctive conceptions of the "bureaucracy problem" presupposed by each model, and how might administrative law address these problems?

III

The Administrative Procedure Act

The Administrative Procedure Act of 1946 (APA) is the fundamental charter of the administrative state, providing a structure of procedural protection to interests affected by agency decisions. The APA has been called a quasi-constitutional statute, with good reason. If there were no APA, the courts—like the Attorney General's Committee on Administrative Procedures that spawned the APA—would certainly have invented something like it in order to implement the constitutional safeguards of the Fifth Amendment's due process clause. (Martin Shapiro, in the article excerpted below, has suggested that the APA is also constitution-like in the sense that it "today means all kinds of things that its drafter could not possibly have intended it to mean.")

The excerpts from Cass Sunstein, a law professor, and Martin Shapiro, a political scientist, are part of a symposium on the APA held to mark its fortieth anniversary in 1986. Sunstein offers a "deliberative model" of administrative governance under the APA, a model that he grounds in the public norms, the "constitutional themes," that he believes animated the APA's draftsmen. Shapiro sharply disputes Sunstein's position on the original intent of the APA and the theoretical merits and practicability of Sunstein's deliberative model. He also advances a provocative thesis about the ideological character and the political, even partisan, stakes in Sunstein's theory of the APA, especially the theory's serviceability in rationalizing and cementing a political alliance between "public interest" lawyers, agencies, and courts against both Congress and the president. (Shapiro's comments also refer to other symposium contributions by Walter Gellhorn, who directed the Attorney General's Committee, and by Alan Morrison, who is chief litigator for Ralph Nader's citizen lobby in Washington, which are not excerpted here. Shapiro describes their points [especially Morrison's] tendentiously but not, I think, in a seriously misleading way. Readers, however, are urged to make up their minds by reviewing all of the articles in their entirety.)

I noted in the introduction that *state* administrative law is routinely neglected in courses and casebooks. This book seeks to rectify that neglect to a small degree by excerpting an article by Arthur Bonfield, a law professor specializing in that subject. Bonfield's discussion underscores the most distinctive features of the Model State Administrative Procedure Act, which more than half of the states have adopted.

Factions, Self-Interest, and the APA: Four Lessons Since 1946

CASS SUNSTEIN

In attempting to control administrative processes, the drafters of the APA responded to two quite general constitutional themes, both of which have played a central role in administrative law since its inception. The first concerns the usurpation of government by powerful private groups. The second involves the danger of self-interested representation: the pursuit by political actors of interests that diverge from those of the citizenry. The relative insulation of administrators from electoral control has given rise to particularly intense fears about these two risks. But the forty years since the enactment of the APA have brought significant advances in knowledge about regulation and administrative law—advances that have important implications for the question of how the legal system should respond to the risks of factional tyranny and self-interested representation.

Four of these advances have been of particular importance. First, it is now possible to identify, with far more sophistication, the functions and malfunctions of regulation. Second, it is clear that dangers to statutory standards lie in unlawful inaction and deregulation as well as in overzealous regulation. Third, faith in administrative expertise has diminished in the wake of a more refined understanding of the complementary roles of technical sophistication on one hand and "politics" on the other. That understanding has brought forth what might be called a deliberative conception of the role of the administrator—a conception that is, however, threatened by a still-tentative trend in the Supreme Court. Fourth, courts have shown themselves to suffer from serious limitations as supervisors of agency discretion; there is a corresponding need for nonjudicial mechanisms of supervision. . . .

The Functions and Malfunctions of Regulation

First, regulation may protect interests that are considered entitlements—with the class of entitlements being defined by reference to values apart from positive law. The best example is protection against discrimination on the basis of race and sex. At least for some observers, protection against discrimination has assumed the role played by protection of private contract and property in the common law era.

Second, regulation may redistribute wealth to people who are thought to

Reprinted by permission from 72 *Virginia Law Review* 271 (1986).

be badly off. The minimum wage is a good example. Whether and how regulation will redistribute resources is a complex question, but improving the lot of the unfortunate is often an intended effect of regulation.

Third, regulation may promote economic "efficiency." Occupational safety and health standards, for example, may be a response to lack of information or to externalities, and regulation of air pollution may be a product of a collective good problem.

Fourth, regulation may be designed to shape or discourage certain preferences or—a related point—to reflect the public's "preferences about preferences." Like Ulysses confronted with the Sirens, the public may decide to bind itself with laws that prohibit the gratification of short-term consumption choices. Prevention of discrimination illustrates the phenomenon of shaping preferences, as do the laws prohibiting sexual harassment. Seat-belt laws, social security provisions, and laws prohibiting cigarette advertisements may be efforts to foreclose misguided consumption choices.

Fifth, regulation may be paternalistic, reflecting a desire to prevent people from making decisions that will harm them. Here, however, the government, and not the public itself, prohibits the satisfaction of consumption choices. Some occupational safety and health regulations may be thus understood.

Finally—and this is an important point—regulation may reflect little or nothing more than interest-group pressures and thus serve no public purpose. Many regulatory programs, including regulation of utility price levels and licensing of interstate motor carriers, have been described in this way. This notion is the "dark side" of the redistribution-of-resources rationale for regulation: the redistribution may turn out to benefit particular classes for no reason other than political power. . . .

The Dangers of Inaction and Deregulation

. . . The original purpose of administrative law was to protect private autonomy—most important, private property—from unauthorized governmental intrusions. Judge-made doctrines primarily sought to enable regulated entities to fend off unlawful government action: rules governing reviewability ensured that members of the regulated class would have access to judicial review, and the principle of standing limited that access to the regulated class, excluding the beneficiaries of regulation. In the face of ambiguity in the APA on the point, however, courts have abandoned the notion that protection of traditional private rights, as defined by the common law, is the sole function of the regulatory process.

This development is a product of three perceptions. The first is that statutes may be undermined through inaction and deregulation as well as through overzealous enforcement. . . . The second perception is that the "rights" created by regulatory schemes are no less deserving of legal protection than those recognized at common law. . . .

Finally, courts have recognized that, as a general rule, political remedies are no more readily available to beneficiaries of regulation than to members of the regulated class. Indeed, regulated class members are often well-organized and may be better able to take advantage of the political process. Members of the beneficiary class, on the other hand, may be quite diffuse and thus unable to overcome transactions costs barriers to the exercise of political influence.

These understandings suggest that the dangers of factional power over governmental processes are no greater in the context of too much regulation than in the context of too little. Administrative law, including doctrines of judicial review, should therefore reflect a concern with unlawful inaction and deregulation as well as with unlawful regulation. This understanding, a significant departure from the original conception of the APA, is to some extent reflected in current law. Even after the recent decision in *Heckler v. Chaney,* inaction is frequently reviewable, and deregulation is subject to the same standards as regulation. Moreover, congressional and executive mechanisms of oversight apply to failure to regulate as well as to regulation.

Expertise and Politics: A Deliberative Conception of Administration

Faction, Deliberation, and Technical Expertise

The debate over the respective roles of "expertise" and "politics" in agency decisionmaking has proved to be one of the most persistent in administrative law. . . . Administrative and judicial efforts to solve this problem have come in the form of a deliberative conception of administration, a conception that amounts to a significant reformulation of previous understandings.

It is useful to begin here with the idea that the role of the administrator is not merely to reflect constituent pressures or to aggregate private interests. Instead, the purpose of the regulatory process is to select and implement the values that underlie the governing statute and that, in the absence of statutory guidance, must be found through a process of deliberation. . . .

The deliberative approach has significant advantages over its major competitors and predecessors, each of which fails to provide an adequate conception of the role of the administrator. Under one view, prominent at the time the APA was enacted, the role of the courts is to require fidelity to statute. Although that function should be uncontroversial, the notion, standing by itself, often fails adquately to constrain agency action because of the open-ended character of many statutory standards—an especially disturbing problem in light of the courts' express authority to invalidate "arbitrary and capricious" decisions. Another view is that, where statutes do not provide to the contrary, courts may constrain the exercise of discretion by requiring administrative agencies to measure and balance the costs and benefits of regulatory action. Although an understanding of costs and benefits is a necessary component of review for arbitrariness, such an assessment is a political rather than a judicial responsibility. A third approach, prominent in the 1970s, sought to ensure participation on the part of all affected interests. That approach foundered in light of four considerations: the fact that the relevant representatives were self-selecting; the weaknesses in the notion that the purpose of administration is to aggregate preferences; the unlikelihood that, even if preference-aggregation were desirable, it would be accomplished by a judicially administered system of interest-representation; and the possibility that such procedures would impose costs not justified by improvements in administrative outcomes.

Although the deliberative approach to administration avoids most of the problems associated with these alternatives, significant difficulties do arise in the effort to implement this approach. The principal question for administrative agencies and reviewing courts is how to define the relevant values. When the values are set out in the governing statute, the answer is easy: the statutory resolution will govern in the absence of a constitutional defect. The major difficulties arise where, as is frequently the case, the statute is ambiguous, and the administrator must ascertain values through a more open-ended process. . . .

To manage this problem of defining relevant values, and to accommodate permissible uses of both expertise and politics, courts have created a four-pronged notion of "reasoned decisionmaking." First, regulatory decisions should be based on a detailed inquiry into the advantages and disadvantages of proposed courses of action—where advantages and disadvantages are not defined merely as "costs" and "benefits" in the economic sense of willingness to pay—and an examination of reasonable alternatives. Second, issues involving value judgments must be resolved consistently with the governing statute. . . .

Third, to the extent that issues of value are to be resolved through an exercise of administrative discretion, the relevant considerations and the actual bases for decision must be explicitly identified and subjected to public scrutiny and review. Finally, the agency's resolution must reflect a reasonable weighing of the relevant factors. . . .

In this light, it should not be surprising that the notion of "capture" has proved of special importance in the cases. The [Motor Vehicle Manufacturers Association v.] State Farm [Mutual Automobile Insurance Co.] Court referred to industry influence, and in Public Citizen v. Steed, the court emphasized the agency's reliance on the treadwear grading practices of tire manufacturers. To be sure, both regulation and deregulation are frequently attacked as the product of faction. In general, however, courts define the phenomenon of "capture" not by reference to outcomes, but by reference to the process the agency has followed in implementing its program: has it reacted to pressure, or has it deliberated? Of course, such an approach creates difficulties for judicial review of administrative processes, and these difficulties are aggravated when administrators act on the basis of mixed motivations or when numerous administrators are responsible for the ultimate decision. This approach also creates a danger of judicial usurpation in the form of decisions that invalidate agency action because of substantive views on the part of courts that cannot be tied to underlying statutory policies.

Moreover, the notion of mechanical-reaction-to-pressure must sometimes be understood as a metaphor for a complex process in which administrators come to share the values of particular affected parties and their approaches to regulatory issues. Regardless of whether the agency's response is mechanical or more complex, however, it is inconsistent with an approach to administration that would properly accommodate expertise and politics in the regulatory process. The various components of the deliberative approach may be understood as a means of simplifying the judicial inquiry. Those components are designed to help "flush out" impermissible motivations without looking into subjective states of mind, and at the same time to minimize judicial intrusions on the merits. The goal, imperfectly realized in practice, is to guard against the dangers of self-interested representation and of factional tyranny in the regulatory process. . . .

Formalism and the Chevron *Approach*

. . . The "clear statement" principle could have dramatic implications, for Congress has rarely "directly addressed the precise question at issue" in any particular case. But for three reasons, the *Chevron* [*USA, Inc. v. Natural Resources Defense Council, Inc.*] approach is an unacceptable basis for judicial review: the approach is unlikely to serve Congress' own goals and expectations; it undervalues the possibility of extrapolating principles from statutory standards; and its conception of separation of powers fails in light of the uneasy constitutional position of the administrative agency. The *Chevron* understanding cannot, therefore, provide a shield against the twin dangers of self-interested representation and factional power, and its failure to do so does not serve the legislative will. . . .

The Limitations of Judicial Remedies and the Need for Alternative Mechanisms of Control

The considerations that support a deliberative approach to administration argue in favor of a moderately aggressive judicial role, but it has become increasingly clear in the last decade that reviewing courts have difficulty handling many of the malfunctions of the administrative process. Although the drafters of the APA believed such malfunctions to be one-shot deviations subject to judicial correction, the most important problems are structural or systemic in character. . . .

The principal problem is that courts have difficulty engaging in the kind of managerial tasks that are essential to successful systemic change. Continuous supervision of the administrative process is impossible to achieve through adjudication, yet such supervision is of special importance in light of the "polycentric" character of administrative decisions, where action in one setting may have unanticipated adverse consequences for other forms of regulatory intervention. These problems are aggravated by the inability of courts to impose a coordinated or hierarchical structure, by their lack of familiarity with the often technically complex issues at hand, and by their lack of political accountability.

Courts should, however, be available to hear complaints of structural or systemic illegalities. . . . Nevertheless, experience has demonstrated that nonjudicial review mechanisms are necessary to supplement and, if they are successful, to displace judicial solutions.

A useful example is the increasing authority of the Office of Management and Budget (OMB) over the regulatory process. . . . There is a common ground in the view that OMB's function is to improve deliberative processes, accommodating both expertise and politics, in the way suggested above. The rhetoric of the debate thus reveals a similar understanding of the function of the regulatory process.

APA: Past, Present, Future

MARTIN SHAPIRO

Professor Gellhorn was one of a cohort of New Deal lawyers who did precisely what [Roscoe] Pound accused them of doing. They created a body of administrative law that rationalized and legitimated the administrative state that the New Deal created and that the New Deal ideology defended.

One basic element of the New Deal ideology was a dedication to that most American cluster of political ideas—the pragmatism of James and Dewey that engendered and became combined with the Progressives' notion that powerful central political authorities guided by technical expertise could develop good working solutions to major social and economic problems. . . .

One of Professor Gellhorn's footnotes nicely reveals another major element of New Deal ideology. Gellhorn makes fun of the anti-New Dealers' Red-baiting, and adds a footnote illustrating his opponents' further odd tendency to accuse those constructing the new administrative law of seeking to establish parliamentary government in America. Their charge, however, was not bizarre. The New Deal and New Deal administrative law indeed were on a parliamentary track, a track customarily labeled as the Democratic theory, the strong theory, or more recently the imperial theory of the presidency. Like many other New Deal ideas, this one is easily traced to the Progressives. . . .

The administrative law invented by Professor Gellhorn and company adopted the executive delegation aspect of parliamentary government by rejecting the nondelegation doctrine. (Of course, it rejected the nondelegation doctrine only after a New Deal Congress rejected the doctrine.) The new

Reprinted by permission from 72 *Virginia Law Review* 447 (1986).

administrative law also managed the Wilsonians' trick of having the sweet without the bitter because the executive got the delegated lawmaking power but was not directly answerable to Congress for the laws it made.

Parliamentary theory, however, has relatively little place for judicial review. One reason executive lawmaking in Britain is so powerful is that British courts do little review of it. The United States was different at the time of the New Deal: American courts did a lot of review, both constitutional and statutory. The New Deal lawyers, however, had two solutions to this problem. On the constitutional side, they developed the theory of judicial self-restraint, which directed unelected judges to defer to elected legislatures. Meanwhile, New Deal historians and political scientists were developing the theory of the strong presidency, which directed the legislature to defer to the president. Put all together, it spelled Roosevelt. Courts should defer to Congress, Congress should defer to the president. So courts really were to defer to the Executive. Q.E.D.

To solve the statutory as opposed to the constitutional review problem, the New Deal lawyers borrowed another Progressive idea, government by experts, which directed nonexpert courts to defer to administrative expertise. The administrative law of the forties and fifties was the culmination of this theory. Reviewing courts routinely deferred to whatever law the agencies made. . . .

Evolution of the APA

The Original Political Bargain

. . . The law of the APA [Administrative Procedure Act] is thus largely a congressional affirmation of the scheme worked out by the executive branch's New Deal lawyers. They formulated a modified and softened version of the prewar vision of Pound and the American Bar Association and fitted it into a basically New Deal plan. It is very important to understand the compromise because it engenders the basic tensions that plague administrative law today.

The APA as originally enacted divided all administrative law into three parts. For matters requiring adjudication, in which government action was directly detrimental to the specific legal interests of particular parties, the compromise was heavily weighted in favor of the conservatives. The Pound-ABA demand for totally separate tribunals was ignored: the agencies themselves adjudicated these matters. But the agencies' processes were to be considered quasi-adjudication and were to be governed by adjudicative-style procedures, presided over by a relatively independent hearing officer, and freely subject to relatively strict judicial review.

The second part, rulemaking, constituted an almost total victory for the liberal New Deal forces. Congress' delegation of vast lawmaking power to the agencies was acknowledged and legitimated. Rulemaking was to be quasi-legislative, not quasi-judicial. No adjudicatory-style hearings or hearing officers were required. Those not directly and immediately affected by the rule could not easily obtain judicial review. Under the APA rulemaking generated no record to be reviewed, and the standard of review made an agency's decisions irreversible unless it had acted insanely. Although the agencies were acting in a quasi-legislative capacity, they were not required to jump through as many procedural hoops as Congress typically did in legislating. Congress normally held oral hearings on pending legislation, a full draft of which was already on the docket, and issued a rather elaborate committee report to explain a bill as it went to the floor of the House or Senate. In contrast, the APA simply required an agency to give notice only of its intention to make a rule. It did not have to submit a draft. It had to receive written comments, but no hearing was required. It merely had to provide a "concise" and "general" statement accompanying its rule.

In the absence of a rulemaking record, reviewing courts were forced to presume that the agency had the facts to support its rule. Given the extremely broad and standardless delegations in most of the New Deal legislation of the thirties and forties, courts rarely found that a rule violated the terms of its parent statute. And it was rare indeed for a New Deal-appointed judge reviewing the work of a New Deal-staffed agency to find that the agency had acted like a lunatic, that is that it had been, in the words of the APA, "arbitrary" and "capricious." Therefore, in the political bargain judicial review of rulemaking is about all the conservatives got, and they got very little of that.

The third part of administrative law originally conceived by the APA included everything that government did that was neither adjudication nor rulemaking. On this point, the liberal New Dealers won almost complete victory, labeling agency action in this area as "committed to agency discretion." No procedures were prescribed. The liberals made only one small mistake: although one section of the APA precluded judicial review of matters committed to agency discretion, another subjected agency actions to review for "abuse of direction."

Since the APA: The Evolution of Administrative Law and the Bonding Between Agencies and Courts

. . . In [Alan] Morrison's words, "the Act, as construed by the courts and implemented by the agencies, has produced workable solutions" and institutionalized due process in the administrative setting.

In Mr. Morrison's administrative law, courts "construe" and agencies "implement," those wonderful code words of the common law that disguise lawmaking by judges and bureaucrats. Furthermore, Mr. Morrison's world contains Republicans but no Democrats. The Republicans are among the several outsiders trying to disrupt the orderly progress of the insiders (namely, we lawyers in our capacity as litigators, agency counsel, and judges) toward an ideal due process.

Mr. Morrison displays the perfect conflation of good lawyering and New

Deal politics that is one of the strongest features of the New Deal consensus. The consensus rests on the following balanced elements: big government, high levels of regulation, welfare state minimums, collective bargaining, and private property rather than government ownership. . . . The regulatory system must be staffed on both the government side and on the private side by lawyers (and lawyer-judges) who so firmly believe in the New Deal ideology that they take it as an integral part of law itself. If one surveys Washington lawyers and judges, no matter whom their clients or employers, they are overwhelmingly liberal Democrats, and they are even more overwhelmingly the products of law schools that for the past fifty years have deliberately infused New Deal ideology into what they teach their students is "the Law." . . . Then and now, New Deal lawyering is the central feature of government regulation, so from his point of view, all is right with the world.

Professor Sunstein's contribution to this symposium discusses the enormous changes in administrative law that have occurred since the APA's passage, and clearly few people understand those changes better than Mr. Morrison. Mr. Morrison, however, interprets these changes not as a reversal of the APA compromises, or as a fundamental rejection of the structure of the APA, but instead simply as an orderly development of its constitutional qualities. . . . All of these changes were merely incremental adjustments designed to facilitate higher levels of New Deal-style regulation. Thus, Mr. Morrison can gently fold them into his constitutional APA. . . .

At first glance, Mr. Morrison's analysis seems strange because it endorses an activist role for courts, whereas the New Deal ideology originally required courts to defer to administrative expertise. To understand Morrison's strategy, one must return to the New Deal's political evolution. Enthusiasm for the strong-presidency theory waned among New Dealers once it became clear that Eisenhower was not simply a postwar aberration and that Republicans might often gain the presidency. The presidency became the imperial presidency to be feared rather than the strong presidency to be loved, once it became the Republican presidency. The president, initially praised by New Dealers as chief legislator, and his Office of Management and Budget, a much-lauded tool designed to make the president truly the chief administrator, are for Mr. Morrison "outside interference" seeking to pressure agencies to produce different results from what would otherwise have occurred.

The presidency was the first thing the New Dealers realized they could lose. Next came Congress. Today the House is overwhelmingly Democratic and the Senate is split about even. But since the heydays of the sixties and seventies, Congress has sporadically taken on an antiregulation, anti–big government mood. It can no longer be trusted completely. The New Dealers of the sixties and seventies, facing an overwhelmingly liberal Congress and a partially captured and lethargic old bureaucracy, trusted Congress to enact detailed agency-forcing statutes. Today, facing the new agencies staffed with their friends and with a sporadically conservative Congress, the same New Dealers want to preserve the spirit of the Congress of the sixties and seventies and get the new

Congress out of the way. They love the sixties and seventies' statutes, so long as their friends "construe" and "implement" them, but they do not want the new Congress to impose its own "construction" and "implementation."

This hostility to congressional control explains Mr. Morrison's active role in and self-congratulation about destroying the legislative veto. It is not good form in academic writing to use the word "stupid." Nevertheless, stupid is the right word for the Reagan administration's opposition to the legislative veto and for various conservative academic commentators' joy at its demise. . . . The offspring of the New Deal, such as Mr. Morrison, correctly understood the issues and knew that destruction of the veto would protect the federal bureaucracy against Congress, thus bolstering the independence of the bureaucratic branch. That, of course, is why Mr. Morrison rightly couples an attack on the legislative veto with an attack on the Office of Management and Budget. He is trying to protect the bureaucracy from both congressional and presidential control.

New Dealers were always fond of bureaucracy, but they were fond of it initially because it was an army of experts commanded by the strong president. Today they are fond of it for quite a different reason—because it is the last bastion of New Deal big government, big regulation ideology in Washington. New Deal lawyers are very busy assembling the constitutional and legal theory of the independent fourth branch of government. . . .

. . . If the new New Dealers are working so hard to shield the bureaucracy from both president and Congress and to create an independent fourth branch of government, why not be equally zealous in shielding the bureaucracy from the third branch, the courts? Originally, the New Deal ideology advocated a united executive branch, president and bureaucracy bonded together with the president in the driver's seat. Having lost the presidency, however, modern New Dealers are attempting nothing less than a fundamental reconstruction of the branches. The bureaucracy is to be torn loose from the presidency and bound to the courts to create a new branch—the due process, rule of law, construing and implementing branch. The new New Dealers do not view judicial activism with the alarm of the old New Dealers because the new New Dealers are trying to cement this agency-court partnership. For the new New Dealers, increased judicial review does not raise the specter of third branch incursion on the strong presidency but of increased activity by one of the partners in their new and favored due process branch.

. . . The bureaucracy was central to the eighties' construction of the sixties' and seventies' statutes because of its rulemaking authority. And glory be, it was in the new bureaucracies of the agencies, created and staffed in the sixties and seventies, that the last great proregulatory fervor in official Washington survived. If the regulatory bureaucracies could establish themselves as an independent branch of government, then they could preserve the sixties' and seventies' style of regulation against the inattentiveness of Congress and against the deregulators who infested the presidency and certain congressional committees.

Law, particularly complex, procedural, lawyer's law, is a wonderful weapon if one is losing in the less esoteric realms of politics. Professor Gellhorn vividly describes the conservatives resorting to this weapon in the thirties, and now the regulators have used it in the seventies and eighties. They have done so in two ways, one broadly institutional, the other more narrowly doctrinal.

On the institutional front, . . . courts have greatly reduced the scope of the discretion category. More importantly, they changed rulemaking from quasi-legislative to quasi-judicial. They invented a host of procedural requirements that turned rulemaking into a multiparty paper trial. They also imposed a rulemaking record requirement that allowed courts to review minutely every aspect of that trial. They invented a "dialogue" requirement and a "hard look" requirement that turned the agency from a legislative rulemaker into a party at its own proceedings. They converted the "arbitrary and capricious" test specified by the APA as the standard of judicial review of rulemaking from a lunacy test into the "clear error" standard that empowers a court to quash a rule not only when it is crazy but also when the judges simply believe it is wrong. The courts did all these things to reduce the independence and discretionary scope of a mistrusted bureaucracy and to subordinate it to more control by the regulated, the beneficiaries of regulation, and the public at large. Mr. Morrison applauds all of this.

. . . But how did it come about that a legal movement originally motivated by distrust of agencies ultimately resulted in confidence in the agency and a distrust of Congress? . . .

With the benefit of hindsight, the answer is clear. Deep-seated in American legal ideology is an identification of legislatures and legislating with the arbitrary, and courts and judging with the reasoned, the principled, and (although this is ultimately self-contradictory) the pragmatically correct. Put another way, we openly acknowledge that legislators do and ought to have broad discretion to enact into law any one of a wide range of policy alternatives. We do not, however, acknowledge similar discretion for judges. . . .

By making the agencies courtlike and by tying themselves closely to the agencies in a "hard look" partnership, the courts clothe the agencies in their own independence mythology and create a bureaucratic-judicial entity whose "constructions" and "implementations" are "institutionalized due process in the administrative setting" and thus beyond reproach. The administrative state has been transmuted from the ravening monster of Pound and the ABA, through the servant of peerless New Deal presidents, and on to something far better and nobler than politics.

In addition to the institutional changes discussed above, the regulators of the seventies and eighties used procedural law to effect a doctrinal change, a point that is relevant to Professor Sunstein's contribution. This doctrinal change involved the concept of "statutory duty." . . . Although the Constitution gives the president the general duty of enforcing all the laws, congressional statutes give particular agencies the particular duty of enforcing particular laws. So to whom are the agencies answerable in implementing the laws—the president or Congress? Of course, no clear answer exists under our constitutional system of checks and balances. The emphasis merely falls one way or the other at various times.

In the heyday of the New Deal, the emphasis fell toward the president, and because Congress wrote general statutes delegating power to the agencies so broadly, few direct clashes could occur between the desires of the chief executive and the explicit commands of Congress. In contrast, the statutes of the sixties and seventies are more explicit and clash frequently with the desires of a deregulation-minded chief executive of the eighties. Thus, emphasis on statutory duty rather than duty to the chief executive bolsters the agencies' quasi-judicial independence against presidential intervention. . . . Today, with the new regulatory enthusiasts embedded in the bureaucracies of the new agencies, I expect a return to enthusiasm for agency expertise. Courts will use the statutory duty doctrine to ward off deregulatory moves by the president and his appointees and to force the agencies to make rules. The rules must be synoptic and so must be made by the only people who can meet synoptic demands for information gathering and processing: the proregulation agency experts. Confronted by enormous synoptic rulemaking records packed with technical data and analysis they cannot understand, both New Deal and Reaganite judges will find deferring to administrative expertise inevitable and so will celebrate it again as a judicial virtue. The judicial shackle will be broken, except for court orders to the agencies to do their statutory duty. The new regulatory bureaucracies of health, safety, and environment then will be very much the senior partner in the new due process, judicial, independent, apolitical, correct, and noble agency-court branch. Just as John Marshall's branch preserved federalism in the dark night of Jeffersonianism, so the new branch will preserve regulation in the dark night of deregulation—the brief faltering of enthusiasm for big government that the inheritors of the New Deal hope, and not without good reason, will soon pass away.

Professor Sunstein's "Deliberative" Review as Economic Substantive Due Process Review

. . . A peculiar utopian flavor . . . marks the school of administrative law commentary to which Professor Sunstein belongs and which is exhibited in the writings of his sometime collaborator, Richard Stewart, who Sunstein cites liberally throughout his own work. Sunstein and Stewart want neither low regulation nor high regulation; they want the *right* level of regulation. Neither of them are concerned with maximizing the outcomes for the beneficiaries of regulatory statutes. Both want regulation that maximizes the public interest, which they do not equate with the beneficiaries' interests. . . .

The tendency of judges to read their own policy preferences into statutes does not discourage Sunstein from proposing activist statutory construction. Nor does he wish to reduce interest group access to agencies simply because, as he notes, maximizing access sometimes facilitates capture. Nor is he suspicious of the people imposing duties on themselves simply because paternalistic governments always impose their paternalisms in the name of the people. Nor does he dismiss OMB regulatory analysis and presidential control simply because the Reagan administration may do them badly. The wise bureaucrat and the wise judge can make the necessary distinctions.

The centerpiece of Sunstein's idealism is his notion of the "deliberation" that should be central to rulemaking. Although if pressed he might come up with some differences in the way agencies and reviewing courts deliberate, he pretty much envisions both of them as carrying on the same deliberative activity. . . . I argued twenty years ago that because courts simply make the same decision that the agency has already made and in the same way, courts need some special excuse for doing so. Sunstein's excuse is that courts are politically more insulated and less subject to capture, so they are good overseers of whether agencies have truly deliberated. I do not find that position very realistic. It returns to an idealistic belief in very good judges who do not confuse their own policy preferences with those of Congress or with the self-evident good.

A second interesting feature of Sunstein's "deliberation" is that it purports to incorporate political factors into administrative and judicial rulemaking. Sunstein rejects the Progressive tradition of administration as wholly neutral, objective expertise. At the same time he decries the contemporary tendency to describe administration as nothing but politics. . . .

Recent Supreme Court cases such as "*Benzene*" [*Industrial Union Dept. AFL-CIO v. American Petroleum Institute*], "*Cotton Dust*" [*American Textile Manufacturers Institute v. Donovan*], and "*State Farm*" [*Motor Vehicle Manufacturers Association v. State Farm Mutual Automobile Insurance Co.*], and even much of the Chicago-style, laissez-faire commentary admit that when courts evaluate the reasonableness question, they must acknowledge that Congress is legitimately entitled to favor some interests over others and that courts should defer to congressional choices. . . . I think Professor Sunstein is saying that courts ought to bar absolutely any rule exhibiting such a malfunction, even when the congressional statute commands it.

. . . This is not the old laissez-faire reasonableness in which entrepreneurial freedom is the rule and regulation is the exception to be justified by good reasons. It is a more demanding standard of reasonableness: the right rule to achieve the right level of a congressionally mandated public interest. It requires a grant of freedom to the entrepreneur to be justified just as much as an infringement on that freedom. Deliberation is full-scale, complete, objective regulatory analysis conducted first by an agency and then by a reviewing court. In the jargon of decision theory, it is synoptic analysis: it requires an explicitly ordered list of all relevant values, a knowledge of all facts, a canvassing of all alternative policies and their possible consequences, and a choice of the alternative that maximizes the values at the least cost. Only if a correct reading of the statute and a full-scale regulatory analysis lead to the conclusion that a number of alternative rules would be *equally* legitimate may the courts grant the agency some discretion to choose among these alternatives. That discretion may be informed by political considerations, such as interest group desires and presidential policies.

Deliberation allows no exception to the rule that no interest group may be favored at the expense of others without a public interest rationale for doing so. Or perhaps Professor Sunstein would prefer to say that there is a special public interest rationale for some groups. He writes, "regulation may protect interests that are now considered entitlements—with the class of entitlements being defined by reference to values apart from positive law. The best example is protection against discrimination on the basis of race and sex." In short, rights are trumps. . . .

The deliberative process's favoritism for those endowed with extra-statutory entitlements or rights is one of three places where it is particularly clear that Professor Sunstein's deliberation comes down to helping the good guys. He argues that redistribution of wealth is a legitimate goal of regulation. If, however, redistribution "benefit[s] particular classes for no reason other than political power," it is not legitimate. And, as noted earlier, the public may bind itself through laws that prohibit the satisfaction of individual consumption choices, but for the government to do so is paternalistic and highly controversial. . . . The fact is ignored that the structure of the APA, its sketchy language, the political and administrative ideologies of the times, the compromise nature of the statute, the compromisers' interests, and the contemporaneous statutory interpretation support nothing like today's shift from the quasi-legislative to the quasi-judicial model of rules, reduction in administrative discretion, and judicially enforced demands for synopticism. . . .

. . . In my view the structure, language, New Deal ideology, contemporaneous interpretation, and political compromise background of the APA point in quite a different direction. The basic and in my view insurmountable objection to the argument that the APA requires a very aggressive judicial review of rulemaking, however, rests on the ground that the essential arrangements for such review did not come into existence until at least twenty-five years after its enactment.

. . . The reviewing courts of the sixties were not prepared to hold trials. They could not conduct de novo-style independent fact determination unless they had an agency rulemaking fact record to review. So they invented the rulemaking record requirement. The APA of 1946 could not function as a legitimator of quasi-de novo review until this invention of the sixties, an invention that was clearly seen at the time as an innovation and was recognized as such by Congress when it began to use special language requiring such a record in some regulatory statutes.

. . . It is a complete anachronism, however, to read the 1946 words "arbitrary and capricious" as requiring a level of review that did not and could not have existed until the judicial innovations of the sixties. To do so completely reverses the traditional understanding of the APA compromise by insisting that agency rulemaking is subject to more judicial review than agency adjudication. . . . Were the courts right in moving to deliberative review? Should the

APA be amended to approve this judicial lawmaking? . . . Professor Sunstein's deliberative review is economic substantive due process reviews. . . .

Professor Sunstein is an idealist. He believes that two deliberative heads, an agency and a court, are better than one. He does not see either of the dark sides to deliberation that I see. The first has already been noted—that the demand for synoptic deliberation encourages agencies to disguise exercises of discretion as exercises of objective synopticism. The second is the obvious one that judges doing economic substantive due process review historically have tended to confuse their own policy preferences with deliberative truth. Indeed, to the degree that judges are encouraged to deliberate, presidents are surely going to be inclined to appoint right-thinking deliberators, particularly to the D.C. Circuit. Thus, preaching deliberation as a judicial role is likely to be self-defeating because it undermines the apparent political neutrality of judges, which is the rationale Professor Sunstein offers for giving reviewing courts a second round of deliberation after the agencies have conducted a first round.

I emphasize, however, that no matter how little the proponents of judicial restraint may be convinced by Professor Sunstein's reading of the APA or by his proposals for a deliberative style of judicial review, much of Sunstein's program is now being carried out in the courts and applauded by many influential commentators. Courts reviewing rulemaking have forced agencies at least to pretend they are deliberative in Sunstein's sense, and the courts themselves are trying to deliberate in Sunstein's sense. Economic substantive due process as substantive review of the reasonableness (read correctness) of rules is alive and well in the D.C. Circuit's clear error/hard look, synopticism-demanding version of "arbitrary and capricious," a version the Supreme Court ringingly endorsed in the *State Farm* decision.

State Law in the Teaching of Administrative Law: A Critical Analysis of the Status Quo
ARTHUR BONFIELD

There are Significant Differences Between State and Federal Administrative Law

Agency Discretion to Make Law by Rule or Order

Some states require agencies to elaborate the major contours of their policy by rule and preclude them from doing so wholly by ad hoc adjudication. . . .

Published originally in 61 *Texas Law Review* 95–137 (1982). Copyright 1982 by the Texas Law Review Association. Reprinted by permission.

The 1981 Model State APA [Administrative Procedure Act] partly codifies this state innovation. It provides that "as soon as feasible, and to the extent practicable, [each agency must] adopt rules, in addition to those otherwise required . . . embodying appropriate standards, principles, and procedural safeguards that the agency will apply to the law it administers." . . .

These state administrative law materials are different in their approach and their conclusion from most federal materials on this subject. . . . That difference may, in part, result from the fact that while both state and federal agency rules are published and therefore are generally available, the same is not true with respect to agency case law. . . .

Rulemaking Procedures

A second area of divergence between state and federal administrative law concerns the scope and coverage of agency rulemaking procedures. Several state statutes, unlike the federal APA, exclude certain narrowly defined categories of rules from usual rulemaking procedures. On the other hand, other broad categories of rules which state statutes subject to usual rulemaking procedures are completely exempt from such requirements by the federal APA. . . .

Executive Review of Agency Rulemaking

The differences between the law of some states and the federal government on the subject of executive review of agency rulemaking is an especially good illustration of how an examination of state law can enrich the discussion of administrative law issues. . . .

The 1981 Model State APA also vests governors with authority to review the rules of their state's agencies, and to disapprove any of them for whatever reasons they deem proper. But it goes further than the schemes previously described because it also empowers governors to "rescind or suspend all or a severable portion of a rule of an agency" at any time. Consequently, a governor may not only veto a rule at the time it is first adopted, but may also do so years later. Under the 1981 Model Act proposal, governors may also summarily terminate any ongoing rulemaking proceeding by issuing an executive order that states the reasons for such action. . . .

The system for executive review of federal agency rules differs in many significant ways from those of the states. No federal statute explicitly vests in the president, or in any other federal executive official, general review authority over agency rules. . . .

Legislative Review of Agency Rulemaking

. . . The discussion above establishes that in many states, legislative rules review mechanisms that allow the suspension or permanent invalidation of particular agency rules by nonstatutory means, or that otherwise allow the legislature to affect the validity of an agency rule by nonstatutory means, are well-accepted, functioning parts of state government. Both the structure and actual operation of these general state legislative rules review mechanisms with "bite" are worthy of serious consideration by students of administrative law because no equivalent exists on the federal level. While in recent years Congress has discussed the desirability of creating some such formal mechanism, existing federal materials do not illuminate all of the possible approaches of their problems nearly as well as the available state materials. . . .

Classes of Agency Adjudication

Another important difference between some recent state administrative law and federal administrative law deals with adjudication. The federal APA expressly regulates the conduct of only one class of adjudication: "formal adjudication." That class, defined by Section 554(a), consists only of those adjudications "required by statute to be determined on the record after the opportunity for an agency hearing. . . ." . . . But if a case does not fit within the Section 554(a) definition, none of the adjudication provisions expressly governs the proceeding; consequently, as long as the agency does not abuse its discretion, it is free to use whatever procedure it chooses in such cases.

. . . The adjudication provisions of the federal act are overbroad . . . and are also underinclusive. . . .

A number of recent state statutes and the 1981 Model State APA take an entirely different approach in their efforts to remedy this problem. They create several distinct classes of agency adjudication, each subject to procedural requirements specially tailored to the needs and circumstances of that particular class. . . .

The Circumstances in Which State and Federal Administrative Processes Operate Are Different

. . . State administrative processes operate under different circumstances than does the federal administrative process; consequently, some of the problems presented in the various state administrative processes differ either in degree or kind from those presented in the federal process. . . .

Compared to the federal process, state administrative processes are responsible to a far smaller constituency and operate through agencies that are substantially smaller in size. As a result, the state processes tend to be more manageable by, in closer contact with, and more visible and accessible to, those they govern than the federal process. The practical politics and interest group diversity surrounding the operation of the various state processes may also be different in some respects from that surrounding the federal administrative process. While federal agencies are usually staffed by full-time officials dedicated only to executing the functions of those agencies, many state agencies are run by part-timers whose primary allegiance lies elsewhere. Further, most federal agencies are not structured to ensure direct and formal representation of interest groups affected by their actions. State agencies, on the other hand, are often structured to accomplish that result. Boards charged with occupational licensing and regulation are a recurrent example. . . .

. . . Because they are smaller, more poorly financed, and less technically competent, state agency staffs are often characterized by somewhat less professionalism than the staffs of most federal agencies. In addition, the staffs of state agencies tend to be more closely associated with partisan politics and the spoils system than the operational staffs of the federal bureaucracy.

It also may be true that, on the whole, the state administrative processes collectively spend more of their time dealing directly with less affluent, less influential, and less well-educated people than does the federal process. . . . In addition, persons dealing with the state administrative processes are less likely to be represented by a lawyer than are persons dealing with a federal agency. Moreover, far more often than in the federal process, the state agency itself will not be represented by a lawyer until the matter goes to court. Lastly, the matters in controversy dealt with by state agencies are, on the average, probably of smaller economic value than the matters handled by such federal bodies. . . .

State and Federal Administrative Processes Operate Within Different Legal Frameworks

. . . Consider, for example, the following characteristics of state administrative process: there are several different levels of government in the states; many agencies are expressly created by state constitutions; and agency heads often are directly elected by the people. In no case can the same be said of the federal process.

NOTES AND QUESTIONS

1. In the same symposium issue in which Sunstein's and Shapiro's articles appear, William H. Allen, a prominent administrative lawyer in Washington, D.C., takes note of the important fact that the APA was amended substantively only thrice between 1946 and 1986. The Freedom of Information Act and its companions, the Privacy and Government in the Sunshine Acts, were added in the 1960s and 1970s; a prohibition on ex parte communications in formal evidentiary proceedings was added in 1976; and the sovereign immunity defense in nondamage suits against the federal government was eliminated that same year. Allen does not mention in this connection two 1980 enactments—the Paperwork Reduction Act and the Regulatory Flexibility Act— presumably because they are not codified as part of the APA, although both affect agency proceedings and rulemaking.

Allen attributes the APA's durability to four major factors. Regulated industries, "public interest" and labor groups, the agencies, and Congress could not resolve differences over possible reforms. As administrative law's focus changed from formal evidentiary proceedings to rulemaking and informal adjudication, the courts (especially the lower courts) interpreted the APA innovatively to accommodate these changes. Congress enacted many agency-specific procedural schemes to which the APA was irrelevant. Finally, a series of executive orders beginning in 1974 went beyond the requirements of Section 553 to impose government-wide, extrastatutory restrictions on agency rulemaking. Allen, "The Durability of the Administrative Procedure Act," 72 *Virginia Law Review* 235 (1986). In 1990, Congress amended the APA for a fourth time to provide for negotiated rulemaking, a subject discussed in Chapter VI, section 6. Negotiated Rulemaking Act of 1990, Public Law 101-648, 104 Statutes at Large 4969.

2. As administrative law and policy become more specialized and technical, does the case for a unified, trans-substantive procedural code like the APA become stronger or weaker? To put the question somewhat differently, should Congress's tendency to adopt agency-specific procedures outside the APA framework be applauded or resisted? Professor Elliott's article in Chapter VIII addresses this issue.

3. As has often been noted, the APA does not prescribe procedures for informal adjudication. Yet the vast majority of decisions that the government makes directly affecting individuals and firms fall into this category, and the adequacy of the procedures (or lack thereof) that agencies employ in making these decisions is often called into question. What are the arguments for and against including this category? If you think the arguments for inclusion are strong, how would you draft such a provision?

4. Can Sunstein's deliberative model be fairly inferred from the text and legislative history of the APA, or would it require amendment? Is Shapiro correct that changes that only occurred during the 1960s and 1970s cannot be said to have been intended by the enacting Congress? How persuasive are Sunstein's efforts to meet this objection? On the merits of Sunstein's proposal, how desirable would a deliberative model be as a policymaking system? Is Shapiro's characterization of Sunstein's model as "economic substantive due process review" (72 *Virginia Law Review* 478) accurate?

5. Shapiro contends that an alliance of activist courts, liberal "public interest" groups, and agencies seeking freedom from political controls have used exotic interpretations of the APA to hijack the policymaking process away from Congress and the president, creating what is essentially a bureaucratic "fourth branch" independent of elected officials. Does the frequency with which public interest groups challenge agency action in Congress and in the courts tend to refute Shapiro's description of the alliance? Does Congress's failure to amend the APA to overturn or modify the judicial decisions bespeak a kind of ratification of those decisions? Or might Shapiro respond that those decisions, quite apart from their merits, enabled the alliance to shift the burden of legislative inertia, always a crucial political resource, in ways that favored the alliance's interests?

In some but not all respects, Shapiro's general account is confirmed by case studies of the effects of judicial review on agencies. A useful review of these studies is presented in the Melnick article in Chapter VI, section 2. In contrast, however, consider the following summary of his analysis of this question by Robert A. Katzmann in *Institutional Disability: The Saga of Transportation Policy for the Disabled* (Washington, D.C.: Brookings Institution, 1986):

This is not a story of judicial imperialism. Courts were the instruments used by all the parties—sometimes bolstering one side, then the other. The judiciary did not seek to impose procedural requirements on agencies. It did not substitute its views for the technical judgments of the agencies, but generally deferred to the administrators. Judges did not involve themselves in the details of public administration, as they have in some other areas of social policy—for example, civil rights, education, environmental regulation, and penal reform. Moreover, the record of decisions presented here belies the commonly asserted view that the politics of rights is synonymous with judicial activism. On balance, courts were less attached to a rights-based approach to policymaking than was either Congress or the bureaucracy (p. 187).

6. For a variety of reasons—conservative administrations in Washington, reformed state governments, complex policy implementation problems, a shrinking federal civil service, and growing political fragmentation—the 1980s and 1990s have seen the devolution of much administrative power to state and local governments. If this trend continues (and even if it does not), the role of state administrative law, discussed in the Bonfield article, will be important. Should federal statutes prescribing joint federal and state administration, such as the Clean Air Act, require uniform—or at least minimal—administrative procedures? How apt is the analogy to the Occupational Safety and Health Act and other federal laws that impose federal regulation unless the corresponding state program is as effective as the federal one? The differences between state and federal administrative law are treated in a recent casebook: Arthur E. Bonfield and Michael R. Asimow, *State and Federal Administrative Law* (St. Paul, Minn.: West Publishers, 1989).

IV

The Determinants of Agency Behavior

As the readings in the earlier chapters indicate, various thinkers at various times have viewed agencies as essential instruments of government and social reform, and as major sources of misgovernment and social pathology. Indeed, many commentators (including your editor) hold both of these views simultaneously. The premise of this chapter is that these radically different appraisals of agencies are perfectly consistent with one another and with the evidence on how agencies behave. The condition that justifies this otherwise paradoxical premise is the extraordinary complexity of agency behavior. What forces shape it? This chapter addresses that question.

Each of the authors represented here emphasizes a somewhat different factor, although all of them surely agree that other factors also matter. Political scientist James Q. Wilson, reflecting on thirty years of deep immersion in the subject, expresses "grave doubts that anything worth calling 'organization theory' will ever exist." Jerry Mashaw tends to confirm this theory-skepticism with respect to administrative process. He elaborates three explanatory paradigms—two of them idealist (public interest and critical theory) and the third realist (rational choice or interest-group theory)—thus underscoring the field's daunting richness of interpretive possibilities. Mashaw's critique of the rational choice model should be kept in mind when the reader turns to the article by Terry Moe, another political scientist. Although this model is often used to explain an agency's policy choices, Moe uses it to explain why the agency's *formal,* statutorily prescribed structure is what it is, a more interesting and encompassing application of the model. (An article by Thomas McGarity, not excerpted here but summarized in note 3 at the end of this chapter, analyzes the variety of *informal* structures for policy development that an agency can create for itself within the formal constraints imposed by statute.)

The subsequent selections offer theories that emphasize causal factors other than interest-group politics. (Some of these, however, are arguably consistent with a rational choice account). Wilson, for example, stresses two related factors that significantly affect agency behavior: the agency's *organizational culture* (its patterned way of thinking) and its *task structure* (how it monitors and measures its employees' performance). Charles Lindblom focuses on how the need to economize on scarce *information* and political resources causes agencies and other political actors to pursue incremental strategies of policy analysis and implementation. (Note that this is quite different than the rational choice theorists' point that information asymmetries between the legislature and agencies account for the structure of administrative law).

Explaining Administrative Process: Normative, Positive, and Critical Stories of Legal Development

JERRY MASHAW

What explains the shape of the administrative decision processes that we observe? . . .

. . . On [one] view, call it the "idealist's vision," administrative processes are part of the general fabric of American public and constitutional law. The law of administrative procedure contributes, as does all such law, to the construction of an operationally effective and symbolically appropriate normative regime. To put the matter slightly differently, administrative procedural requirements embedded in law shape administrative decisionmaking in accordance with our fundamental (but perhaps malleable) images of the legitimacy of state action. That is administrative procedure's purpose and its explanation.

Not all administrative lawyers subscribe to this idealist vision of a general (and largely judge-made) law of administrative procedure that is based on normative premises having a constitutional or quasi-constitutional status. . . .

The explanatory notion implicit in the "legal dissenters" literature, however, seems to be one of decision processes developed to meet the "needs" of particular substantive regimes. Thus, in one plausible interpretation of its behavioral assumptions, the dissenting, like the mainstream, view presumes that process forms emerge as part of a normative program. The dissenters, too, are functionalist, procedural idealists who see process as constitutive of and supplementary to the elaboration of a normatively appropriate legal order. For them that legal order is merely the more particularistic one found in specific statutory regimes, rather than in some overarching normative structure of American public law.

Realist Critiques

. . . In the current intellectual climate, critical legal studies (CLS) and positive political theory (or "political economy" or "social choice theory") are the principal contenders for the realist throne. And so it has come to pass that Gerry Frug, in an elegant CLS critique, has explained how the normative structure of administrative (and corporate) law serves to maintain an ideology of bureaucracy that both legitimates and masks coercion. Administrative procedure assures us of the objectivity of administration even as it subjects us to the discretionary dominion of administrators. As such, the law's stories of bureaucratic legitimacy are preeminently "mechanism[s] of deception."

Positive political theorists, in particular, Mathew McCubbins, Roger Noll and Barry Weingast, have weighed in with a similarly bleak view. The argue that administrative processes can be understood as the means by which political victors maintain the gains from successful interest-group struggle at the legislative level. Administrative decision structures are the devices through which legislative principals control the actions of potentially deviant administrative agents. Legislatures can thereby deliver on the electoral deals that maintain them in office. In both the CLS and positive political theory (PPT) accounts, the normative rhetoric of the law, the crucial data analyzed in both the legal literature and judicial opinions, is largely epiphenomenal—a product or constituent of more fundamental underlying material processes that are obscured, if not misrepresented, by lawyers' talk. . . .

Toward a Comparative Test

To a degree, idealist and realist explanations of administrative procedure are noncomparable perspectives of the same phenomena. Lawyers and legal scholars provide an internal interpretation of process purposes and goals articulated in the normative vernacular of American political and constitutional ideology. Realists are "external" or "critical" observers looking past the law's internally prescribed rhetoric to explain process phenomena in terms of material interests and political power. Idealists, by ignoring issues of behavioral motivation, seem to view expressed intentions as relatively nonproblematic guides to what is being done and why. Realists tend to combine rhetoric, objective interests, and concrete behaviors to construct both their whats and their whys, either through the development and testing of behavioral hypotheses (PPT) or through the reflexive or dialectical examination of rhetoric and practice (CLS). These two perspectives not only look at different data— idealists at expressed purposes; realist at implicit (sometimes hidden) interests or ideology, but they have dramatically different methodologies for the interpretation of behavior, and conflicting ideas of what could count as a reason for action.

Given this radical disjunction between the two approaches (to say nothing of the intellectual gulf that divides CLS and PPT realists), mediation of their rival claims may be impossible. If methodology determines both evidence and its interpretation, then to ask who has the best explanation of American administrative processes may simply pose an issue of taste. Each explanation may be adequaate within the terms of its own methodology, but hopelessly inadequate, or indeed obviously false, from the perspective of the alternative visions. The "best" explanation then becomes the one that best fits the analyst's own preferences concerning the style of explanatory stories.

But perhaps methodological bias might also provide a way past methodological relativism. If what we see depends upon what we look for and how we look for it, then the choice of evidentiary base and interpretive stance should privilege the theory whose methodology is chosen. Precisely because of this privileging, there emerges a possible test of the power of administrative process (and perhaps other sorts of legal) storytelling. For, if we privilege one style by adopting its methods, and it still fails to beat its competitors, then we should at least be very skeptical of its adoption as the primary account of the phenomena to be explained. Can such a comparative test be constructed? . . .

I choose to privilege the positive political theory's behavioral approach for several reasons. First, a positive model has already been developed from this perspective by others, and that makes the task of specifying procedural implications much easier. Second, that methodology is familiar and well-adapted, at least in its own domain, to adjudication among competing explanatory hypotheses. Third, legal-normative and CLS approaches do have empirical presuppositions that should, in principle, be translatable into positive theory terms. Fourth, I believe that the PPT model, at least as currently constructed, needs refinement. Choosing to pursue the analysis from the perspective of positive political theory, thus, at least holds out the prospect of contributing to the further elaboration of that perspective. And, finally, the attempt to force other theories into the "if–then" mode of positive theory may illuminate and sharpen our understanding of them, even if we admit that to "translate" is in some sense to falsify.

Models of Administrative Procedure

Idealist Visions of Legislation, Administration, and Administrative Law

Let me oversimplify. During the crucial middle decades of this century (say, the 1930s through the early 1970s), the vision of legislation and administration that seems to have dominated legal consciousness was a vision of government as a well-ordered input/output machine. Into the machine went social problems and political values; out of the machine came legislative programs that would make social reality conform to social ideals. The new activist state was purposeful and pragmatic. Collective purposes—visions of the public interest— were shaped by the macropolitical processes of electoral and legislative debates. Concrete realizations of those purposes were achieved through the practical application of the expertise possessed by administrators to whom various policy domains were delegated. . . .

Yet administrative law was not all flexible accommodation to the administrative enthusiasms of the moment. Judicial review was designed to keep administrators within their jurisdictions and harnessed to the values and purposes expressed in the macropolitical processes of legislation and electoral accountability. Moreover, procedural protections for both individuals and groups, backed again by judicial enforcement, reinforced the image of citizens as rights holders in the new administrative state, and supported participation in the micropolitical processes of administration.

Mediation of the tension between the ideals of governmental efficacy and of self-, or participatory, governance defined the central issues for administrative law and administrative lawyers. . . .

While these developments can be described in a terminology reminiscent of liberal legality—as attempts to limit administrative discretion by rendering administrators accountable to the substantive demands of statutory regimes and to the procedural requirements of the Constitution and APA [Administrative Procedure Act]—the particular form of control employed (call it "proceduralism") has other normative bases. The proceduralist project in administrative law is the micropolitical analogue of Ely's currently popular vision of the function of constitutional adjudication. In this view, judicial review is justified by the necessity to maintain reasonable access for all groups to a political process whose democratic character consists precisely in its responsiveness or potential responsiveness to the wishes of those groups. Similarly, the normative, pluralist vision of administrative processes sees both process design and judicial policing of the implementation of those designs primarily as devices for providing policy access to relevant political forces or interests. Administrative legality entails policymaking that accommodates these interests procedurally by giving them serious consideration in the development of substantive norms.

Given the conventional "new wine in old bottles" techniques of legal development, it is perhaps not surprising that the pluralist and liberal projects can be carried on using essentially the same legal concepts. Demands for process transparency and decision rationality can be used both to protect rights and limit government (liberalism) and to assure access and an appropriate accommodation of interests (pluralism). This may be a happy feature of administrative law: new arrangements can be accommodated without the stress of developing a new politicolegal vocabulary. However, from the vantage point of our current project—the description of the procedural implications of a normative account of administrative law—the capaciousness of the legal categories is something of a problem. It suggests that legal structures and processes may be given multiple normative interpretations. Their purposes, therefore, remain ambiguous.

These interpretive uncertainties suggest that there may be major difficulties in developing a set of determinate and specific process characteristics that would implement the major visions of constitutionalism and administrative legality to which legal idealists have traditionally subscribed. It may be possible to interpret every instance of administrative structure or process as furthering the goals of liberal legality or normative pluralism, or of both simultaneously. This is particularly true if we allow procedures also to be compromises between, or syntheses of, these often simultaneously held visions. . . .

The liberal-pluralist compromise that has shaped the legal idealist vision of American administrative procedure thus seems to require courts capable of structuring and controlling a process of legislative and agency procedural innovation that is directed at interest accommodation, but subject to conditions of widespread procedural rights-holding. The idealist conception of administrative law presumes, indeed features, control of administrative procedure by courts, rather than by agencies or legislatures. Furthermore, on this dimension, if no other, it differs from the models to which we now turn.

The Positive Political Theory of Administrative Process

A group sometimes designated by the awkward label "rational choice political economists" has, in recent years, asked us to jettison our vision of goverance as a benign input/output machine for the definition and effectuation of the public interest. It its place, one branch of this literature has described governmental action as nothing more than self-interested political bargaining in the pursuit of individual or group material interests. The black box of macropolitics and bureaucratic decisionmaking has been pried open to reveal copious opportunities for "rent-seeking" behavior both by the people's representatives and by the "experts" in charge of public programs. Moreover, if ubiquitous free-rider difficulties induce rational ignorance in voters and radically skew the structure of pluralist, interest-group activity, then the self-interested manipulation of both legislative and administrative processes can go on largely unchecked by electoral restraints.

This revised, prinicipal-agent perspective has reenergized a focus on institutional design, and has generated explanations of the activist, administrative state radically different from those embedded in Legal Idealism. Rather than effectuating the public good while maintaining liberty and democratic control, public institutional arrangements are virtually all explained as devices to facilitate private gain at public expense. On this view, broad delegations of power and limitations on executive control structure politics in the interest of "iron triangles" of self-aggrandizing representatives, bureaucrats, and interest groups. Judicial review cements the gains from the private deals for the use of public power, and agency processes are designed primarily to permit "capture" by the already advantaged. Pursuit of "the public interest" as a guide to an understanding of the structure and behavior of public institutions is thus replaced root and branch by hypotheses featuring pursuit of private material gain. By 1984 an article in the *American Economic Review* could plausibly claim that "the economic theory of regulation long ago put public interest theories of politics to rest."

To be sure, the empirical record of this branch of PPT is one that should induce the utmost caution in its practitioners. However, PPT practitioners are not all wedded to interest-group explanations. Indeed, they work with a host of models, having quite different assumptions and emerging out of differing "public choice" traditions. Their familial relationship is difficult to capture save in a core general presumption that political behavior is to be explained as the outcome of rational (and often strategic) action by relevantly situated individuals within some set of defined institutionalized boundaries. This variety, however, makes PPT a nonstarter as a basis for constructing a good explanation of administrative process. What is needed is a more detailed theory with determinate procedural implications for testing. In the field of administrative process, the most ambitious effort to date is what I will call the "McCubbins-Noll-Weingast (MNW) Hypothesis."

The MNW thesis is relatively straightforward and, in the abstract, quite plausible. Electorally accountable officials, the president and the Congress, have a difficulty. They must often put the implementation of public programs in the hands of other nonaccountable officials who may have their own designs on the programs. Monitoring is always costly, as is the application of sometimes cumbersome sanctions. Moreover, the deviation of bureaucrats from politicians' desires may be inherently uncorrectable so long as the bureaucrats' actions remain within that set of options that might have been approved ex ante by some winning coalition. How can politicians control bureaucrats? The MNW answer is "through administrative process."

MNW posit that the two major control issues facing the Congress are problems (1) of information asymmetry and (2) of the erosion of an original legislative coalition over time. We will take the latter problem first because it illustrates a serious issue of vagueness in the model that is provided. The problem with the MNW analysis, as it currently stands, is that it may be nontestable.

To solve the problem or eroding legislative coalitions, administrative procedures would need to provide an opportunity whereby the constituencies that motivated the original legislative coalition could themselves act on implementing agencies to preserve the bargain that was struck at the legislative level. MNW hypothesize, therefore, that legislators will "stack the [administrative] procedural deck" in favor of the winning coalition. So far so good, but there are some troublesome loose ends in this scenario.

For example, why would the political controllers want to preserve the original policy position in the face of the erosion of the original legislative coalition? If the legislative coalition is one that is derivative from the demands of constituencies, then a change in the preferences of the legislative coalition should signal a change in the preferences of the constituencies. Since the desire for political control in this model seems to be to service relevant constituencies, rather than to preserve some preferred policy on other grounds, it is difficult to understand why the political principals want to preserve the old coalition at the administrative level when a new winning coalition of voters or interest groups desires a different policy. This just seems a good way for legislators to lose the next election.

MNW's refinement of the idea of "erosion" in their most recent article does not solve this problem. Even if the legislature's inability to rectify administrative deviations is entirely a function of a change in the status quo point at which legislative bargaining begins, the status quo point for constituent interests has also changed. Hence, the politically effective winning coalition for constituents will also have changed. Why it is politically advantageous to thwart this coalition must still be explained. The answer that legislators would want to prevent deviation ex ante in order to increase the value of the original deal merely raises further questions. The most important is this: If so, why were the administrators given the discretion to deviate from an ex ante known preferred position?

At another point in their discussion, MNW seem to address the issue of why administrators are given sufficient discretion to deviate from what appear to be the legislators' original objectives by suggesting that political controllers want to use administrative procedure as an "auto pilot." By this they mean the administrative processes should allow dominant coalitions of constituents to work their will at the administrative level as their preferences change, without further activity on the part of legislative or presidential politicians. One can certainly imagine this being a desire of political principals. However, it would presumably be accomplished by quite different processes than those that "stack the deck." Hence, no matter what procedures we find (either procedures that enfranchise original coalitions to resist alteration of legislative policy or procedures that permit flexible adaptation of policy pursuant to the coalition's new demands), we should presumably find the theory "proved." On this reading, there is no null hypothesis here.

It might be, or course, that MNW are claiming that the same processes can be used both to stack the deck in favor of policies adopted at the behest of particular constituencies and simultaneously to provide those same constituents with the opportunity to force policy shifts as their preferences change. If so, then the theory suggests that we should be looking for processes that somehow enfranchise particular constituencies or interests independently of the relationship of procedural requirements to possible substantive outcomes. This might be evidenced, for example, by statutory provisions creating special personal access of decisional control for identifiable constituencies. But these sorts of statutes are quite rare at the federal level. Certain regulated agricultural marketing arrangements might qualify (see Block v. Community Nutrition Institute, 467 U.S. 340 [1984]). However, we should not forget that it was the specter of balkanized constitute-producer control of economic policy that energized the nondelegation doctrine in the 1930s.

If statutes fail to identify specially enfranchised constituents explicitly, then the determination of whether procedures satisfy the MNW hypothesis becomes difficult indeed. That some interests turn out to have an advantage in employing particular procedures is not very good evidence that they were the constituencies meant to be served by the legislation. Over time, one would expect some such groups to emerge with respect to *any* statutory scheme. And historical inquiry into whether those benefited were really members of the dominant coalition for passage of a particular statute is almost certain to founder on the inadequacy of the written record, if not on prior issues of how to operationalize the idea of "coalition" or what should count as evidence of "membership." Much modern legislation passes by lopsided majorities after extensive compromise that includes widely divergent viewpoints within the "winning coalition." . . .

For our "comparative testing" purposes, we need then only specify some implication of the MNW perspective that points toward different procedural implications than does an idealist vision. Here again, the obvious suggestion is the location of procedural control. While idealist legality seems to presume judicial control over process, the MNW thesis, both originally and as modified here, posits legislative control. After all, it is the heed for low-cost legislative monitoring through control over procedural or structural design that motivates the behavioral hypothesis.

The requirement of retained or operative *legislative* control over process is particularly pointed with respect to the second (and as yet undiscussed) monitoring difficulty that MNW posit as ubiquitous in the congressional-bureaucratic principal-agent game: the bureaucracy's control over information. Thus, for example, if we find the Congress structuring administrative processes in ways that require agencies to divulge politically relevant information to congressional monitors or their surrogates, then we may have some confirmation for the MNW thesis. To be sure, there is some overlap here between this information-divulging idea and the openness and transparency demands of idealists' view of administrative law. However, from the principal-agent perspective, the important element ought to be not just that information is made available, but, what kind, to whom, and at whose request. If administrative law makes most information available to others, for example, in forms not particularly useful for congressional or legislative monitoring, and through processes outside legislative control, then the MNW information thesis would not be confirmed.

It may be objected that this description oversimplifies the PPT vision. Legislators might employ courts and constituents as continuous monitors. Hence, provisions for judicial review or revelation of information to citizens through FOIA requests could satisfy legislative monitoring requirements without any active oversight by the legislature itself. This is, indeed, possible. But if monitoring or access by anyone is to count as legislative control, so long as it is in any way facilitated by legislation, then we are back to the problem of nonfalsifiability. Everything will count as legislative control. It remains to be seen, therefore, whether some more discriminating criteria can be developed to discern whether processes evidence legislative control or control primarily by others.

A Critical Approach to Administrative Procedure

The CLS description of administrative law is not substantially different from the description given by the idealist legal literature. The critical project is, after all, one that works from within the conceptual structure of legal regimes to expose the ways in which that structure fails in legitimation that it has set for itself. Moreover, the CLS approach recognizes that the task of legitimation has a relationship both to the concrete exercise of power and to the intellectual stories that describe the way in which power is structured, checked, and controlled. The normative task of administrative law is simultaneously to explain and to contruct a legal regime that will effectively control state power. Conceptualization and implementation are separated only with difficulty, and then at the cost of a certain artificiality.

Hence, when a critique of a legal domain is proposed from a perspective that takes a "critical" stance, that enterprise, too, has tightly connected empirical and conceptual dimensions. The critique is one of how the concepts put in place by an existing legal regime fail in practice to exercise the constraints or controls that they purport to establish. But critical analysis goes further to demonstrate the ways in which the conceptual structure itself contains contradictions, or unresolvable tensions, that presumably defeat its practical program. There is implicit in critical analysis a causal linkage between the inadequacy of conceptual structures and their failure to provide the forms of legal constraint or empowerment that they advertise.

This causal linkage between conceptualization and practical operation is obviously problematic. Many would assert that it is just the case that individual consciousness and social life are both deeply compromised, perhaps inherently contradictory. Yet both individuals and societies cope, even thrive. There is, thus, no necessary connection between the adequacy of the conceptualization of a political or legal ideology and its stability, legitimacy, or success. This latter view is, of course, also controversial. For present purposes, however, it is not necessary to focus on the linkage between conceptualization and action—the praxis problem—that haunts critical theory and energizes its critics. Nor need we worry too much about the reflexsiveness of critical method or its liberating objective. It is, for now, enough to recognize that there is embedded in critical analysis of legal institutions a set of empirical presuppostions about the structure of legal concepts and their effects in action.

The critical story of administrative process is, in this "translated" view, relatively straightforward. Administrative law, like corporate law and perhaps other fields, confronts a basic conflict between individual freedom and bureaucratic control. The function of the law is the maintenance of freedom within an apparently "necessary" bureaucratization of social, and particularly economic and political, life. This story is, as we have said, highly reminiscent of the conventional lawyers' and legal academics' story of administrative process. However, it takes that story to a more fundamental ideological level. To put the matter somewhat abstractly, bureaucratic domination is avoided in the legal idealists' vision of legality through legal constraints or legal structures that rely on either decisional impersonality or rational consent. These two basic approaches work themselves out in four models or paradigms of explanation, which can be developed historically or conceptually.

The crucial claim of critical legal studies for our purposes is that none of these legitimation stories work. Although they attempt to render bureaucratic power nonthreatening, and in operation to check and structure such power so that human freedom is maintained, they fail. Moreover, they all fail for the same reason: the impossibility of separating objective and subjective spheres of decisionmaking and, therefore, of confining bureaucratic discretion to objective judgments that preserve the citizens' (or employees') subjective freedom.

The plausibility of this critique can be appreciated by looking at a single example. One way of maintaining objectivity in administration is through a "formalist" explanation of the exercise of administrative power. In this view, bureaucratic implementation is a form of pure instrumental rationality. The goals of administration are prespecified in statutes. Those goals are objective in the sense that they are socially common and specified through a process of democratic choice. The task of administration then is itself liberating. It allows us to pursue our predetermined goals efficiently, while preserving social resources for the pursuit of other social ends. The coerciveness of any exercise of bureaucratic power is but an act of self-paternalism; it imposes on us decisions that we have already made—and which represent our best collective judgment of what our social purposes should be.

To state the formalist theory in this way is virtually to reveal its defects. No selection of statutory goals can perfectly represent our collective desires even if the underlying legislative process was itself an exercise in ideal democratic choice. Not only are there gaps and "unprovided for cases," the purposes stated in any one bit of legislation are likely to work at cross purposes with those articulated in others. Bureaucratic coordination and gap-filling will, at the very least, provide occasions for exercises of judgment and discretion which are in no strong sense "controlled" by statutory language.

The doctrinal articulation of this formalist approach to legitimation, the "nondelegation doctrine," itself contains contradictions which recognize the impossibility of formalist legality. After all, one of the purposes that we may choose by legislation is to delegate flexible authority to administrators. Thus, the administrators' subsequent and better understanding of the problems giving rise to collective action can produce better and more sophisticated policies than we could have constructed at the time of legislating. Hence, the notion that legislators *may not* delegate legislative authority to administrators must also contain within it the idea that they *should* delegate legislative powers to administrators. Nondelegation must permit delegation. Furthermore, in order to accommodate both ideas, the nondelegation doctrine must be conceptualized as a set of standards that are sufficiently abstract to permit the continuing dialectic of constraint and flexibility in the actual practice of administrative judgment.

This is, of course, to say that the legal "rules" that maintain formalist legality are inadequate precisely to the task of maintaining formalist legality. They contain contradictions that can be managed only through the exercise of a decisional competence that cannot be explained by, or controlled by, the Janus-faced legal rule. We noted this double aspect to administrative law in our earlier discussion of the conventional normative approach. There we described mediating the competition between efficacy and legality in the benign rhetoric of "a continuing task" for administrative law. From the critical perspective this task is viewed, not only as continuous, but as impossible, and as simultaneously masking its impossibility through the rhetoric of legal standards.

Thus, there is at least one straightforward empirical implication of the critical approach. The structure of administrative law doctrine, legal institutions, and legal processes should be one that, whatever its normative rhetoric, maintains the power of bureaucratic officials to exercise state power, constrained largely only by bureaucratic imperatives. Therefore, although critical theorists are committed to the seriousness of intellectual or conceptual dialectic, and to its imminent relation to practice, their critique seems to lead in a thoroughly "realist" direction. Like Karl Llewellyn, their analysis implies that the power of the law is in the hands of those "who have the doing in charge."

We have emerged from our excursion through normative, positive, and critical theories with three divergent implications about the structure of administrative decisionmaking. Legal idealist theory describes control over administrative process as if it were preeminently a judicial task. Positive political theory posits an administrative process constructed by the legislature as a solution to its principal-agent problems concerning bureaucrats. Critical theory suggests that because of the contradictions of formalist legality, administrative process will remain firmly within the control of the bureaucracies themselves. . . .

Assessing the Evidence

. . . What then is to count as evidence of the plausibility or persuasiveness of one or another view? At one level the answer to the question is obvious. We are looking here for evidence of who has control over administrative procedures. . . . To simplify matters, we will consider only . . . PPT, and focus primarily on the sorts of evidence that have been marshaled in support of the MNW hypothesis.

The World According to PPT. One would expect to find a Congress that is using administrative process to insure the fidelity of agencies to the congressional will to be intensely involved at all the levels of procedural design that we have mentioned. In addition, one would expect the legislature to draft the procedural elements of statutes in great detail in order to control procedural meddling by the president (or the Executive Office of the President), the courts, and agencies themselves. Procedures should be specific and hand-tailored, not general and off-the-rack.

There is, indeed, some evidence for this view of legislative procedural design. . . .

Yet these detailed and process-specific incursions into administrative process seemed dwarfed by the degree to which the Congress acts generically and leaves the crucial details of procedural implementation to agencies, courts, and perhaps the president. Indeed, from the viewpoint of administrative lawyers who work with and observe the changes in agency processes, agency strategic control of process would seem to be the rule. . . .

But there is a more fundamental difficulty with viewing the Administrative Procedure Act, even as interpreted by the courts, as providing an opportunity for the Congress to surmont information asymmetries that inhibit political control. The same procedures that to some degree promote information revelation simultaneously disable political intervention. Congressional intervention, other than through additional amendatory legislation, must take place on the administrative record (Sierra Club v. Costle, 657 F.2d 298 [D.C. Cir, 1981]). Moreover, the requirements of rationalization of decisions in terms of that administrative record mean that "political" concerns cannot be given weight unless the factual record and statutory purposes justify it (Motor Vehicle Manufacturers Association v. State Farm Mutual Insurance Co., 463 U.S. 29 [1983]).

Similar points can be made about the Freedom of Information Act. . . .

On this account, the MNW, principal-agent, legislative-control model seems, at best, radically incomplete. Moreover, the fact that the bureaucracy must take account of at least two political principals is glossed over as if it were of little importance. Yet, in the life of agencies and in the design of administrative structures, the Congress and the president seem to view this issue as of fundamental importance. . . . Competition between or among principals may eliminate, at least reduce, the power of all political principals. A model that takes the principal-agent paradigm seriously must confront this issue. . . .

. . . The Congress also has the peculiar habit of delegating huge amounts of responsiblity for the development of administrative processes to states and localities. . . . Why would a congress desiring to control the implementation of federal programs (or federal spending) through administrative process make these vast delegations to agencies whose structure within state government and procedural requirements within state administrative law are almost completely outside its control? . . .

I can only conclude that major insights into the structure and processes of federal administrative agencies as they actually operate are unlikely to flow from viewing agency structure and process primarily in terms of the monitoring and sanctioning problems that political controllers have with federal bureaucracies. That the MNW story has something to tell us I have little doubt. But the convincing parts of that story will not persuade aficionados of idealist or critical explanation that their modes of analysis should be abandoned.

Conclusion

Given the methodology employed here, we cannot assess the power of our other competitors without another act of translation into a newly privileged approach. Nevertheless, I am suspicious that administrative law and administrative processes present a much more complex interaction (if not systematic confusion) of principals and agents than any of our three story-telling traditions suggests. It seems perfectly plausible, depending on context, to put the legislature, courts, or agencies themselves at the forefront of stories about the design and operation of administrative processes. Indeed, the history of many administrative regimes is likely to bear all three interpretations. Moreover, the president (or the executive branch) is largely missing from all of these tales. Yet because the effects of presidential action on administrative agencies are themselves complex and obscure, adding the president to the mix hardly simplifies matters. Attempts to articulate a presidential-control hypothesis are not likely to provide a more convincing central heuristic than the three analytic traditions we have, in part, explored. So where does that leave us?

When in a cheerful mood, the inadequacy of simple stories concerning the origins and evolution of administrative processes suggests to me that the modern administrative state is a construct of which Madison, Hamilton, and Jefferson might justly be proud. It is checked and balanced, motivated and constrained in ways so complex and continuous that there are no final victories in the political competition to control the exercise of administrative power. On more somber days, however, I am troubled by the possibility that the inadequacy of our theoretical perspectives may simply mean that we cannot recognize the degree of which the founders' hopes for combining individual liberty with effective governance have been disappointed by the growth of modern, bureaucratic administration.

"The Politics of Bureaucratic Structure"
TERRY MOE

American public bureaucracy is not designed to be effective. . . .

The central question boils down to this: what sorts of structures do the various political actors—interest groups, presidents, members of Congress, bureaucrats—find conducive to their own interests, and what kind of bureaucracy is therefore likely to emerge from their efforts to exercise political power? In other words, why do they build the bureaucracy they do? . . .

Most citizens do not get terribly excited about the arcane details of public administration. . . . Organized interest groups are another matter. They are . . . normally the only source of political pressure when structural issues are at stake. Structural politics is interest-group politics. . . .

Interest Groups: The Technical Problem of Structural Choice

. . . Although the group has the political power to impose its will on everyone, it almost surely lacks the knowledge to do it well. It does not know what to tell people to do. . . .

A group with the political power to tell everyone what to do, then, will typically not find it worthwhile to try. A more attractive option is to write legislation in general terms, put experts on the public payroll, and grant them the authority to "fill in the details" and make whatever adjustments are necessary over time. This compensates nicely for the group's formidable knowledge problems, allowing it to pursue its own interests without knowing exactly how to implement its policies and without having to grapple with future contingencies. The experts do what the group is unable to do for itself. And because they are public officials on the public payroll, the arrangement economizes greatly on the group's resources and time.

It does, however, raise a new worry: there is no guarantee the experts will always act in the group's best interests. Experts have their own interests—in career, in autonomy—that may conflict with those of the group. And, due largely to experts' specialized knowledge and the often intangible nature of their outputs, the group cannot know exactly what its expert agents are doing or why. These are problems of conflict of interest and asymmetric information, and they are unavoidable. Because of them, control will be imperfect. . . .

. . . A politically powerful group, acting under uncertainty and concerned with solving a complex policy problem, is normally best off if it resists using its power to tell bureaucrats exactly what to do. It can use its power more productively by selecting the right types of bureaucrats and designing a structure that affords them reasonable autonomy. Through the judicious allocation of bureaucratic roles and responsibilities, incentive systems, and structural checks on bureaucratic choice, a select set of bureaucrats can be unleashed to follow their expert judgment, free from detailed formal instructions.

Interest Groups: The Political Problem of Structural Choice

Political dominance is an extreme case for purposes of illustration. In the real world of democratic politics, interest groups cannot lay claim to unchallenged legal authority. Because this is so, they face two fundamental problems that a dominant group does not. The first I will call political uncertainty, the second political compromise. Both have enormous consequences for the strategic design of public bureaucracy—consequences that entail substantial departures from effective organization. . . .

There are various structural means by which the group can try to protect and nurture its bureaucratic agents. They include the following.

- It can write detailed legislation that imposes rigid constraints on the agency's mandate and decision procedures. While these constraints will tend to be flawed, cumbersome, and costly, they serve to remove important types of decisions from future political control. The reason they are so attractive is rooted in the American separation-of-powers system, which sets up obstacles that make formal legislation extremely difficult to achieve—and, if achieved, extremely difficult to overturn. Should the group's opponents gain in political power, there is a good chance they would still not be able to pass corrective legislation of their own.
- It can place even greater emphasis on professionalism than is technically justified, since professionals will generally act to protect their own auton-

omy and resist political interference. For similar reasons, the group can be a strong supporter of the career civil service and other personnel systems that insulate bureaucratic jobs, promotion, and pay from political intervention. And it can try to minimize the power and number of political appointees, since these too are routes by which opponents may exercise influence. . . .

- It can oppose formal provisions that enhance political oversight and involvement. The legislative veto, for example, is bad because it gives opponents a direct mechanism for reversing agency decisions. Sunset provisions, which require reauthorization of the agency after some period of time, are also dangerous because they give opponents opportunities to overturn the group's legislative achievements.
- It can see that the agency is given a safe location in the scheme of government. Most obviously, it might try to place the agency in a friendly executive department, where it can be sheltered by the group's allies. Or it may favor formal independence, which provides special protection from presidential removal and managerial powers.
- It can favor judicialization of agency decisionmaking as a way of insulating policy choices from outside interference. It can also favor making various types of agency actions—or inactions—appealable to the courts. It must take care to design these procedures and checks, however, so that they disproportionately favor the group over its opponents.

The driving force of political uncertainty, then, causes the winning group to favor structural designs it would never favor on technical grounds alone: designs that place detailed formal restrictions on bureaucratic discretion, impose complex procedures for agency decisionmaking, minimize opportunities for oversight, and otherwise insulate the agency from politics. The group has to protect itself and its agency from the dangers of democracy, and it does so by imposing structures that appear strange and incongruous indeed when judged by almost any reasonable standards of what an effective organization ought to look like.

But this is only part of the story. The departure from technical rationality is still greater because of a second basic feature of American democratic politics: legislative victory of any consequence almost always requires compromise. This means that opposing groups will have a direct say in how the agency and its mandate are constructed. One form that this can take, of course, is the classic compromise over policy that is written about endlessly in textbooks and newspapers. But there is no real disjunction between policy and structure, and many of the opponents' interests will also be pursued through demands for structural concessions. What sorts of arrangements should they tend to favor? . . .

- Opponents want structures that work against effective performance. They fear strong, coherent, centralized organization. They like fragmented authority, decentralization, federalism, checks and balances, and other structural means of promoting weakness, confusion, and delay.
- They want structures that allow politicians to get at the agency. They do not want to see the agency placed within a friendly department, nor do

they favor formal independence. They are enthusiastic supporters of legislative veto and reauthorization provisions. They favor onerous requirements for the collection and reporting of information, the monitoring of agency operations, and the review of agency decisions—thus laying the basis for active, interventionist oversight by politicians.

- They want appointment and personnel arrangements that allow for political direction of the agency. They also want more active and influential roles for political appointees and less extensive reliance on professionalism and the civil service.
- They favor agency decisionmaking procedures that allow them to participate, to present evidence and arguments, to appeal adverse agency decisions, to delay, and, in general, to protect their own interests and inhibit effective agency action through formal, legally sanctioned rules. This means that they will tend to push for cumbersome, heavily judicialized decision processes, and that they will favor an active, easily triggered role for the courts in reviewing agency decisions.
- They want agency decisions to be accompanied by, and partially justified in terms of, "objective" assessments of their consequences: environmental impact statements, inflation impact statements, cost-benefit analysis. These are costly, time-consuming and disruptive. Even better, their methods and conclusions can be challenged in the courts, providing new opportunities for delaying or quashing agency decisions. . . .

In short, democratic government gives rise to two major forces that cause the structure of public bureaucracy to depart from technical rationality. First, those currently in a position to exercise public authority will often face uncertainty about their own grip on political power in the years ahead, and this will prompt them to favor structures that insulate their achievements from politics. Second, opponents will also tend to have a say in structural design, and, to the degree they do, they will impose structures that subvert effective performance and politicize agency decisions.

Legislators and Structural Choice

. . . [L]egislators tend not to invest in general policy control. Instead, they value "particularized" control: they want to be able to intervene quickly, inexpensively, and in ad hoc ways to protect or advance the interests of particular clients in particular matters. . . .

Presidents and Structural Choice

Presidents are motivated differently. Governance is the driving force behind the modern presidency. . . .

This raises two basic problems for interest groups. The first is that presidents are not very susceptible to the appeals of special interests. . . .

The second problem is that presidents want to control the bureaucracy.

While legislators eagerly delegate their powers to administrative agencies, presidents are driven to take charge. . . . They are the only participants who are directly concerned with how the bureaucracy as a whole should be organized. And they are the only ones who actually want to run it through hands-on management and control. . . .

This is just what the winning group and its legislative allies do not want. They want to protect their agencies and policy achievements by insulating them from politics, and presidents threaten to ruin everything by trying to control these agencies from above. The opposing groups are delighted with this, but they cannot always take comfort in the presidential approach to bureaucracy either. For presidents will tend to resist complex procedural protections, excessive judicial review, legislative veto provisions, and many other means by which the losers try to protect themselves and cripple bureaucratic performance.

Bureaucracy

. . . Once an agency is created, the political world becomes a different place. Agency bureaucrats are now political actors in their own right. . . .

Thus, however active the agency is in forming alliances, insulating itself from politics, and otherwise shaping political outcomes, it would be a mistake to regard the agency as a truly independent force. It is literally manufactured by the other players as a vehicle for advancing and protecting their own interests, and their structural designs are premised on anticipations about the roles the agency and its bureaucrats will play in future politics. The whole point of structural choice is to anticipate, program, and engineer bureaucratic behavior.

Conclusion

The Consumer Product Safety Commission [CPSC], the Occupational Safety and Health Administration [OSHA], and the Environmental Protection Agency [EPA] are all prime examples of the new social regulation. In each case . . . the real battles over policy took place within an arcane realm of politics remote from the concerns of ordinary citizens: the politics of structural choice. . . . Each is a grotesque combination of organizational features that clearly are not conducive to effective performance. . . .

. . . At the risk of oversimplifying, there are three major reasons that help explain why this is so and thus why public bureaucracy cannot be organized for effective performance. First, even the group that successfully pressures for the creation of a public agency—consumers for the CPSC, labor for OSHA, environmentalists for the EPA—will not demand an effectively designed organization. While it certainly wants a bureaucracy that will do the best job possible, it must also reckon with political uncertainty: its political enemies may soon gain sufficient power to exercise legitimate authority over the winning group's agency. Something must therefore be done to protect the group's accomplishments from being captured or destroyed.

Consumers were responding to political uncertainty when they chose an independent commission, a form they had loudly damned in earlier days. Perhaps the most pervasive examples, though, are the host of agency-forcing mechanisms that all the winning groups employed in imposing detailed and onerous requirements on their agencies, the EPA being the extreme case. There is little doubt that these formal constraints were debilitating and a direct cause of ineffective performance. There is also little doubt, especially in the EPA's case, that many of these agency-forcing requirements were technically unjustified. The experts clearly could have done a much more competent job if they had been granted the discretion to put their expertise to proper use. But the groups had no intention of granting them discretion. By directing bureaucratic behavior themselves via detailed formal requirements—even if these requirements were technically ill advised and took a toll on agency performance—the groups were removing crucial decisions from the realm of future influence by business. This was tremendously valuable, and they were willing to pay a price for it. As a result, they purposely created bizarre administrative arrangements that were not well suited to effective regulation.

Second, the winning group must usually compromise with the losing group when structural choices are being made. This is democracy in action. Unfortunately, the losing group is dedicated to crippling the agency in whatever ways it can and gaining a measure of control over agency decisions. Thus it will pressure for fragmented authority, labyrinthine procedures, mechanisms of presidential and congressional intervention, and other structures that impede vigorous agency performance. This is precisely why OSHA is hobbled by the OSHRC [Occupational Safety and Health Review Commission], why OSHA's and the EPA's regulatory arrangements are so heavily reliant upon the states, why the CPSC was slapped with a congressional veto, and why all these agencies must give business procedural protections that go well beyond those required by the Administrative Procedure Act and the courts. In the private sector, structures are generally designed by participants who want the organization to succeed. In the public sector, bureaucracies are designed in no small measure by participants who explicitly want them to fail.

Third, presidents have the power and incentive to impose their own layer of structure on top of the one that the legislative process has already produced. Although presidents have a direct stake in how well government performs, they are a constant threat to impose structures that undermine the policy goals of individual agencies and their group supporters. Most obviously, presidents have electoral coalitions that may dispose them to favor an agency's opponents. Clearly, business has successfully worked through Republican presidents in putting a structural rein on the regulatory agencies, particularly OSHA and the EPA.

In addition, though, all presidents have broader social concerns that thrust them into the enemy's role. For agencies are single-minded in pursuit of their mandates, and presidents do not want them to be. Presidents want them to give balanced consideration to economic growth, inflation, and other issues of national concern. This means they must prevent agencies from simply doing a good job at what their designers intended for them to do. This may or may not be in the best interests of society. But whatever the case may be, the structures imposed on agencies will be even more byzantine and convoluted because of the president's sumperimposed layer, and they will be even less well suited to the achievement of congressionally ordained goals. . . .

There is some reason to believe, in fact, that the current administrative tangle may actually get worse over time. Consider the following.

- The kinds of socioeconomic problems government is called upon to address seem to be growing increasingly interdependent and complex. . . . The greater the technical requirements of society's problems, the more poorly designed American bureaucracy is likely to prove as it struggles to address them.
- Politics has become much more competitive in the last decade or two, as an interest-group system heavily weighted in favor of business has been transformed by groups representing consumers, women, blacks, environmentalists, and other broad social interests. . . .

    It is well known that public agencies have become increasingly formalized and proceduralized during the 1970s and 1980s, in part because of restrictions imposed by the courts. Group competition may well have been at least as consequential. And the more vital and competitive American democracy becomes in the future, the worse its "bureaucracy problem" will get.
- The intrusive structural designs of presidents are only going to become more intrusive over time. The development of an institutional capacity to control the bureaucracy is not something peculiar to Republican presidents, nor is the aggressive use of this capacity going to fade into the past with the Reagan presidency.

"Beliefs"

JAMES Q. WILSON

Defining Tasks: Professional Norms

During a typical week, the Federal Trade Commission (FTC) in downtonwn Washington, D.C., receives several dozen letters from individuals and firms complaining about what the writers believe are the unfair business practices of certain enterprises. These letters are screened by an Office of Evaluation consisting of several attorneys. Many of the letters are simply gripes about behavior that does not violate any federal law; others contain accusations that may involve a violation; and still others call attention to what appear to be clear cases of wrongdoing. While these letters are being reviewed, the office also learns from newspaper stories about planned corporate mergers. If the merger is a large one, the firms involved must notify the FTC directly. Meanwhile, economists in the FTC make recommendations for the investigation of certain industries that seem to be dominated by a small number of large firms or that appear to be making abnormally high profits. All of these sources of information go to an Evaluation Committee that recommends to the top officials of the FTC which cases should be investigated and which should not.

In reaching these decisions, the FTC gets rather little guidance from the statutes. These are the Federal Trade Commission Act and the Clayton Act. They make it illegal for firms to engage in "unfair" or "deceptive" methods of competition, practice certain forms of price discrimination, or buy (or merge with) other firms if the effect of this "may be to substantially lessen competition or tend to create a monopoly." As political scientist Robert A. Katzmann and others have pointed out, these words are too vague to provide clear guidelines for deciding what to do. The FTC has produced an *Operating Manual,* but it does not tell the Evaluation Committee exactly what these statutory words are supposed to mean or how to decide what cases to bring. Yet cases are brought and lawyers go to work on them. Somehow, the tasks of the FTC get defined with sufficient clarity to enable its staff to know what to do.

What does tell the employees of the FTC what to do is, to a significant degree, the professional norms that they have learned and the career opportunities those professions hold out to them.

Medicine, nursing, engineering, economics, the law—these and countless other occupations are popularly referred to as professions. People engaged in a profession often get more respect, income, and deference than do people who work at "ordinary" jobs. But a popular label is not a useful definition. What is distinctive about members of a profession, at least for purposes of explaining organizational behavior, is not how much status, income, or deference they receive but the sources of these rewards. A professional is someone who receives important occupational rewards from a reference group whose membership is limited to people who have undergone specialized formal education and have accepted a group-defined code of proper conduct. The more the individual allows his or her behavior to be influenced by the desire to obtain rewards from this reference group, the more professional is his or her orientation. Thus, not every member of an occupation whom we think of as a professional may in fact be a professional. An attorney who serves a client even when that service is likely to incur the displeasure of the bar association and law school professors is not a professional (though he or she may be an excellent attorney); by the same token, a physician who follows the procedures recommended by other physicians and by medical school professors even when it is not in the best interests of the doctor is highly professional.

In a bureaucracy, professionals are those employees who receive some significant portion of their incentives from organized groups of fellow practitioners located outside the agency. Thus, the behavior of a professional in a bureaucracy is not wholly determined by incentives controlled by the agency. (We might define a bureaucrat as someone whose occupational incentives come entirely from within the agency.) Because the behavior of a professional is not entirely shaped by organizational incentives, the way such a person defines his or her task may reflect more the standards of the external reference group than the preferences of the internal management.

The tasks performed by employees of the Federal Trade Commission crucially depend on whether they are lawyers or economists. As Robert Katzmann has shown, attorneys prefer to pursue a case in which there is clear evidence that the conduct of some firm violates the law. They are prosecution-minded. This orientation reflects both their training in the law and court procedures and (in many cases) their desire to use the skills they develop at the FTC to land a job with a well-paying, prestigious law firm. They evaluate potential cases on the basis of the strength of the evidence and the reigning interpretation of antitrust law. Economists, by contrast, are trained to evaluate potential cases in terms of their likely impact on consumer welfare. To them, an easily prosecuted case of business misconduct may have no impact on the prices consumers pay or the quality of goods they receive, and so economists argue against taking on these cases and in favor of finding cases that involve large concentrations of economic power that if broken up might unleash competitive forces which will make consumers substantially better off. Moreover, the FTC economists are disproportionately drawn from among those who studied at or were influenced by the University of Chicago. "Chicago economics" emphasizes the social benefits of free markets.

The lawyers reply that economists, especially Chicago ones, are prone to pick cases on the basis of economic theory rather than legal reality, cases that are difficult to investigate and may find little support in the courts. At the extremes, a lawyer is happiest with an allegation that two companies have illegally conspired to fix prices, while an economist is happiest with the discovery that the size and market power of the Exxon corporation may reduce competition in the oil industry.

Differences within the FTC over what the agency's job should be are not simply professional preferences about policy choices; they are competing visions, sometimes invested with a good deal of emotion, as to how best to discharge a public responsibility. Lawyers sometimes refer to economists as "case killers," "dogmatic" people engaged in "God-playing." For their part, economists see themselves as social scientists—as "dispassionate searchers for truth"; lawyers, on the other hand, are seen as people less interested in truth than in finding facts to support a predetermined case.

The government lawyers' professional orientation toward prosecution was also noted by Arthur Maass in his study of federal investigations of state and local political corruption. Federal law provides no clear justification for the many heavily publicized cases brought by U.S. attorneys against mayors, judges, city councilmen, and state legislators. Nor can the desire for partisan advantage explain the sharp growth in such cases since the mid-1970s: prosecutors in both Republican and Democratic administrations have pressed these investigations. What drives the program, Maass concludes, "is the prosecutorial mentality" coupled with the personal career benefits that come from having sent some corrupt local official to prison.

But lawyers are not wholly professional in their outlook. They are, after all, trained to serve the interests of a client and may do this with little regard for what other lawyers think. Even in the FTC, some lawyers endorsed the shift toward investigating large structural cases of the sort that, beginning in the 1970s, many FTC commissioners favored. And personal advantage, such as getting a job in a big firm or laying the groundwork for a campaign for governor, reinforces their occupational bias toward vigorous prosecution.

Engineers, on the other hand, rarely benefit personally from bringing the norms of the engineering profession into government service. When they define vague or ambiguous tasks in ways that conform to engineering predilections, it is usually without the added incentive of enhanced opportunities for winning elective office. This strong professional orientation helps explain why the National Highway Traffic Safety Administration (NHTSA) has acted as it has in discharging its responsibility to reduce death and injury on the highways. By law, NHTSA could have chosen to improve the skills and habits of drivers, enhance the design of roads, or eliminate the defects in the autos; to some extent, it has done all three. But to a remarkable degree, it has favored changing the automobile rather than the driver or the highway. Charles Pruitt explains this as the result of the decision to staff the agency predominantly with engineers. In part this decision was made knowing—and wanting—its consequences.

Born in a crisis atmosphere after a bruising congressional attack on the auto maufacturers, NHTSA felt it had to take immediate, highly visible action to improve auto safety. Educating people to drive more safely was a policy that would have taken years to produce results (if in fact it would have produced any results at all); redesigning the car to meet federal standards could be done much more quickly. The redesign that could be made the fastest was one intended to make the car more "crashworthy"—for example, less likely to impale a driver on a steering column or light switch. No doubt cars have become more crashworthy in part as a result of NHTSA rules, thereby vindicating the initial hiring decisions. But the decision to staff the agency with crash engineers (instead of specialists in driver education, alcohol abuse, or highway design) had long-term consequencees that may not heve been intended. These engineers worked persistently and enthusiastically to require the installation of air bags (devices that, in the event of a crash, deploy automatically to protect the occupants), despite the intense political controversy engendered by this proposal. Moreover, NHTSA resisted demands from other federal agencies that it devote more attention to human and environmental factors in highway safety.

The impact of the engineering profession on the work of the National Aeronautics and Space Administration (NASA) has been frequently remarked. When the space-shuttle Challenger exploded, killing its crew, there was a review to find out what happened. The engineers in NASA and the firms supplying NASA were a gifted and dedicated group that had accomplished extraordinary feats. Buth their approach to their work may have created blind spots. Engineers value quantitative data and distrust personal opinions. Thus, the anxieties that some may have felt about the readiness of the shuttle to be launched were not as forcefully expressed as they might have been or attended to as seriously as they should have been because they were perceived as hunches, not facts or numbers. One engineer working on Challenger put it this way: "I have been personally chastised in flight readiness reviews at Marshall [Space Flight Center] from using the words 'I feel' or 'I think,' and I have been crucified . . . because 'I feel' and 'I think' are not engineering-supported statements, but they are just judgmental."

Even in a highly bureaucratic welfare office, some of the intake workers have acquired, by virtue of their training and experience, a quasi-professional attitude toward their clients. As a consequence, though the situation presses them to treat clients equally and distantly (as explained previously), a few operators whom Prottas observed tried to think of themselves not as intake clerks but as social workers. Contrary to the views of those who believe that welfare workers seek to degrade or punish applicants, many of these bureaucrats wished to use what little discretion they had to help clients, or at least those who seem to be "deserving." To the extent circumstances permitted (which given the case loads and organizational requirements was not much), "they look[ed] forward to opportunities to behave as social workers, which means an opportunity to do something more than the routine processing of simple cases."

Nowhere is the influence of professional norms more evident than in the contrasting histories of two otherwise similar agencies: the Forest Service and the Park Service. The former, created in 1905, and the latter, created in 1916, each manage vast tracts of public lands, many of which are so similar as to be indistinguishable. The two agencies, with roughly comparable functions, often have been bitter bureaucratic rivals. But the Forest Service from the first committed itself to develop and be guided by a doctrine of "professional" forestry, by which was meant the scientific management of forests in order to produce a sustained yield of timber and other natural resources. Though this utilitarian focus was later modified, the commitment to professional education and research never changed. By the time Herbert Kaufman studied it in the 1950s, 90 percent of all rangers were trained foresters, meaning that they had at least a bachelor's and sometimes a master's degree in forestry or some related subject. Most are members of the Society of American Foresters. The Forest Service provided the impetus for much of this professional education and, in turn, is now molded by the doctrines developed in those schools and societies. In recent years, many Forest Service practices, such as clear-cutting and below-market timber sales, have become controversial. However, right or wrong, these policies are not the result of having caved in to economic interests; they reflect beliefs about what good forest management requires. To change the way Forest Service tasks are performed today would require equivalent changes in forestry doctrine that is probably beyond the capacity of the Forest Service to dictate: the parent has become hostage of the child.

By contrast, the Park Service inherited many of its lands, including Yellowstone, from the army, and saw its job as balancing the need to preserve the wilderness with the popular desire to enjoy it. These goals seemed to require law-enforcement and engineering skills—law enforcement to set down and apply rules for campers, engineering to build facilities for them. The task became defined as managing *people* more than managing parks. There was never an effort to develop a park ranger profession that was rooted in science and higher education. Even today, the main career track in the Park Service is the "ranger" track, by which is meant experience in visitor protection and safety, and not the naturalist track, which includes research. The ranger track, reports Alston Chase, a Park Service critic, is officially classified as nonprofessional, with no educational prerequisites beyond a high-school diploma. The few scientists in the service have little influence. In 1985 the Park Service, which manages over three hundred parks, spent about $16 million on research, less than the budget of one Forest Service experiment station.

One consequence of these differences in operator values can be found in how the two services manage visitors. Craig W. Allin found that even in comparable areas attracting comparable visitors, the Forest Service gives less emphasis to regulating visitor use than does the Park Service. The Park Service, like the Forest Service, has been deeply embroiled in controversy, especially since a vast brush fire ravaged Yellowstone Park in 1988. At issue are the age-old questions of popular access versus wilderness preservation; managing the wildlife versus deferring to natural forces; and fighting fires by "letting them burn" versus practicing controlled burning. Each agency handles its controversies in characteristic ways: the Forest Service engages in extraordinarily detailed planning and research activities designed to defend its view of correct forest management; the Park Service searches for external allies and engages in periodic changes of direction.*

Politicians and interest groups know that professionals can define tasks in ways that are hard for administrators to alter, and so one strategy for changing an organization is to induce it to recruit a professional cadre whose values are congenial to those desiring the change. We have seen already how the supporters of highway safety worked to insure that crash engineers were hired by NHTSA, thereby giving that agency at birth a distinctive focus. By the same token, the tendency . . . of the Occupational Safety and Health Administration (OSHA) to work harder on improving safety than on health standards began to give way to a greater concern for health matters as OSHA began to hire more and more research scientists and public health experts. These professionals were less concerned with finding the bureaucratically easiest task

---

*That I have described the Forest Service as a professional organization and the Park Service as a nonprofessional one does not mean that I think the former is good and the latter bad. Professionalism and nonprofessionalism each can lead to good or bad outcomes.

and more concerned with focusing agency efforts on what their professional techniques identified as hazards and their professional norms taught them to take seriously.

Even established bureaucracies with strong professional traditions already in place can be altered by inserting a new profession. The Army Corps of Engineers and the Forest Service both have changed the way in which they approach certain tasks because they were obliged to hire a large number of persons who identify with the emerging environmentalist professions. The National Environmental Policy Act (NEPA) requires federal agencies to take into account the environmental impact of their projects. To do this they must prepare (among other things) environmental impact statements (EIS), and therefore must hire people skilled at preparing such reports—biologists, ecologists, biochemists, and the like. In 1984, when Serge Taylor published his account of the EIS process, the two agencies employed between them several hundred of these new professionals. In this way, attention to different values was institutionalized in ongoing organizations, but at a price: having tasks defined by rival professions weakened the ability of the agencies to develop and maintain a shared sense of mission.

The Forest Service, for example, was an organization dominated by a single professional culture when Herbert Kaufman studied it in the 1950s. By the 1980s it had become divided into many different professional cultures. Today foresters have to contend with engineers, biologists, and economists, among others. The foresters dislike the tendency of engineers to elevate mechanical soundness over natural beauty, of biologists to worry more about endangered species than about big game, and of economists to put a price on things foresters regard as priceless.

Political Ideology

In the United States the higher levels of the federal government are staffed by bureaucrats who are more liberal than the population at large and certainly more liberal than business executives. Stanley Rothman and S. Robert Lichter interviewed 200 top-level career administrators in a variety of federal agencies. Over half described themselves as politically liberal and said they had voted for George McGovern for president in 1972; nearly two-thirds voted for Jimmy Carter in 1976. Overwhelmingly they supported a woman's right to an abortion and agreed that environmental problems are serious. On the other hand, they displayed no generalized antipathy toward American society: the great majority believed that private enterprise is fair to workers and that less regulation of business would be good for the country; only a tiny fraction thought the nation should move toward socialism.

There were important differences among senior bureaucrats depending on the kind of agency for which they worked. Those employed by "activist" agencies (such as the Environmental Protection Agency, the Consumer Products Safety Commission, and the Department of Health and Human Services) were significantly more liberal than those employed by "traditional" agencies (such as the Departments of Agriculure, Commerce, and the Treasury). In the activist agencies, bureaucrats were more likely than their counterparts in traditional agencies to say that women and blacks should get preference in hiring, that poor people are the victims of circumstances, and that U.S. foreign policy has aimed at protecting business.

Essentially the same conclusions were reached by Joel Aberbach and Bert Rockman in their survey of senior bureaucrats holding office at the time Richard Nixon was president. They found evidence that Nixon was correct in his perception of bureaucratic ideology—that the officials were more liberal than he—and that the most liberal of all were in the social service agencies. Over 90 percent of the Democrats in these activist agencies were classified by Aberbach and Rockman as on the left. By contrast, only 25 percent of the Democrats in traditional agencies were leftists.

These findings show that liberals who worry that high-level bureaucrats will have conservative opinions are wrong. In retrospect, it is clear that Kingsley was mistaken to think that because they are unrepresentative of the public at large, bureaucrats will prefer the status quo. Conservatives who fear the liberalism of bureaucrats do have grounds for concern, but exactly how much concern is not at all clear. We really don't care what bureaucrats think, we care what they do. Does ideology determine behavior? There is no systematic evidence bearing on this question.

From what already had been said about attitudes and behavior, we would expect ideology to make little difference in routinized or higly structured roles. There is no liberal or conservative way to deliver the mail or issue a driver's license. The example of police arrest and shooting practices suggests that there are even limits to how much difference personal beliefs can make in relatively ambiguous roles. But that still leaves a lot of ground to cover.

What some of that ground may look like is suggested by Jeremy Rabkin's account of the development of the Office for Civil Rights (OCR). Created in 1965 to implement various civil rights laws, the OCR had a great deal of latitude in defining its own task. It is charged with insuring that no "program or activity receiving federal financial assistance" shall practice discrimination with regard to race, color, sex, handicap, or national origin. If discrimination were shown, the guilty organization could lose its federal funding. It was left to OCR to decide what constituted a "program or activity," whether the federal assistance had to be received directly or need only be received indirectly in order to bring the entity under the law, and what constituted evidence of discrimination.

After some initial fumbling and uncertainty, the OCR staff began to issue regulations that took an expansive view of the reach and force of the law. Discrimination was forbidden in any part of a school or other organization receiving federal aid, not merely in those parts that were directly supported. For example, if a school library bought books with federal aid, the cafeteria staff and the sports programs could not discriminate, even if they received no money. Discrimination was broadly defined to include any case in which a school district employed (for example) blacks in a lesser proportion than their representation in the community, whether or not anyone had produced evidence of discrimination. In perhaps its most controversial decision, OCR issued orders banning father-son and mother-daughter banquets at schools, orders later reversed by Congress in response to a public outcry.

There were several reasons for OCR taking this expansive view of its powers. Civil rights and feminist organizations often attacked it for "dragging its feet." Schools and universities were not effective in lobbying against the regulations. Cabinet officials and congressional committees did rather little to restrain OCR. But in part the agency acted as it did because it attracted people who had "considerable enthusiasm for its mission."

The OCR is not alone in this regard. When the National Labor Relations Board (NLRB) was created in 1935, it attracted pro-labor lawyers to its staff who gave that agency a distinctive ideology until later events muted or ended the ideological enthusiasms of the staff. The Office of Surface Mining Reclamation and Enforcement (OSM), an agency created to regulate strip-mining in the coal fields, attracted dedicated environmentalists to its staff who went well beyond the letter of the law in drawing up mining regulations.

Why did ideology seem to influence the behavior of the staff of OCR, the lawyers in the NLRB, and the environmentalists in OSM, but not the arrest decisions of police officers? The answer, I conjecture, is to be found in the differing environments in which these bureaucrats work. The OCR, NLRB, and OSM were newly created policymaking bodies operating in a political environment shaped (if not dominated) by like-minded partisans in Congress and the supporting interest groups. Attitudes—whether ideology in the OCR, prior experiences in the ECA [Economic Cooperation Administration], or professional norms in the FTC—are most likely to influence the performance of weakly defined roles, especially when the attitudes are reinforced by other incentives. The jobs in a new agency charged with designing new policies are about as vaguely defined as one can imagine; moreover, a newborn agency is surrounded by its political parents—people and groups eager to applaud behavior that is consistent with the zeal of those who won the fight to create the agency.

Police officers have a great deal of discretion, but that does not mean that their tasks are vaguely defined. It only means that what they do is not clearly defined by formal organizational rules. Their work is shaped by informal understandings that are the product of daily, street-level contact with clients and years of on-the-job experience. These contacts and experiences tell police officers how to handle unruly citizens, when to use deadly force, what demeanor will win the approval of more senior colleagues, and which arrests will both stand up in court and be rewarded by promotion boards. The outside observer may not notice these sources of task definition and thus suppose that officers are free to do whatever they want; but that conclusion is wrong—and when it leads to ill-considered efforts to change behavior by writing rules, it is michievous.

This perspective on ideology suggests that the formative years of a policy-making agency are of crucial importance in determining its behavior. As with people, so with organizations: Childhood experiences affect adult conduct. This means . . . that one must study history.

Agencies, like adults, learn from experience, and so the formative childhood years are only part of the story. Bureaucracies will in time acquire a distinctive personality or culture that will shape the attitudes of people who join these organizations. . . . When critics of an older agency complain that bureaucrats with the "wrong attitude" are determining its behavior, they are often reversing the causal process. The agency is in fact producing certain attitudes in its members.

There is a good deal of evidence that the political views of bureaucrats tend to correspond to their agency affiliation more than they reflect their social status. Kenneth Meier and Lloyd Nigro looked at the beliefs of a sample of higher-level federal civil servants and found that the social origins of these officials explained only about 5 percent of variance in their opinions; agency affiliations, on the other hand, explained much more. Bernard Mennis compared foreign service officers and military officers performing similar jobs, mostly having to do with managing foreign political and international-security affairs. He found that their policy views differed sharply in the way one would expect: The foreign service officers were more liberal, the military officers more conservative. But these differences existed despite the fact that the social backgrounds of the two groups of officials were about the same. As Charles Goodsell summarized these studies, "the agency positions in which bureaucrats 'sit' have much to do with where they 'stand.' "

The Science of "Muddling Through"

CHARLES LINDBLOM

[In formulating administrative policy, one possible approach] might be described as the method of *successive limited comparisons*. I will contrast it with [another] approach, which might be called the rational-comprehensive method.* More impressionistically and briefly—and therefore generally used in this article—they could be characterized as the branch method and

Reprinted by permission from 19 *Public Administration Review* 79 (1959).
*I am assuming that administrators often make policy and advise in the making of policy and am treating decisionmaking and policymaking as synonymous for purposes of this article.

root method, the former continually building out from the current situation, step-by-step and by small degrees; the latter starting from fundamentals anew each time, building on the past only as experience is embodied in a theory, and always prepared to start completely from the ground up.

Let us put the characteristics of the two methods side by side in simplest terms.

Rational-Comprehensive (Root)

1a. Clarification of values or objectives distinct from and usually prerequisite to empirical analysis of alternative policies.
2a. Policy formulation is therefore approached through means-end analysis: First the ends are isolated, then the means to achieve them are sought.
3a. The test of a "good" policy is that it can be shown to be the most appropriate means to desired ends.
4a. Analysis is comprehensive; every important relevant factor is taken into account.
5a. Theory is often heavily relied upon.

Successive Limited Comparisons (Branch)

1b. Selection of value goals and empirical analysis of the needed action are not distinct from one another but are closely intertwined.
2b. Since means and ends are not distinct, means-end analysis is often inappropriate or limited.
3b. The test of a "good" policy it typically that various analysts find themselves directly agreeing on a policy (without their agreeing that it is the most appropriate means to an agreed objective).
4b. Analysis is drastically limited: i. Important possible outcomes are neglected. ii. important alternative potential policies are neglected. iii. Important affected values are neglected.
5b. A succession of comparisons greatly reduces or eliminates reliance on theory.

• • •

Assuming that the root method is familiar and understandable, we proceed directly to clarification of its alternative by contrast. In explaining the second, we shall be describing how most administrators do in fact approach complex questions, for the root method, the "best" way as a blueprint or model, is in fact not workable for complex policy questions, and administrators are forced to use the method of successive limited comparisons.

Intertwining Evaluation and Empirical Analysis (1b)

The quickest way to understand how values are handled in the method of successive limited comparisons is to see how the root method often breaks down in *its* handling of values or objectives. . . .

In summary, two aspects of the process by which values are actually handled can be distinguished. The first is clear: evaluation and empirical analysis are intertwined; that is, one chooses among values and among policies at one and the same time. Put a little more elaborately, one simultaneously chooses a policy to attain certain objectives and chooses the objectives themselves. The second aspect is related but distinct: the administrator focuses his attention on marginal or incremental values. Whether he is aware of it or not, he does not find general formulations of objectives very helpful and in fact makes specific marginal or incremental comparisons. Two policies, X and Y, confront him. Both promise the same degree of attainment of objectives *a, b, c, d,* and *e.* But X promises him somewhat more of *f* than does Y, while Y promises him somewhat more of *g* than does X. In choosing between them, he is in fact offered the alternative of a marginal or incremental amount of *f* at the expense of a marginal or incremental amount of *g.* The only values that are relevant to his choice are these increments by which the two policies differ; and, when he finally chooses between the two marginal values, he does so by making a choice between policies.

As to whether the attempt to clarify objectives in advance of policy selection is more or less rational than the close intertwining of marginal evaluation and empirical analysis, the principal difference established is that for complex problems the first is impossible and irrelevant, and the second is both possible and relevant. The second is possible because the administrator need not try to analyze any values except the values by which alternative policies differ and need not be concerned with them except as they differ marginally. His need for information on values or objectives is drastically reduced as compared with the root method; and his capacity for grasping, comprehending, and relating values to one another is not strained beyond the breaking point.

Relations Between Means and Ends (2b)

Decisionmaking is ordinarily formalized as a means-end relationship: means are conceived to be evaluated and chosen in the light of ends finally selected independently of and prior to the choice of means. This is the means-ends relationship of the root method. But it follows from all that has just been said that such a means-end relationship is possible only to the extent that values are agreed upon, are reconcilable, and are stable at the margin. Typically, therefore, such a means-ends relationship is absent from the branch method, where means and ends are simultaneously chosen. . . .

The Test of "Good" Policy (3b)

. . . For the root method, there is no test. Agreement on objectives failing, there is no standard of "correctness." For the method of successive limited comparisons, the test is agreement on policy itself, which remains possible even when agreement on values is not. . . .

In an important sense, therefore, it is not irrational for an administrator to defend a policy as good without being able to specify what it is good for.

Noncomprehensive Analysis (4b)

. . . In the method of successive limited comparisons, simplification is systematically achieved in two principal ways. First, it is achieved through limitation of policy comparisons to those policies that differ in relatively small degree from policies presently in effect. Such a limitation immediately reduces the number of alternatives to be investigated and also drastically simplifies the character of the investigation of each. . . .

The second method of simplification of analysis is the practice of ignoring important possible consequences of possible policies, as well as the values attached to the neglected consequences. If this appears to disclose a shocking shortcoming of successive limited comparisons, it can be replied that, even if the exclusions are random, policies may nevertheless be more intelligently formulated than through futile attempts to achieve a comprehensiveness beyond human capacity. Actually, however, the exclusions, seeming arbitrary or random from one point of view, need be neither. . . .

Even partisanship and narrowness, to use pejorative terms, will sometimes be assets to rational decisionmaking, for they can doubly insure that what one agency neglects, another will not; they specialize personnel to distinct points of view. . . .

Succession of Comparisons (5b)

The final distinctive element in the branch method is that the comparisons, together with the policy choice, proceed in a chronological series. Policy is not made once and for all; it is made and remade endlessly. Policymaking is a process of successive approximation to some desired objectives in which what is desired itself continues to change under reconsideration. . . . [T]heory is sometimes of extremely limited helpfulness in policymaking for at least two rather different reasons. It is greedy for facts; it can be constructed only through a great collection of observations. And it is typically insufficiently precise for application to a policy process that moves through small changes. In contrast, the comparative method both economizes on the need for facts and directs the analyst's attention to just those facts that are relevant to the fine choices faced by the decisionmaker. . . .

Successive Comparison as a System

Successive limited comparisons is, then, indeed a method or system; it is not a failure of method for which administrators ought to apologize. None the less, its imperfections, which have not been explored in this article, are many. For example, the method is without a built-in safeguard for all relevant values, and it also may lead the decisionmaker to overlook excellent policies for no other reason than that they are not suggested by the chain of successive policy steps leading up to the present. Hence, it ought to be said that under this method, as well as under some of the most sophisticated variants of the root method—operations research, for example—policies will continue to be as foolish as they are wise. . . .

. . . While much of organization theory argues the virtues of common values and agreed organizational objectives, for complex problems in which the root method is inapplicable, agencies will want among their own personnel two types of diversification: administrators whose thinking is organized by reference to policy chains other than those familiar to most members of the organization and, even more commonly, administrators whose professional or personal values or interests create diversity of view (perhaps coming from different specialties, social classes, geographical areas) so that, even within a single agency, decisionmaking can be fragmented and parts of the agency can serve as watchdogs for other parts.

NOTES AND QUESTIONS

1. Mashaw discusses three models: idealist, positive political theory (PPT), and critical legal studies (CLS). These models, he claims, seek to "explain[s] the shape of the administrative decision processes . . . ," but he fails to make clear precisely what he means by this. Does he think that the models are designed to explain the motivations behind decisions, the formal processes and rules through which they are reached, the informal processes, or their outcomes? Do all these models purport to explain the same thing(s)?

Having elaborated the three models, Marshaw appraises the evidence bearing on a major implication of the PPT model (that administrative process is designed to assure legislative control of outcomes) and finds that the evidence does not bear it out. He concludes by expressing his suspicion that a rigorous testing of the other two models would invalidate them as well. Is it possible that even Mashaw, who is at considerable pains to emphasize the difficulty of testing hypotheses about administrative process, has still understated it? Consider, for example, how many cases one would need to examine in detail in order to test an idealist hypothesis of judicial control, how difficult it is to define outcomes, and how the notion of "anticipatory reactions" by one or another branch to the others could complicate any causal analysis.

In their response to Mashaw, published in 6 *Journal of Law, Economics & Organization* 203 (1990), McNollgast concede most of Mashaw's argument about the importance of idealist norms in the actual design of administrative process but contend that this fact is not inconsistent with the PPT model because rational politicians concerned only with their own reelection could nevertheless support procedures, such as those in the Administrative Procedure Act, that limit their ability to intervene in administration. They also acknowledge courts as a source of procedures that embody normative goals. Given these concessions, how robust is PPT as a predicative explanatory model?

2. Compare Moe's description of bureaucratic incentives and dependence on "the other players" with Shapiro's analysis in Chapter III of the bureaucratic "fourth branch." Which is more consistent with the evidence? With a plausible theory of democratic legitimacy? How plausible is Moe's differentiation among the incentives that he claims drive legislators, the president, and the bureaucracy? Why does Moe think that "presidents are not very susceptible to the appeals of special interests"? Doesn't that depend in part upon the nature and fragility of his electoral coalition?

Moe, who like PPT theorists analyzes politics from a rational actor perspective, nevertheless criticizes the PPT approach for its preoccupation with legislatures and disinterest in bureaucracy and for its emphasis on political institutions (including agencies) as solutions to legislators' collective-action problems rather than on their role as "weapons of coercion and redistribution" in which there are winners and losers. Moe, "Political Institutions: The Neglected Side of the Story," 6 *Journal of Law, Economics, & Organization* 213 (1990). Is Moe's failure in this excerpt to discuss the role of courts in shaping agency structure a serious omission? From Moe's perspective, what are their incentives? Is Moe's analysis consistent with Robert Kagan's discussion in Chapter V of the sources of "adversarial legalism"? In a commentary on another rational choice analysis of how legislators attempt to control agency decisionmaking through structural design, Glen Robinson argues that this approach ignores "the difficulty of successful political manipulation of process or structure without agreement on substantive policy choices," that the public choice model's description of processes and structures is too general to support inferences about how they can be used to "stack the deck" in favor of certain interests, that administrative programs in fact do not exhibit the kind of political manipulation that the model predicts, and that the institution of judicial review compounds the legislator's principal-agent problem. 75 *Virginia Law Review* 483 (1989). See also Robinson, *American Bureaucracy: Public Choice and Public Law* (Ann Arbor: University of Michigan Press, 1991). Do these criticisms apply to Moe's argument as well? The public choice approach to explaining agency behavior is also discussed in the readings and notes and questions in Chapter I above.

3. In "The Internal Structure of EPA Rulemaking," 54 *Law and Contemporary Problems* 57 (1991), Thomas McGarity explores the phenomenon of "institutional expertise"—an agency's ability to integrate the contributions of widely varying professional perspectives into a single coherent policy product. He argues that the mobilization of such expertise is a special imperative for the modern social regulatory agency, as contrasted with the New Deal agency, which focused on a single industry or narrow set of economic problems. His example is the Environmental Protection Agency, which must somehow synthesize at least the following different perspectives: scientific (both pure and applied), engineering, management, enforcement, "economic-analytical," legal, and political. Drawing upon the rule-development process in EPA and other social regulation agencies, McGarity distinguishes five different decisionmaking models, each with its advantages and disadvantages: the team (used in EPA), the hierarchical (used by EPA when the agency is under time pressures), the outside advisor (used by the Occupational Safety and Health Administration), the adversarial (used by the National Highway Traffic Safety Administration), and the hybrid (formerly used by the EPA).

McGarity's taxonomy of informal decisionmaking structures, of course, is not exhaustive. Another model, employed by Joseph Califano when he was Secretary of

Health, Education, and Welfare, attempted to maintain close secretarial control over all phases of the policy development process, especially for rules involving issues cutting across program jurisdictions, a common situation in the department. The executive secretariat in the secretary's office would designate an agency, often a staff unit in the Office of the Secretary such as Planning and Evaluation or General Counsel, to lead and coordinate a task force consisting of representatives of all affected programs and staff agencies within the department. A member of the executive secretariat would attend all task force meetings on behalf of the secretary, infusing secretarial perspectives into the task force's deliberations and keeping the secretary apprised at an early stage of the process about all developments in which he might have an interest or want to intervene.

4. How does professionalism in agencies affect other social values relevant to administrative government such as hierarchical control, public accountability, efficiency, civil service morale, and citizen participation? In this connection, compare Wilson's brief discussion of professionalism in social work bureaucracies to William Simon's strikingly different view in Chapter VI, section 4.

5. Lindblom's incrementalist model accurately describes much agency decision-making. It not only corresponds to the popular view of bureaucracy as hidebound and unimaginative, but it is also borne out by many political science studies of particular agencies. An illuminating study of the difficulties created by imposing new tasks on an agency with a traditional mission is Martha Derthick, *Agency Under Stress: The Social Security Administration in American Government* (Washington, D.C.: Brookings Institution, 1990). Perhaps more instructive, then, are the many cases in which agencies have adopted innovative policies, rejecting the familiar for the unknown. Many examples are reviewed in chapter 12 of James Q. Wilson, *Bureaucracy: What Government Agencies Do and Why They Do It* (New York: Basic Books, 1989).

Does "muddling through" have an inescapably conservative bias? Lindblom's more intriguing claim for this model (coming as it does from one who is no conservative) is normative. Ironically, he argues, incrementalism produces superior results than the more synoptic or "rational-comprehensive" method. What light does this comparision between incrementalism and synopticism shed on the choice betwen adjudication and rulemaking as agency policymaking modes? On the choice between common law and legislation? On this point, see Colin Diver, "Policymaking Paradigms in Administrative Law," 95 *Harvard Law Review* 393 (1981).

V

Models of Procedural Justice and Effective Governance

The administrative process is not simply a machine for churning out regulatory policies and adjudicating claims, although it is certainly that. It is also an institutional embodiment of at least two kinds of normative visions of that process—one of procedural justice and one of effective governance. These visions, of course, are closely related. Only a system perceived as procedurally just is likely to command the respect of the governed, but even a fair procedure will not be tolerated for very long if it is too cumbersome or unreliable to produce effective decisions. For that reason, normative analyses of administrative process generally integrate the justice and governance criteria into each model.

What are those models? Which visions of justice and governance do and should guide us when we evaluate the structures of the administrative state and design new ones? These questions are of fundamental importance and have engaged the intense interest of lawyers, philosophers, political scientists, and organization theorists. Individual commentators employ different taxonomies and even different normative frameworks, but all are concerned with how bureaucratic power, especially the exercise of discretion, is to be legitimated.

In an excerpt from Jerry Mashaw's book, *Due Process in the Administrative State,* he elaborates three models of bureaucratic legitimation—"transmission belt" (a formalist model, derived from Richard Stewart's article excerpted in Chapter VI, section 5, which views the agency as strictly implementing the legislature's goals), expertise (another Stewart category, emphasizing the agency's technocratic capacities), and participation (a post–New Deal model but one increasingly beleaguered in the courts). The article by psychologist John Thibaut and law professor Laurens Walker presents an empirically based theory that emphasizes disputant control as the most attractive feature in a process capable of independently conferring legitimacy on distributional outcomes.

Charles Reich offers a rather different, more substantive approach. In his celebrated article, "The New Property," Reich defends a model of procedural justice that derives more from a strong normative view concerning how entitlements to public benefits work to protect and nourish individual freedom than from abstract notions of fair and reliable process. By assimilating government largess to the traditional categories of entitlement and property rights, Reich can extend to the former the more extensive procedural rights associated with the latter. Robert Rabin casts an interesting light on the magisterial, timeless quality of Reich's essay. Reviewing it at a quarter-century's remove, Rabin finds it curiously parochial and time bound in the way it assesses and balances the relevant social values.

My own study of the use of an exceptions process to ensure "regulatory equity" in a system dominated by rules presents yet another model for harmonizing the claims of procedural justice and effective governance. The excerpt distinguishes this model from several cognates—agency adjudication, pure discretion, discretion-within-rules, and judicial equity—which are also designed to respond to the various justice and efficiency limitations of rules.

Finally, Robert Kagan's case study of a particular port-dredging project in Oakland emphasizes the social costs of the highly participatory, formalized, legalistic system that he calls "adversarial legalism." Although this system certainly reflects long-standing legal and political structures and traditions, Kagan argues that it now poses a far more serious threat to the effectiveness of the activist state than ever before. This is a threat, moreover, that other liberal democracies have managed largely to avoid.

Due Process in the Administrative State

JERRY MASHAW

The difficulties with the transmission belt and expertise justifications for the exercise of administrative authority might . . . be ameliorated by the third model of administrative legitimacy—a model that concentrates on the administrative process itself. If administrative decisionmaking is too divorced from the macropolitics of legislative action to satisfy democratic ideals, those ideals might be embodied, at least partially, in administrative procedures that ensure the participation of affected interests in the process of administrative policy formation. Similarly, if rational analysis is inadequate to justify choices among competing values, those values might be represented in a pluralist, bargaining approach to administration. Presumably the outcomes of that process would combine administrative expertise with the political legitimacy of the traditional legislative compromise. Attempts to overcome the limitations of either the transmission belt or the expertise models thus suggest a participatory approach. Moreover, the notion of direct participation in administrative governance responds to deep strains of individualism and political egalitarianism in the American character. It rekindles the nostalgic image of the town meeting.

On closer examination, however, the participatory approach also has serious limitations. For one thing the micropolitics of a participatory administrative structure may undermine rather than support what remains of the attachment to legislative control or to rational decisionmaking. There is no obvious reason to expect that the outcome of pluralist bargaining at the micropolitical, administrative level should reproduce the results of bargaining in the macropolitical, legislative process. And a participatory, bargaining process will complicate, perhaps stifle, the application of those rational modes of thought that are necessary for the solution of many of our more complex and subtle problems.

Participation might, of course, be urged on other grounds. It might, for example, be argued that administration that cannot be shown to be democratically controlled or expert should at least be "fair," in the sense of providing equal opportunities to affected parties to influence the ultimate administrative judgment. But even were this principle accepted, it is not at all clear what a fair structure would look like. The immediately appealing analogy is to a judicial trial—open, neutral, and with strict equality of participation. Yet for questions that are broad and complex, such procedure is terribly cumbersome. Moreover, although the participatory approach seems to presume that the process of decision can be made open and neutral, it seems rather more likely that, with respect to many areas of administrative policy formation, certain interests, because of their intensity, resources, and organization, will come to dominate even an open decisionmaking process. Or, to put the point somewhat differently, interests that are substantially affected might, because of lack of resources or organization, fail to participate effectively in administrative forums. Any attempt to prevent these inequalities from skewing policy in "unfair" directions, by fine-tailoring the decision processes through the use of evidentiary presumptions, burdens of proof, required disclosure of information, and the like, reveals the need for some criteria for determining the importance of the interests represented by the various potential participants. The participatory model itself provides no such guidelines.

The Path of the Law Revisited

. . . The question of appropriate forms of participation in administrative proceedings does not present itself to a court (or to a legislative committee or to an administrator) as a question of competing theoretical models of administrative due process. Instead, a court will confront a claim by a particular party that it has a right to participate, or to participate in a particular way at a particular time, in an agency decision process. In making that process claim the suitor may well advance arguments that relate to all three of the models of administrative legitimacy discussed here. Indeed, it is commonplace to find litigants urging: (1) that faithful implementation of the statute requires that the administrator attend to their evidence attend to their evidence and arguments; (2) that decisionmaking absent their contribution is unreliable, uninformed, and therefore inexpert; and (3) that, given the impact of the decision on their interests, the agency process excluding their participation is unfair.

These arguments demand some reformulation or reinterpretation of the transmission belt and expertise models. The legislature must be viewed as both initiating a policy *and* creating or recognizing the interests that will be benefited or constrained by the policy. The transmission belt is thus seen as carrying both policies and rights. Expertise must also be extended or generalized to imply rationality as its goal. Participants are then viewed as increasing the stock of information and analysis available to an expert agency that is not omniscient, merely skilled. Yet, given the intuitive appeal of these reinterpretations—their tighter fit with contemporary conceptions of pluralistic politics and bounded rationality—the process lens through which claims are refracted increases the power and appeal of the due process paradigm that such claims invoke. Such a paradigm synthesizes the historic foundations of administrative legitimacy into an ideal of delegated, rational, and fair administration, realizable through decision processes designed to guarantee appropriate participation. It is this "access," "participation," "some kind of hearing," that the plaintiffs in thousands of administrative due process cases demand.

An Emerging Counterrevolution?

The attractiveness of this vision notwithstanding, a contemporary counter-tendency is discernible. Participation has costs as well as benefits. As recognition of those costs mounts, particularly the costs of intense participation by affected interests via trial-type procedures, the political appeal of participatory legitimation has declined. The Supreme Court has begun, directly and indirectly, to discourage the procedurally revisionist stance of the lower federal courts that provided much of the legal underpinning for the participatory approach. Both at the legislative and judicial levels there is in a sense a turning back to reexplore the possibilities of substantive legitimation either through legislative review of the transmission belt or through rational discourse. This return to a substantive focus increasingly features flirtation with the contemporary incarnations of rationality, cost-benefit and risk-benefit analysis.

A Theory of Procedure

JOHN THIBAUT AND LAURENS WALKER

We wish to emphasize that our proposal is general in the sense that it is intended to apply to all instances of interpersonal conflict. In the realm of the law, we have intentionally disregarded such traditional subject matter categories as . . . "administrative procedure." . . . The proposal is largely derived from systematically collected empirical evidence. . . .

The theory begins with the distinction between the two conflict resolution objectives of "justice" and "truth." We contend that in most instances one or the other of these objectives is dictated by the subject matter of the dispute, or more specifically by the outcome relationship that exists between the individual parties to the conflict.

At one extreme, the relationship of the parties may be such that one resolution of the dispute will uniformly enhance the outcomes of all interested parties, while a different resolution will uniformly reduce these outcomes. ("Will this path lead us out of the forest, or that one?") In this situation, it may be said that the interests of the parties are coincident, and if any dispute arises it will entail simply a "cognitive" conflict—a dispute as to which resolution is to the common advantage. . . .

Disputes that develop in scientific inquiry are the prototype of cognitive conflict in a setting of common interest. . . .

At the other extreme of the subject matter continuum is the situation of maximum conflict of interest. In this case the respective interests of the parties are perfectly opposed because a particular solution will maximize the outcome of one of the parties only at the expense of the other. Here the ultimate test of any particular solution is the character of the distribution of outcomes among the interested parties, and no solution will ultimately be recognized as "correct" by all of them. Hence, the objective of resolving conflicts of interest must frankly be seen as something other than finding the "true" or scientifically valid result. From the time of Aristotle the objective in resolving this kind of dispute has been characterized as "justice."

The legal process is concerned, for the most part, with the resolution of conflicts of interest. . . .

It is true, of course, that the legal process is often concerned with resolving disputes about "facts," and factfinding can be determinative when the conflicting justice claims are controlled by established legal rules. But typically determinations of fact are subordinate to the justice objective. The purpose of factfinding in the legal process contrasts sharply with the purpose of factfinding in scientific inquiry. In science the facts found have an enduring significance because they guide future conduct. . . . In contrast, the significance of factual determinations in a legal proceeding generally ends with the division of outcomes and there is no future reliance on the cognitive decision. . . .

The distribution of control among the procedural group participants is the most significant factor in characterizing a procedural system. "Control" involves at least two elements: control over the decision and control over the process. . . . "When the ratio of their rewards is equal, as perceived by *all* the parties, to the ratio of their contributions, then the distribution of reward is said to be fair, just, or equitable." This conception of distributive justice is empirical and relativistic; it is not a philosophical ethic. It is grounded in the *perceptions* that justice has or has not been done, according to the cultural or subcultural standards that define the values of contributions or inputs and of rewards or outcomes.

The foregoing analysis demonstrates that the procedural model best suited to the attainment of distributive justice in disputes entailing high conflict of interest is arbitration, or more specifically in legal settings, the Anglo-American adversary model. Most of the process control rests with the disputants, who are able to present their claims from their own perspectives, with full particularities and contexts. The impartial decisionmaker hears the contending presentations, evaluates the relative weights of the input claims, and renders the decision that distributes the outcomes. The freedom of the disputants to control the statement of their claims constitutes the best assurance that they will subsequently believe that justice has been done regardless of the verdict. And though they must exercise this kind of process control, they are in no position to evaluate the relative weights of the rival claims. Decision control must, therefore, be in the hands of a third party who applies a normative standard to the conflict between idiographic claims.

Our endorsement of the adversary model as likely to obtain distributive justice is strictly limited to our definition of that model as a system of legal decisionmaking that assigns virtually all process control to the disputing parties but that reserves decision control to the judge or jury.

Some Practical Constraints

Our argument concerning the optimal distribution of process and decision control in conflict resolution groups is meant to apply generally to the legal process, though we recognize limitations on its applicability in certain situations. One such situation arises when the outcome of a dispute is controlled by a precise substantive or normative rule. . . . Not only does this reduce the disputants' process control, but it also transforms the nature of the third party's decision control. The decisionmaker becomes a scholar of the code, a sophisticated factfinder whose task is to match the disputants' actions with the relevant statutory provision. The decisionmaker's function is directed less toward justice and more toward truth.

Of course, specificity in rules may enhance perceptions of the fairness of the legal system to the extent that it increases evenhandedness in the administration of justice. . . . Some erosion of justice in the courts may be viewed as a necessary trade-off against the benefits of codification as caveat.

The substitution of legislative judgments for case-by-case determinations of distributive justice in the courts will hinder the achievement of justice unless the legislative process compensates the disputants for the loss of process control in the courts. . . . The inherent problems in anticipating circumstances and context in future disputes about the distribution of outcomes suggest that citizen process control in the legislature cannot be an adequate substitute for disputant process control in the decisionmaking forum.

A second limitation on our argument arises when the disputants have insufficient resources to present their input claims fully. . . .

Resolution of these mixed disputes presents particularly difficult problems. When disputes involve an admixture of intense conflicts of interest and strongly divergent claims about matters of fact, neither an autocratic nor an unrestrainedly adversarial procedure is satisfactory.

A Two-Tier Solution

These two examples of the class of mixed disputes are typical. How can these intricately compounded disputes be properly resolved? Application of straightforward autocratic methods is likely to violate the concerns of justice, while the unfettered use of adversarial methods will impede the attainment of truth. Although there appears to be no ideal solution, our theory suggests that a two-stage procedure will best reconcile the two objectives in such disputes. The first stage should resolve issues of fact with the objective of determining truth; the second stage should resolve policy questions in a wholly separate procedure.

Our proposal requires the separation of questions of fact from those of justice or policy. Many difficulties are entailed in making this separation. Yet the distinction is clear enough in principle, and, in psychological research, it has proved feasible to make the separation both conceptually and empirically.

Once the separation has been made, the next step is to determine the appropriate procedure for resolving the issues of fact. Since this procedure must be capable of yielding a provisional resolution of the cognitive conflict in spite of a strong conflict of interest, it must allocate total decision control to a third party. Furthermore, since the primary objective of the procedure at this stage is to determine the truth, it must also allocate to the decisionmaker a degree of process control. Inevitably, the conflict of interest inherent in the relationship between the disputants remains as a latent force underlying the proceedings, even though the only explicit aim at this point is to resolve the cognitive elements of the truncated dispute. In deference to the real, if unofficial, presence of the interest conflict, a measure of process control must be retained by the disputants themselves. But the decisionmaker should excise statements of the values, preferences, and special interests of the disputants from the documents of the case and should proscribe all references to them. In general, then, the procedure suggested for this factual dispute extracted from its context of conflicting interests lies between the autocratic and the adversarial models.

According to this proposal, the decision control of the third party must be strong. If questions of science and technology are at issue, the decisionmaker should be a scientist (or a panel of scientists) who is fully capable of evaluating the particular claims in dispute. If the cognitive conflict derives in any part from contending "paradigms," the decisionmaker should be able to understand and assess them, and perhaps to assist in translating each paradigm into terms that are understood by the other side. Even if the scientist (or the panel) has the authority to decide the dispute, a lack of understanding will impair effective decision control by forcing him to decide on the basis of considerations other than the scientific merit of the claims. And even though the nature of science and technology makes any such decision provisional, an uninformed decision is unnecessarily inaccurate and thus subject to immediate, and wasteful, revision.

We have suggested that the disputants should exercise less process control, and the decisionmaker more, than in an adversarial model. The disputants' process control could be reduced by requiring that the contending representatives reach agreement in advance on the rules of evidence to apply in the proceedings, including rules of relevance. The representatives should be scientists, not lawyers. Therefore, the rules of evidence would probably conform to the norms that constrain presentations at a scientific meeting. A referee might be necessary to monitor compliance with the agreed rules of procedure. This close control over the procedure would not only reduce the process control of the disputants but would also increase the process control of the decisionmaker, particularly if the referee worked closely with the decisionmaker or if both functions were performed by a single judge. . . . The analysis developed in the preceding pages suggests that the proposal for the Science Court is the most promising method for achieving the objectives of truth and just policy.

The New Property

CHARLES REICH

[In this article, one of the most influential and frequently cited in the entire legal literature, Reich explores the threat that the expansion of government largess—in the form of welfare benefits, job opportunities, grants, contracts, licenses, franchises and the like—posed to individuality. Reich documents the growth of public lawyers and examines the legal regime that permitted the government to treat this largess as revocable while conferring more secure protection on traditional forms of property. Reich calls this regime, "the new feudalisms" because it possesses the following features:

> The comparison of the general outlines of the feudal system may best be seen by recapitulating some of the chief features of government largess. (1) Increasingly we turn over wealth and rights to government, which reallocates and redistributes them in the many forms of largess; (2) there is a merging of public and private, in which lines of private ownership are blurred; (3) the administration of the system has given rise to special law and special tribunals, outside the ordinary structure of government; (4) the right to possess and use government largess is bound up with the recipient's legal status; status is both the basis for receiving largess and a consequence of receiving it; hence the new wealth is not readily transferable; (5) individuals hold the wealth conditionally rather than absolutely; the conditions are usually obligations owed to the government or to the public, and may include the obligation of loyalty to the government; the obligations may be changed or increased at the will of the state; (6) for breach of condition the wealth may be forfeited or escheated back to the government; (7) the sovereign power is shared with large private interests; (8) the object of the whole system is to enforce "the public interest"—the interest of the state or society or the lord paramount—by means of the distribution and use of wealth in such a way as to create and maintain dependence.

In the article's concluding section, Reich urges that we convert this largess into a new form of property.]

Reprinted by permission of The Yale Law Journal Company and Fred B. Rothman & Company from *The Yale Law Journal*, Vol. 73, pp. 773–87.

Toward Individual Stakes in the Commonwealth

Ahead there stretches—to the farthest horizon—the joyless landscape of the public interest state. The life it promises will be comfortable and comforting. It will be well planned—with suitable areas for work and play. But there will be no precincts sacred to the spirit of individual man.

There can be no retreat from the public interest state. It is the inevitable outgrowth of an interdependent world. . . . As we move toward a welfare state, largess will be an ever more important form of wealth. And largess is a vital link in the relationship between the government and private sides of society. It is necessary, then, that largess begin to do the work of property.

The chief obstacle to the creation of private rights in largess has been the fact that it is originally public property, comes from the state, and may be withheld completely. But this need not be an obstacle. Traditional property also comes from the state, and in much the same way. . . . Hence, all property might be described as government largess, given on condition and subject to loss.

If all property is government largess, why is it not regulated to the same degree as present-day largess? Regulation of property has been limited, not because society had no interest in property, but because it was in the interest of society that property be free. Once property is seen not as a natural right but as a construction designed to serve certain functions, then its origin ceases to be decisive in determining how much regulation should be imposed. The conditions that can be attached to receipt, ownership, and use depend not on where property came from, but on what job it should be expected to perform. . . . Our primary focus must be those forms of largess which chiefly control the rights and status of the individual.

Constitutional Limits

The most clearly defined problem posed by government largess is the way it can be used to apply pressure against the exercise of constitutional rights. A first principle should be that government must have no power to "buy up" rights guaranteed by the Constitution. It should not be able to impose any condition on largess that would be invalid if imposed on something other than a "gratuity." . . .

The problem becomes more complicated when a court attempts, as current doctrine seems to require, to "balance" the deterrence of a constitutional right against some opposing interest. In any balancing process, no weight should be given to the contention that what is at stake is a mere gratuity. It should be recognized that pressure against constitutional rights from denial of a "gratuity" may be as great or greater than pressure from criminal punishment. And the concept of the public interest should be given a meaning broad enough to include general injury to independence and constitutional rights. It is not possible to consider detailed problems here. It is enough to say that government should gain no power, as against constitutional limitations, by reason of its role as a dispenser of wealth.

Substantive Limits

Beyond the limits deriving from the Constitution, what limits should be imposed on governmental power over largess? Such limits, whatever they may be, must be largely self-imposed and self-policed by legislatures; the Constitution sets only a bare minimum of limitations on legislative policy. The first type of limit should be on relevance. It has proven possible to argue that practically anything in the way of regulation is relevant to some legitimate legislative purpose. . . . But legislatures should strive for a meaningful, judicious concept of relevance if regulation of largess is not to become a handle for regulating everything else.

Besides relevance, a second important limit on substantive power might be concerned with discretion. To the extent possible, delegated power to make rules ought to be confined within ascertainable limits, and regulating agencies should not be assigned the task of enforcing conflicting policies. Also, agencies should be enjoined to use their powers only for the purposes for which they were designed. . . .

A final limit on substantive power, one that should be of growing importance, might be a principle that policymaking authority ought not to be delegated to essentially private organizations. The increasing practice of giving professional associations and occupational organizations authority in areas of government largess tends to make an individual subject to a guild of his fellows. A guild system, when attached to government largess, adds to the feudal characteristics of the system.

Procedural Safeguards

Because it is so hard to confine relevance and discretion, procedure offers a valuable means for restraining arbitrary action.

The grant, denial, revocation, and administration of all types of government largess should be subject to scrupulous observance of fair procedures. Action should be open to hearing and contest, and based upon a record subject to judicial review. The denial of any form of privilege or benefit on the basis of undisclosed reasons should no longer be tolerated. Nor should the same person sit as legislator, prosecutor, judge, and jury, combining all the functions of government in such a way as to make fairness virtually impossible. There is no justification for the survival of arbitrary methods where valuable rights are at stake.

. . . Today many administrative agencies take action which is penal in all but name. The penal nature of these actions should be recognized by appropriate procedures.

Even if no sanction is involved, the proceedings associated with government largess must not be used to undertake adjudications of facts that normally should be made by a court after a trial. . . .

From Largess to Right

The proposals discussed above, however salutary, are by themselves far from adequate to assure the status of individual man with respect to largess. The problems go deeper. First, the growth of government power based on the dispensing of wealth must be kept within bounds. Second, there must be a zone of privacy for each individual beyond which neither government nor private power can push—a hiding place from the all-pervasive system of regulation and control. Finally, it must be recognized that we are becoming a society based upon relationship and status—status deriving primarily from source of livelihood. Status is so closely linked to personality that destruction of one may well destroy the other. Status must therefore be surrounded with the kind of safeguards once reserved for personality.

Eventually those forms of largess which are closely linked to status must be deemed to be held as of right. Like property, such largess could be governed by a system of regulation plus civil or criminal sanctions, rather than a system based upon denial, suspension, and revocation. As things now stand, violations lead to forfeitures—outright confiscation of wealth and status. But there is surely no need for these drastic results. Confiscation, if used at all, should be the ultimate, not the most common and convenient penalty. The presumption should be that the professional man will keep his license, and the welfare recipient his pension. These interests should be "vested." If revocation is necessary, not by reason of the fault of the individual holder, but by reason of overriding demands of public policy, perhaps payment of just compensation would be appropriate. The individual should not bear the entire loss for a remedy primarily intended to benefit the community.

The concept of right is most urgently needed with respect to benefits like unemployment compensation, public assistance, and old age insurance. These benefits are based upon a recognition that misfortune and deprivation are often caused by forces far beyond the control of the individual, such as technological change, variations in demand for goods, depressions, or wars. The aim of these benefits is to preserve the self-sufficiency of the individual, to rehabilitate him where necessary, and to allow him to be a valuable member of a family and a community; in theory they represent part of the individual's rightful share in the commonwealth. Only by making such benefits into rights can the welfare state achieve its goal of providing a secure minimum basis for individual well-being and dignity in a society where each man cannot be wholly the master of his own destiny.

Reflections on "The New Property"

ROBERT RABIN

"The New Property," like any forceful messianic tract, is written from a perspective that tnds to lose sight of the countervailing considerations that make more intelligible the salutary intentions underlying developments which otherwise would appear to be sheer Nemesis. In context, the Administrative State is not just the latest version of state-initiated witch hunts or pogroms designed to root out nonconformists and heretics (although, in fairness to Reich, McCarthyism itself could certainly have been so regarded). Instead, the legacy of Progressive and New Deal ideology is a highly complex, mixcd public-private system of regulation and benefit distribution in which retaining the "precincts sacred to the spirit of individual man" becomes a necessarily complicated endeavor.

The crux of the problem is very simply stated. Those who would exercise their individuality without constraint often would do so at the expense of others. This was obvious in the fifties, in fact, in the context of a political struggle quite apart from the benighted anticommunist crusade, namely, the civil rights movement. At lunch counters, department stores, and bus stations, proprietors justified their racial bigotry as an expression of individualism. . . .

In its 1990s guise, the question is whether public educational institutions can adopt a "moral character" condition on educational opportunities—a form of government largess—that sanctions incivility in the face of claims of free expression. In my view, "The New Property" provides virtually no assistance in framing a focused answer to this question—whether the offending student is marching around in a Nazi uniform screaming racial epithets, publishing racially stereotypical cartoons in a student newspaper, or engaging in one of the variants discussed above. Without a morc fine-tuned balancing of the limits of free expression and the protections of minority status than "relevance" (to the educational enterprise) provides, any resolution is simply an exercise in circular reasoning.

The same difficulties are present in the welfare rights area. If one puts aside blatant fifties practices, such as midnight raids to implement a state-enforced moral code prohibiting a man from spending the night in the house of an AFDC [Aide to Families with Dependent Children] recipient, the precincts of protected individualism again become murky. Suppose the social worker's home visit was to insure that a special grant for the dependent child's winter clothing had not, in fact, been used to buy liquor. Or

"The Administrative State and Its Excesses: Reflections on *The New Property*. Copyright © 1990 by University of San Francisco Law Review. Reprinted by permission from 25 *University of San Francisco Law Review* 273 (1990).

consider the monitoring—again through home visits—of a crack-addicted parent, living alone with a dependent infant under borderline economic circumstances.

There was plenty of life-style supervision going on here, and the parent could have been coercively subjected to a regime of parenting classes and treatment programs at risk of losing custody of her child. But anyone who observed the consequences of these life-styles unaltered—observed the children at a later point in county shelters, juvenile halls, or foster homes—would have been hard-pressed to promote the virtues of a parental right to be left alone. Again, "relevance" as a limitation on the state's intrusion into the life style of the welfare recipient fails to provide sufficient guidance to the appropriate measure of official scrutiny.

The central thrust of these observations is that "The New Property" provided a clarion call to vigilance in an era of egregious intrusions into the sphere of individual freedom, but, in the process of doing so, it failed to take account of a simple but critical distinction: While government largess programs at times have been hedged with conditions the dominant purpose of which is to discourage nonconformist behavior as an end in itself, other programs feature limitations designed to assure that largess will not be used in a fashion that poses risks to the health and safety of others. In the latter kinds of cases, where the administrative program is designed to promote a *conscientious* pattern of conduct, rather than a *conformist* mode of behavior, a more refined conception of due process, as well as a system of internal bureaucratic controls, is essential to safeguard effectively against arbitrary official conduct. . . .

My difficulty with Reich's secondary thesis—in essence, a thesis disparaging expertise and promoting public participation—parallels my reservations about his central theme. In the paradigmatic cases of agency abuse that he discusses, there is no doubt that he is correct. Economic regulatory agencies did abuse their mandates when they systematically stifled economic competition by adopting policies discriminatory against new entrants and price-cutters. Natural resource agencies did ignore their statutory responsibilities when they failed to consider intangible costs of environmental degradation in allocating sites and resources to commercial interests. Nonetheless, a populist demand for greater public participation is of limited utility in resolving more subtle issues of when it is appropriate to defer to agency claims of expertness.

Take Scenic Hudson [Preservation Conference v. FPC] itself. After the case was remanded to the FPC [Federal Power Commission], the agency proceeded to take reams of testimony and documentary evidence on the aesthetic, conservational, historical, and ecological values affected by the proposed hydroelectric power plant at Storm King Mountain. In the final analysis, however, no talismanic process could conclusively determine the weight to be assigned those values and the corresponding economic costs and benefits that the additional electric power would provide for New York residents. . . . A transformed administrative law based on participatory values will almost certainly improve the process of decision through its sensitivity to a wider array of issues, but it has no substantive content.

Once again, however, Reich was a sharp observer of his times. He wrote just as consumer and environmental activism was about to crystalize in the public interest law movement and the social regulation era of the late sixties and early seventies. Almost overnight, agency processes opened up to traditionally unrepresented interests and a new judicial activism arose, animated by demands for broad access to agency officials and expanded justifications of administrative policy. While the new administrative law of the seventies provided no key to a substantive definition of the public interest, it did widen the horizons of government regulators and planners in a salutary fashion.

When the Exception Becomes the Rule: Regulatory Equity and the Formulation of Energy Policy Through an Exceptions Process
PETER SCHUCK

This article examines the pursuit of regulatory equity through an administrative "exceptions process." Such a process relieves a person, firm, or entity subject to a valid statutory or administrative rule from the legal obligation to comply with the rule. It does so by issuing a formal exception, waiver, or similar form of special relief from the rule's application based upon the special features of the applicant's situation. . . .

The Limits of Rules

That rules cannot fully achieve justice is an ancient theme in legal theory. Aristotle distinguished between "legal" justice, based upon general rules, and "equity," which corrects what is merely legally just. That distinction has been reflected in many important institutional and doctrinal developments. Examples include the equitable jurisdiction of courts; the evolution of many rigid common law rules into more flexible standards, such as "reasonableness," in which equity is "built-in"; and the open-ended, indeterminate standards contained in much modern legislation.

Regimes of rules and of equity are based upon fundamentally different conceptions of justice; each conception, taken alone, is radically incomplete. The regime of rules is general and prescriptive. It requires at a minimum that a decisionmaker accurately apply general, clearly articulated norms to fairly

Peter H. Schuck, "When the Exception Becomes the Rule: Regulatory Equity and the Formulation of Energy Policy Through an Exceptions Process," 1984 *Duke Law Journal,* 163. Reprinted with permission.

established facts. In contrast, the regime of equity is specific and ad hoc. It requires that any such norms be subordinated to an overriding standard of contextual fairness—that is, fairness to individuals in light of their particularized situations. Rules respond to the gravitational pulls of backward-looking precedent, forward-looking policy, or outward-looking analogies and notions of legal equality. Equity resonates to the inward-looking concerns of the immediate, to the riveting force of the particular, the situational. Those who value rules emphasize the need for commonality, regularity, continuity, and order. Those who esteem equity stress the importance of individuality (we are unique), spontaneity (we are not bound by the past), responsiveness (our needs change), and freedom (our course is not predetermined). Whereas rules exalt formal categories, equity is profoundly suspicious of abstraction and seeks to free diversity from the delusive discipline of order. Equity speaks not to the typical case—indeed, it may even deny that such a thing exists—but rather to the exotic, the unexpected, the exigent. Its distinctive techniques are neither comprehensive vision, synoptic analysis, nor disembodied principle, but sound judgment, contextual analysis, and an intuitive "feel" for what is fair under the circumstances.

Rules glorify precisely those technocratic values—predictability, stability, uniformity, and control—about which equity is most skeptical; they systematically overlook or override certain particularities of time, place, and context. Our sense of justice, however, often demands that these particularities receive greater recognition and weight than rules ordinarily accord them. We can readily imagine sets of facts that come well within a rule's terms yet seem to call for a different disposition. When such an instance occurs, justice seems to call for suspending the rule in that case, without necessarily calling its general validity into question. Equity, too, entails certain risks, but they are the risks of uncertainty, irrationality, favoritism, diversity, and incoherence—precisely those that an increasingly rationalistic, centralized, control-oriented society wishes to minimize. The tension in our law between rule and equity, then, can never be eliminated. It will persist, as Thurman Arnold noted, "[s]o long as men require a moral and logical ideal to satisfy their impulses toward mercy and common sense." The quest for regulatory equity is a search for an appropriate means to harmonize these eternally conflicting values.

Any complex system of rules that aspires to justice must make some provision for regulatory equity. Although classes of cases always exist in which rules create injustice, this does not explain what it is about a case that places it within those classes. It fails to reveal which characteristics demand that the rule be suspended rather than applied, and which reasons or motives make an appeal to equity seem appropriate. The following discussion attempts to provide that explanation.

The limits of rules define the boundaries of regulatory equity. Broadly speaking, three general categories of conditions or reasons may mark those limits. The need for regulatory equity may derive from certain features of the rule itself, from the institutional context in which the rule is developed and applied, and from situational needs of the decisionmaker that may have nothing to do with the rule's own features or context.

The Character of Rules

. . . Every rule has at least some limitations. These may often be reduced (at some cost) but can never be wholly eliminated, for they inhere either in the nature of rules or in the complexity of reality. Insofar as these limitations remain, perfect justice solely through a regime of rules is unattainable.

Limitations of form. Three formal attributes of a rule determine the extent to which it can achieve its purposes without equity's assistance: its generality (the number of cases to which it is meant to apply), its transparency (the degree to which it evokes uniform interpretations in different minds), and the number of unweighted factors that it makes relevant to its application. These attributes are empirically as well as analytically distinct.

Any given rule, then, expresses an inescapable tension between the virtues and vices of each of its formal elements. A more general rule controls more cases but only by subjecting more diverse phenomena to its homogenizing prescriptions. A vaguer, more ambiguous rule permits greater flexibility in adapting to new or unanticipated conditions but only by reducing the predictability and uniformity of outcomes. A multifactored rule, especially one that fails to weight the factors, enriches the administrator's ability to discriminate among complex phenomena but diminishes the rulemaker's (or reviewing court's) control over particular decisions. Regulators can only transcend these limitations by looking beyond the regime of rules to the ethos of equity.

Limitations of knowledge. We are extraordinarily ignorant of the intricate web of cause, consequence, and circumstance in which our intentions and actions are shaped and executed. But however little we know of the world as it is, we know even less of the world that will be. A regime of rules, of course, is directed to the future. To Aristotle, equity meant deciding cases as the "legislator himself would have said had he been present, and would have put into his law if he had known." In this view, when cases arise that the rule did not envision and to which it cannot justly be applied, equity must effect what the rulemaker's want of prescience makes necessary; it must suspend or reformulate the rule (Aristotle does not indicate which) to achieve a just outcome in the unforeseen case.

This notion of the relationship between imperfect knowledge and equity seems quite incomplete, at least in the regulatory context. Imagine a perfectly prescient rulemaker. She fully comprehends the implications of all rules that she might issue, foresees all changes that might occur, and anticipates all cases that may arise in the future. Is it clear, as Aristotle implies, that this clairvoyant seer would formulate a rule resolving in advance all cases yet to come? And if she would, would this achieve perfect prescriptive justice, rendering equity superfluous? The answer to both questions, it would seem, is no. She might believe, for example, that by framing a rule today that will anticipate and correctly decide tomorrow's case, she would incur unacceptable political costs in attempting to persuade others to accept that rule. She might therefore prefer to proceed incrementally, issuing a more limited but noncontroversial rule now and deferring the resolution of future cases to a later day. If she also faced high information costs in formulating a rule capable of fully anticipating the future, she might find an incremental strategy even more attractive.

Aristotle also seems to have been wholly concerned with rules formulated by rulemakers who not only are ignorant of the future but also proceed as if they were *unaware* of their ignorance. Yet some rulemakers *are* aware of their ignorance and guard against it in fashioning rules. Both kinds of rules demonstrate the need for regulatory equity, but each implies somewhat different responses to that need.

Limitations of comprehensiveness. A rule does not always govern autonomously, even within its well-defined boundaries. Dominion over its subjects may be hotly contested. . . .

Threats to the comprehensiveness and autonomy of an agency's rules, consequently, come from several quarters—from conflicts between rules, from policies and principles that exemplify other values, and from other institutions. These sources of tension reflect moral, administrative, and political imperatives to which an agency must somehow respond. For example, when the Occupational Safety and Health Administration is challenged to justify an occupational safety standard that imposes heavy costs upon regulated firms, it cannot simply point to its broad statutory authority to develop rules that minimize risks to workers. Even before an agency promulgates a statutorily authorized rule, it must take account of competing policies (e.g., the objective of maintaining a viable industrial sector) and competing principles (e.g., a worker's health is not protected if her employer is forced out of business). Equity is one integrating technique.

Limitations of articulation. In our legal culture, the legitimacy of a rule—and perhaps its validity as well—ordinarily requires that the reasons adduced in its support be justified in terms of some preexisting, appropriate premise of decision, some rule or principle whose generality, by transcending the particular case, avoids the risks of arbitrariness inherent in ad hoc decisionmaking. . . .

. . . The norm of reason-giving is not nearly as well-established for regulators as for judges. Agencies sometimes provide reasons when they are not legally obligated to do so, but their articulation is often cursory, conclusory, and of little use to one seeking to evaluate a decision's substantive rationality.

Although the law often *permits* decisionmakers to refrain from providing reasons, the nature of certain decisions may actually *preclude* them from doing so. . . .

The Institutional Context of Rules

A regime of rules is embedded in a set of institutional arrangements through which general prescriptions are developed, promulgated, applied, evaluated, and perhaps modified. The ability of such a regime to achieve justice without recourse to equity depends upon the institutions charged with performing these rule-sustaining functions. Because certain institutions are especially hospitable to rule-based justice and others to equity, the legal system can affect how equitable considerations shape a rule's application by entrusting it to the ministrations of one kind of institution rather than another.

The organization, ideology, and operating procedures of common law courts, especially those engaged in private adjudication, encourage judges to be more preoccupied with the interests of particular litigants than with those of the larger society. Regulatory agencies, however, tend to weigh policy and programmatic considerations more heavily, even when they adjudicate. As [I later] demonstrate, this distinction reflects differences in the institutional purposes, types of rules, decision procedures, and organizational forms and settings of courts and agencies.

Legislatures present a somewhat more complex pattern. Their most characteristic output, the public bill, ordinarily takes the form of a general rule. This is not surprising. Generality not only minimizes the risk and appearance of legislative favoritism but also captures substantial "economies of scale" in legislative production. Legislatures, however, have also established specialized organs to dispense particularistic justice. Private bills, which typically prescribe results intended for only one or a small number of beneficiaries, are ordinarily processed under unique procedures. Public bills, of course, are sometimes drafted to achieve decidedly particularistic results—for example, so-called "Christmas tree" tax amendments designed to benefit one or relatively few companies. Even there, however, special procedures are needed to minimize abuse.

The relationship between different conceptions of justice and the institutions created to implement them is not simply a matter for institutional designers in the legislative and executive branches. Courts also influence this relationship, most notably by elaborating a common law of administrative process designed to render agency decisionmaking more regularized, generalized, visible, and susceptible to judicial scrutiny and control. To facilitate judicial review, for example, courts have encouraged agencies to use rules. This, in turn, increases the need for regulatory equity to supplement and refine those rules. On the other hand, when adjudicating in a common law mode, courts tend to avoid rule-like decisions that might constrain their future flexibility.

The Situational Needs of Rulemakers

Both the inherent imperfections of rules and the particular institutional settings in which they are developed and applied may create lacunae in the structure of regulatory justice that only equity can fill. But not all demands for equity are so systematic. Some are ad hoc, reflecting the incentives that confront particular rulemakers in particular circumstances. Two of these circumstances are particularly important: the need to win political support for agency decisions and the need to process a large volume of cases with scarce administrative resources.

Regulatory Equity and Discretion

There is an intimate relationship between regulatory equity, in the sense in which that concept is used here, and discretion. Discretion, according to its chief expositor, Professor Kenneth Davis, exists "whenever the effective limits on [an official's] power leave him free to make a choice among possible courses of action or inaction." An equitable standard, by its very nature, creates some discretion in the official who applies it; a number of different outcomes will be consistent with that standard. Discretion, therefore, affords decisionmakers leeway within which they can give equitable considerations some weight when they apply a rule.

But regulatory equity and discretion are by no means identical concepts. Regulatory equity is an objective or goal of the legal system, while discretion is a legal form or technique, a means to that end. Discretion is necessary to achieve regulatory equity but it is not an end in itself. The relationship between discretion and regulatory equity is revealed in the different techniques through which regulatory decisions can assimilate equitable considerations, the role of discretion in each of those regulatory techniques, and its somewhat different role in an exceptions process.

First, regulatory equity may be achieved through *rule formulation*. . . .

Second, regulatory equity may, within certain limits, be achieved through *rule interpretation*. . . .

Third, regulatory equity can be introduced through *rule enforcement*. This is not quite the same as rule interpretation. . . .

Finally, regulatory equity can be introduced by creating a *formal exceptions process* for the explicit purpose of considering applications for relief from particular rules. Whether the relief that this process grants is described as an exception, waiver, special relief, variance, or exemption, the technique is essentially the same. An exceptions process, as we shall see, also employs rule formulation, rule interpretation, and rule enforcement techniques in performing its tasks.

Discretion plays an essential but distinctive role in each of these regulatory equity techniques. . . .

In a formal exceptions process, discretion plays yet another role. There, the process itself assumes that the rule, as formulated and interpreted, does apply to a particular situation and presumably will be enforced as such. In that context, discretion does not operate upon the rule but is only invoked to determine whether that situation qualifies for special dispensation under the legal standards governing exceptions relief.

It is tempting to view this as a distinction without a difference, to conclude that the use of discretion in the exceptions process simply effects a reformulation, reinterpretation, or nonenforcement of the rule. This view, however, ignores several important points. The first concerns the moral meaning of rules. . . .

A second distinction between discretion that acts directly upon a rule and discretion that is used in an exceptions process to decide whether relief from a concededly applicable rule should be granted, relates to the institutional structure within which the decisions are made.

Regulatory Equity and Judicial Equity

The pursuit of equity in the regulatory process has certain obvious parallels to the more familiar, traditional phenomenon of judicial equity through case-by-case adjudications. . . . Nevertheless, the differences between the ways in which courts and agencies deploy equitable values are at least as important as the apparent similarities, underscoring the distinctive character of regulatory equity. They also suggest why a specialized administrative organ or arrangement is often necessary to infuse regulatory equity into the administrative process.

Institutional Purpose

The dominant purposes of courts and agencies differ. Characteristically, courts adjudicate the rights and obligations of disputants on the basis of principled justifications and distinctions derived from previously adopted legal norms. . . .

A regulatory agency, in contrast, is an engine of continuous social policy formulation and implementation. Although obliged to render neutral, principled decisions with respect to individual disputes brought before it, an agency's principal purpose is to effectuate an externally created but bureaucratically internalized legislative purpose, usually the protection of certain collective values or group interests. . . . This orientation encourages agencies systematically to undervalue particularized justice in favor of the social interests and policy goals that they are required to pursue through their regulatory programs. These interests and goals usually transcend those of the particular parties before them.

Types of Rules

Common law adjudication often requires judges to devise new principles or rules where existing ones cannot fairly resolve a dispute or where no plausibly applicable rule exists at all. But the rules that courts elaborate through case-by-case adjudication typically differ in important respects from those that agencies develop in rulemaking—and even from those that they apply in agency adjudication. . . .

There is an intriguing irony here, one highly relevant to the problem of regulatory equity. . . . Agencies tend to spawn hard-edged rules that resist the solvent of further discretion; an equitable capacity must therefore be added. Courts, while eschewing discretion, tend to produce tractable, malleable rules capable of absorbing equitable claims as they arise in individual cases.

Decision Procedures

Given their distinct purposes and products, it is hardly surprising that the procedures and doctrines of courts and regulatory agencies differ significantly; procedures, after all, are intended to reinforce an institution's dominant goals even as they are shaped by those goals. Most court adjudication, for example, is structured to focus attention narrowly upon the claims of individual litigants rather than upon a decision's larger social consequences. . . .

For most regulatory agencies, however, individual dispute resolution is decidedly ancillary to the discretionary, policy development function. . . .

Organizational Forms and Settings

Even in an age of burgeoning judicial caseloads and complex forms of litigation, a court remains a remarkably solitary decisionmaker. . . .

The contrast between this relatively closed, unpopulated milieu and the dense, teeming environment in which regulatory officials make decisions could hardly be more striking. From their perches in the organizational hierarchy, regulators preside over a swarm of bureaucratic, policy-generating activity. Their decisionmaking machinery, unlike that of judges, is highly differentiated—by function (e.g., policy development, enforcement, research), by programmatic subject matter (e.g., water, air, pesticides), by geography (e.g., field operations, state and local programs), and by other dimensions of specialization (e.g., legal counsel, economic analysis).

That courts exhibit little or no functional differentiation, and that agencies display a great deal, directly affects the way in which each goes about reconciling the competing claims of rule and equity. In the courtroom, particularity finds a limited sanctuary from the generalizing impulse of the outside world. . . .

In a court, the trier of fact also integrates the conflicting claims of rule and equity. As ad hoc decisionmakers without continuing life or responsibilities, juries are probably less preoccupied with maintaining a system of legal rules than with dispensing situational justice. To a lesser extent, this is also true of judges, who do not systematically supervise any specialized policy system and who receive little feedback concerning how their equity-infused decisions affect the integrity or effectiveness of rules. . . .

The agency's pursuit of situational justice, in contrast, is a relatively subordinate enterprise. . . . Unlike a judge, who combines these strivings in herself, an agency must somehow find a bureaucratic way to integrate the conflicting conceptions of justice that these specialized structures express.

Regulatory Equity and Policymaking Through Agency Adjudication

When regulatory equity is pursued in the context of an exceptions process, it takes the form of a particularized adjudication of the rights of individual claimants (or a class of similarly situated claimants) by the agency's exceptions tribunal. Not all agency adjudication, however, is designed to achieve regulatory equity in the "situational justice" sense in which I have defined it. Adjudication may also be used to pursue the agency's broader policy goals. Indeed, as noted above, all regulatory agencies use adjudication to elaborate and enforce their policies to some degree, and a few develop policy almost exclusively through the adjudicatory process. . . .

One must be careful, then, not to conflate what are actually two very distinct forms of adjudication that demand very different analytical and normative frameworks. This is difficult, however, for at least two reasons. First, at a formal level, these two types of adjudications resemble one another.

The second reason why regulatory equity and policy development through common law agency adjudication are easy to conflate is that both the exceptions applicant and the agency may be tempted to counterfeit the decision process by passing the latter off as the former. This is an especially promising strategy when, as in the DOE [Department of Energy], the same bureaucratic unit performs both functions. . . .

To an important extent, largely overlooked by scholars, the use of the exceptions process for policymaking purposes can be understood as an administrative response to a set of constraints upon informal rulemaking under the Administrative Procedure Act. Ironically, these constraints have been imposed largely in the name of "regulatory reform." To a policymaker who wishes to change the prevailing regulatory course quickly, informal rulemaking is increasingly a process to be avoided. . . . As pressures for increased formality, more extensive records, and analytical rigor have dissipated many of the much-celebrated virtues of informal rulemaking, certain advantages of case-by-case adjudication in general, and of exceptions adjudication in particular, have come to seem correspondingly great.

Some of these advantages are intrinsic to agency adjudication. By limiting the impact of decisions to their particular facts, agency adjudication facilitates cautious and flexible policy development and exploits incrementalism's considerable political and intellectual virtues. In principle, agency adjudication limits the scope of factual inquiry, demanding fewer analytical resources than rulemaking and consuming less time. But case-by-case adjudication also has extrinsic advantages, peculiar to particular administrative contexts. . . .

First, policymaking by way of adjudication is particularly attractive where an agency's regulatory jurisdiction extends to a broad subject matter or to numerous and diverse regulated firms or individuals. . . .

This jurisdiction-related temptation to proceed by case-by-case adjudication is even stronger when the program is conceived to be emergency in nature. In such situations, rules must be issued quickly, often in wholesale lots.

. . . Again, if relief is to be granted in such cases, adjudication will be necessary to identify the specific situations in which the rule pinches. . . .

Agency adjudication may also seem particularly attractive when the rulemaking process is ineffective for any of a number of intellectual, bureaucratic, or political reasons. . . .

Finally, agency adjudication typically enjoys lower visibility and hence greater freedom from outside control than rulemaking. . . .

Several points about the above discussion should be stressed. First, these particular advantages of the case-by-case approach tend to be distributed in a highly skewed fashion; they accrue almost exclusively to the agency, not to regulated firms or the general public. The disadvantages of adjudication, however, are borne in just the reverse pattern. . . .

Second, many of the considerations that make rulemaking seem inadequate as a regulatory tool may imply not that case-by-case adjudication is a superior mode for implementing regulatory policy, but that the particular market should not be subjected to regulation of a command-and-control variety at all. In this view, regulatory inequity is so pervasive in such a program that an exceptions process attempts to use Band-Aids to staunch a massive hemorrhage.

Regulatory Equity and the Exceptions Process

In this part, I raise some fundamental questions about the role of exceptions in the regulatory process, questions that [another part] of this study . . . addresses. They may be organized around five related problems.

The Problem of Functional Integration

Does the exceptions process simply serve the situation-oriented, particularistic concerns of equity in particular cases, or does it also implicate the agency's larger policymaking functions? If the exceptions process affects both, how does the agency manage to integrate equitable and policy development purposes without sacrificing the distinctive values of each?

The Problem of Organizational Integration

Does the bureaucratic organization of the exceptions and rulemaking processes affect the way in which the agency's rules are formulated, interpreted, and enforced? Does that organizational structure affect the way in which exceptions decisions are actually made? Does it solve the problem of functional integration? What values are served and sacrificed by that particular structure?

The Problems of Equitable Criteria and Legitimacy

How can the exceptions process establish and preserve legitimacy and integrity in a decision context that invites imputations of favoritism, political pressure, ad hoc judgments, and unequal treatment?

The Problem of Procedures

Does the exceptions process employ procedures that are adequate to the kinds of tasks that it undertakes? Are its procedures appropriate both to the pursuit of regulatory equity and to common law-type agency adjudication? In particular, how can it organize what is an essentially adjudicatory process to generate decisions that are not only procedurally fair to individual parties but also expeditious and sensitive to political and policy factors?

The Problem of Accountability and Control

What roles do other institutions—the department, the Congress, the courts, and private groups—play in structuring and influencing the various functions that the exceptions process exercises? How well do these other institutions actually perform them and how should those roles be altered?

Adversarial Legalism and American Government
ROBERT KAGAN

Derek Bok, president of Harvard University, complains that every year America's educational system turns thousands of bright students into lawyers, who work at redistributing pieces of the economic pie, while Japan's best and brightest become engineers, makers of a bigger pie. An administrator of the U.S. Environmental Protection Agency estimated that more than 80 percent of EPA's regulations have been challenged in court and that each year that kind of litigation consumed 150 person-years of EPA program staff and lawyers' time. . . . A waiter in a *New Yorker* cartoon answers an inquiring diner, "You won't catch me recommending *anything,* sir. I have a lawsuit on my hands right now."

Social scientists who carefully study the legal system, on the other hand, tend to denigrate these stories as atypical, alarmist, or politically biased. Statistically, million-dollar judgments are rare. Surveys show that most people confronted with a problem (including victims of medical malpractice) don't sue or raise legal defenses. . . . In America today, the rate of civil lawsuits per capita, viewed cross-nationally and historically, does not appear to be remarkably high. . . . Regulatory inspectors negotiate informal settlements more often than they "go by the book". . . .

Reprinted by permission of John Wiley & Sons, Inc., from 10 *Journal of Policy Analysis & Management* 369 (1991). Copyright © 1991 by the Association for Public Policy Analysis and Management. All rights reserved.

I'm not inclined to dismiss the "too much litigation" perspective so quickly. Yes, I would grant the scholars, law is a good thing, a barrier against official arbitrariness, a check on economic rapaciousness, a force for tolerance and healthy social change. But the scholars, interestingly, seem to have a parochial view of law. Focusing on only the current-day American version, they say, "This is modern law. Take it (with all its excesses) or leave it (and throw away its far more important social benefits)."

When one views the American legal system from a cross-national perspective, however, a different set of possibilities emerges. Other economically advanced democracies, too, have legal systems. Western European polities care about justice, environmental regulation, and preventing professional or governmental malpractice. In some areas, such as workers' rights and land use regulation, many Western European polities have "more law" than the United States. Japan has a more detailed and extensive set of legally mandated product standards and premarketing testing requirements. . . . But according to an accumulating body of studies, the United States, when viewed from a comparative perspective, has a unique "legal style."* For one social problem after another—compensating injured people, regulating pollution and chemicals, ensuring equal educational opportunity, deterring malpractice by policemen, physicians, or presidential aides—the American policymaking system encompasses, on average:

1. more complex legal rules;
2. more formal, adversarial procedures for resolving political and scientific disputes;
3. slower, most costly forms of legal contestation;
4. stronger, more punitive legal sanctions;
5. more frequent judicial review of and intervention into administrative decisions; and
6. more political controversy about (and more frequent change of) legal rules and institutions.

*For some illustrative comparative studies, see Badaracco [for the sake of brevity, cited authors have been given without publication information] on occupational health regulation in Germany, France, England, Japan, and the United States; Bayley on regulation of police in Japan and the United States; Braithwaite on regulation of coal mine safety in several countries; Day and Klein on nursing home regulation in Great Britain and the United States; Brickman et al. and Jasanoff on regulation of carcinogens in several countries; Kelman on occupational safety regulation in Sweden and the United States; Kirp on racial desegregation in British and American schools; Kirp on regulation of education for handicapped children, United Kingdom and United States; Langbein on civil litigation methods in West Germany and the United States; Lundqvist on air pollution regulation in Sweden and the United States; Quam et al. on medical malpractice litigation in Great Britain and the United States; Vogel on environmental regulation in Great Britain and the United States; Bok and Flanagan on selection of labor union representatives; Glendon on abortion policymaking and dispute-resolution related to divorce and child support; Reich on how different bank regulations and labor law affect governmental bailouts of large corporations faced with bankruptcy. Of course, national legal styles are not monolithic. They vary within nations and even within branches of the same legal institution. With respect to regulatory enforcement style, see Kagan.

Searching for a handy summary rubric for these legal propensities, I label them "adversarial legalism"—a method of policymaking and dispute resolution characterized by comparatively high degrees of the following:

- *Formal legal contestation.* Disputants and competing interests frequently invoke legal rights, duties, and procedural requirements, backed by the threat of recourse to judicial review or enforcement.
- *Litigant activism.* The gathering and submission of evidence and the articulation of claims is dominated or profoundly influenced by disputing parties or interests, acting primarily through lawyers.
- *Substantive legal uncertainty.* Official decisions are variable, unpredictable, and reversible; hence adversarial advocacy can have a substantial impact.

This definition might be clarified by suggesting its opposites. Table V.1 displays two dimensions along which legal or administrative decisionmaking processes vary. The horizontal dimension involves the degree of *legal formality*—that is, the extent to which disputing parties or interests invoke formal legal procedures and preexisting legal rights and duties. The vertical dimension concerns the extent to which the decisionmaking process is *hierarchical*—dominated by an authoritative official decisionmaker, applying authoritative norms or standards—as opposed to *participatory,* that is, influenced by disputing parties and their lawyers, their normative arguments, and the evidence they deem relevant. Taking each of these dimensions to their extreme form produces four ideal-types.

1. *Negotiation/mediation.* A decision process in the lower left cell of Table V.1 is adversarial in the sense that it would be dominated by the contending parties, not by an authoritative governmental decisionmaker. And it would be informal, nonlegalistic, in that neither procedures nor normative standards would be dictated by formal law. The purest cases would be dispute resolution through negotiation without lawyers and policymaking through bargaining among legislators representing contending interests. The cell would also include mediation, whereby an "official" third-party attempts to induce contending parties to agree on a policy or settlement, but refrains from imposing a settlement in accordance with law or official policy.
2. *Expert/political judgment.* The more an official third party (as opposed to the contending interests) controls the process and the standards for decision, and the more authoritative and final the third party's decision is, the more "hierarchical" the decision process. Hierarchical processes can be legally informal, as suggested by the upper left cell in Table V.1. Consider, for example, what Jerry Mashaw has called the "professional treatment model"—such as decisions concerning eligibility for disability benefits when made by a panel of government-appointed physicians (or perhaps physicians and social workers), without significant probability of intensive judicial review. (European disability and workers' com-

Table V.1. Modes of Policymaking and Dispute Resolution

Informal to Formal

Hierarchical
to
Party-Influenced

Expert or political judgment	Bureaucratic rationality
Negotiation/ mediation	Adversarial legalism

pensation systems tend to follow this model). Another example of legally informal, hierarchical decisionmaking is provided by Badaracco's description of regulatory rulemaking concerning occupational safety and health in Great Britain, France, West Germany, and Japan. In these cases, a government ministry, with final authority to promulgate a rule, conducted a series of informal, closed-door, consensus-building discussions with a limited number of industry and labor representatives and scientists; in contrast with rulemaking on similar issues in the United States, participation and assessment of evidence was not organized in a "judicialized" manner, and the agency's decision was not subjected to judicial reversal, based on procedural or substantive legal criteria. Rather, faith is placed in the agency's political judgment and its ability to forge acceptable compromises from the expert advice of scientists, engineers, and economic analysts.

3. *Bureaucratic rationality.* A decision process characterized by a high degree of hierarchical authority and legal formality (the upper right cell of Table V.1) would resemble Weber's ideal-typical bureaucracy. The submission and assessment of evidence would be governed by written rules and procedures, as would substantive decisions, made by carefully trained, apolitical civil servants. The more hierarchical the system, the smaller the role for legal representation of and influence by affected citizens or contending interests. In contemporary democracies, this pure case rarely occurs, but it is an ideal systematically pursued, for example, by tax collection agencies, or the U.S. Social Security Administration. Tending toward this cell, too, would be German or French courts, where highly professionalized judges—not (as in the United States) the parties' lawyers, and not lay juries—dominate both the evidentiary and the decisionmaking processes. . . .

4. *Adversarial legalism.* The lower right cell of Table V.1 implies a decision process that is procedurally formalistic but in which hierarchy is weak and party influence on the process is strong. American civil and criminal adjudication provide vivid examples. Complex legal rules govern pleadings, jurisdiction, pretrial discovery and testing of evidence,

and so on, but the gathering of evidence and invocation of legal rules is dominated not by the judge but by contending parties' lawyers. . . . At the same time, as comparisons of American and British "adversarial systems" make clear, hierarchical, authoritative imposition of the law is relatively weak in the United States. . . . From a comparative perspective, American judges are more political . . . , their decisions less uniform. Law is treated as malleable, open to parties' novel legal arguments and pleas of extenuating circumstances. In civil cases, lay jurors still play a large and normatively important role in the U.S., magnifying the importance of skillful advocacy by the parties and reducing legal certainty.

Similarly, when compared to policymaking in European democracies, regulatory decisions in the United States entail more legal formality—more complex legal rules concerning public notice and comment, open hearings, ex parte contacts, evidentiary standards, formal response to interest-group arguments, and so on. But hierarchical authority is weak. Agencies cannot restrict participation by interest groups. Agency decisions are frequently challenged in court by dissatisfied parties and reversed by judges, who dictate further changes in administrative policymaking routines. Lawyers, scientists, and economists hired by contending industry and advocacy groups play a large role in presenting evidence and arguments. The clash of adversarially advanced argument, rather than top-down application of official norms, is the most important influence on decisions.

No legal system falls entirely into any single cell in Table V.1. Different programs tend toward different policymaking and dispute-resolution methods, and variation also occurs within programs. Adversarial legalism can and does occur in reputedly cooperative nations such as the Netherlands . . . and Japan. . . . Americans, conversely, often refrain from adversarial legalism, resorting to negotiation or submitting to bureaucratic or expert judgment. But viewed in the aggregate, adversarial legalism seems more common in the United States than in other democracies, and more common today than in the America of thirty years ago. Adversarial legalism—party-dominated legal contestation—seems to be a barely latent, easily triggered potentiality in virtually all contemporary American political and legal institutions.

The Costs of Adversarial Legalism

Adversarial legalism, of course, is deeply embedded in American ideals and ways of governance. . . .

Without intending to dismiss these social benefits, I suggest it would be useful to attend to the other side of the ledger. The spirit of distrust of authority that underlies adversarial legalism can be used against the trustworthy, too. An equal opportunity weapon, it can be invoked by the misguided, the mendacious, and the malevolent as well as by the mistreated. Its processes enable contending parties to use the extraordinary costs and delays of adversarial litigation in a purely tactical way, to extort unjustified concessions from the other side.

No comprehensive account of the social and economic costs of adversarial legalism is readily available. . . .

Simply because it costs so much and takes so long, adversarial legalism often undermines the law's most basic aspirations for effectiveness and justice. For some enterprises and organizations, avoiding the legal process becomes more salient than fighting for what they believe is the right or just result. As procedures for involuntary commitment of mentally ill individuals have become increasingly demanding and complex, many police and hospital personnel refrain from initiating the commitment process even when they feel it is fully warranted. School administrators not infrequently accede to what they consider educationally unjustified parental demands concerning education of handicapped children, simply to avoid the costs of repetitive hearings and litigation. . . . Liability insurers understandably concentrate costly legal defense efforts on the lawsuits with the most potential for huge damage awards. The consequence is undercompensation, on average, for the most severely injured claimants—and overcompensation, it appears, for claimants in many smaller lawsuits, which the insurance companies find more costly to defend than to pay off. . . . Even more often, injured plaintiffs, especially those with small claims, are deterred from filing lawsuits. In debt collection and criminal adjudication, too, the high litigation costs and delays that flow from adversarial resistance commonly result in the abandonment or compromise of just claims and defenses.

At the policymaking level, adversarial legalism provides citizen watchdog organizations access to the rulemaking process in government agencies and, through the threat of judicial review, helps guard against administrative arbitrariness or "capture." But adversarial legalism also breeds legal deadlock and socially harmful inertia. The implementation of new regulations is often blocked by judicial appeals, sometimes at the behest of regulated entities complaining of unreasonable strictness or inadequate analysis, sometimes at the behest of proregulation advocacy groups complaining of regulatory inaction or laxity. . . . When every decision must be bolstered by legally "bullet-proof" scientific evidence and procedural methods, then vital protective measures concerning workplace health risks, hazardous air pollutants, . . . and motor vehicle safety features . . . remain bogged down in the bureaucracy for years. Virtually every management plan for each of the many National Forests has been held up while waiting for appellate scrutiny. In the last decade, virtually every Department of Interior decision about offshore petroleum exploration has been challenged in court. . . . Operating in the shadow of potential or actual legal appeals, the processing of license applications for nuclear plants increased from an average of twelve months for plants completed in 1960 to thirty-three months for those completed in 1973, to fifty-six months for 1981 plants. . . . While these legal pressures enhanced safety, overseas power plants, "equivalent to American reactors in quality and safety . . . have been built in much less time and at far lower cost". . . .

One further, less tangible cost of adversarial legalism is its corrosive effect on personal and institutional relationships, as when physicians, in a corner of their minds, regard certain patients as potential medical malpractice claimants. Similarly, when a regulatory inspector and a regulated enterprise become locked into a legalistic, adversarial posture, the cooperation and exchange of information so essential to effective regulation is cut off. . . . When regulatory rulemaking is only a prelude to litigation, a National Academy study of the EPA found, contending interest groups are more prone to exaggerate or minimize risks and to suppress or distort information that weakens their position. . . . Adverarial legalism and the distrust it symbolizes is demoralizing to teachers, nurses, architects, police officers, environmental engineers, and other occupants of what Eugene Bardach refers to as the "trustee stratum" of the nation, who are forced by the prospect of legal review to spend hours doing defensive paperwork rather than discharging their professional responsibilities.

Yes, the reader might think, adversarial legalism does entail significant costs in some settings. But isn't it a necessary concomitant of the kind of legal controls needed to achicve justice in American society? Couldn't one argue that sharply defined legal rules, strong penalties, legal rights to challenge administrators, and formal adversary procedures are quite appropriate for a society like ours—where citizens are more mobile, individualistic, and culturally diverse than in Western European nations or Japan; where traditional informal social controls are weaker; where citizens are less deferential to government authority . . . ; and where entrepreneurs denigrate regulatory controls and perhaps need to be dealt with more legalistically. . . . From this perspective, isn't adversarial legalism just an unfortunate but relatively minor side effect of a generally desirable institutional system, like the pollution emitted by an electrical power plant?

It is difficult to answer such questions definitively, just as it is hard to know precisely when the emissions of scores of power plants start adding up to a serious environmental and health problem. What can be done is to look closely at the costs and the causes of adversarial legalism—as in the following case study—and to try to imagine changes in our political and legal institutions that may facilitate responsive public policy implementation and fair dispute resolution, but without so much costly, slow, and divisive litigation.

The Dredging Dilemma: A Case Study of Adversarial Legalism

Increasing trade volumes and larger ships have generated powerful pressures for port expansion. Clogged roads between urban ports, railheads, and intercity highways evoke demands for new docks, linked more closely to highways and rail lines, large enough to store thousands of containers. Port authorities, backed up against crowded urban neighborhoods, have often sought to create new docks by dredging and landfill operations. Similarly, ports are pressured to dredge deeper harbor channels and dock areas to accommodate ever-larger ships. On the other hand, as this section will show, response to these demands is hindered by a complex, politically fragmented, and legalistic governmental process for funding and approving new port expansion projects and for minimizing their adverse impacts on the environment.

• • •

Negotiating the Legal Maze: The Oakland Harbor Case

In November 1984, after years of political deadlock over funding water and harbors projects, the Corps of Engineers completed its cost-benefit analysis and an environmental impact statement (EIS) concerning the Port of Oakland's harbor-deepening proposal, and in early 1987 Congress authorized funding. In 1986, however, California water quality and fish and game agencies raised concerns about adverse effects on San Francisco Bay fisheries and water quality if the 7 million cubic yards of dredged sediments were dumped at the originally planned site—near Alcatraz Island. Although the Corps of Engineers Supplementary EIS disputed their arguments, the state agencies had legal power to block in-bay disposal. This impelled the Corps to select an ocean disposal site (designed 1M) fifteen miles from the Golden Gate, although the added distance would double dredging costs to $39 million. The EPA, however, refused to authorize use of the 1M ocean site. Furthermore, Citizens for a Better Environment (CBE) seemed poised to bring a lawsuit challenging the adequacy of the Corps' Supplementary EIS and demanding use of a disposal site beyond the edge of the Continental Shelf.

In January 1988, anxious Port of Oakland officials convened several meetings of representatives from the various regulatory agencies, environmental groups, and fishing organizations. Compromise was elusive. The Corps argued that in the absence of demonstrated environmental differences that would justify the greatly increased costs, it was legally precluded from endorsing disposal at more remote ocean sites. EPA and U.S. Fish and Wildlife officials said they were legally precluded from accepting 1M without further research showing it was environmentally acceptable compared to more distant, deeper ocean sites.

In March 1988, under pressure from Port of Oakland officials, EPA and the Corps of Engineers convened a "Technical Review Panel" in Washington, D.C. to review the scientific questions. Lacking sufficient data, the panel made a political decision: Since fishery interests were likely to sue and delay the project if 1M were used, a different ocean site (B1B) should be used for the first 500,000 cubic yards of dredged material (except for material from a clearly contaminated area), enabling the port to make the channel just deep enough to accommodate the first larger ships. Meanwhile, further testing and study should precede a decision concerning disposal of the remaining 6.5 million cubic yards. CBE and the Pacific Coast Federation of Fishermen indicated that they endorsed this solution. The Corps hastily "beefed up" the section of its EIS concerning the effects of dredge disposal at B1B. Port of Oakland officials made arrangements for the actual dredging and transport of sediment to B1B, which was twice as deep as 1M but also twice as far—thirty miles from the Golden Gate (but only ten miles off the San Mateo County coast)—and hence 50 percent more costly.

Where access to court is easy, however, compromise is unstable. In mid-April 1988, the Half Moon Bay Fishermen's Marketing Association (HMBF MA), alleging that dumping dredged material at B1B would disrupt a valuable fishing ground, filed suit against the Corps in federal court, arguing that the Supplementary EIS concerning B1B was inadequate under NEPA [National Environmental Policy Act]; that the selection of B1B violated the Marine Protection, Research and Sanctuaries Act; and that HMBFMA had not had adequate notice and opportunity to request a public hearing on the use of B1B. On May 5, the federal trial judge denied HMBFMA's request for a temporary restraining order stopping the dredging, still poised to begin. HMBFMA immediately appealed. The Ninth Circuit Court of Appeals first issued a restraining order pending hearing, but on May 12, despite its expressed misgivings about the quality of the Corps' EIS, dissolved the stop order. Port officials ordered the dredging to commence; harassed by fishing boats, barges dumped the first loads of sediment at B1B.

On May 16, Half Moon Bay fishermen dumped a ton of fish heads at the Port of Oakland. . . . Press reports indicated the dredging company had taken contaminated dredged material to B1B; dredgers claimed they had made a mistake, misreading their charts. Most significantly, on the same day, a San Mateo County Superior court judge, acting in a suit against the Port of Oakland filed by the County of San Mateo (later joined by the HMBFMA), issued a temporary restraining order stopping the dredging. The dredging permit had been issued, the court held, without a requisite certification from the California Coastal Commission that the project was consistent with its coastal development plan, which included enhancement of fisheries. (Actually, the Coastal Commission long before had been given notice of the project, but had not informed the port or the Corps that it thought "consistency review" was legally required.) On July 15, 1988, a state appellant court rejected the port's appeal of the lower court's restraining order.

Port of Oakland officials, under increasing pressure from shipping lines, had become desperate. They were still paying for the rental of expensive dredging equipment, in case the injunction should be lifted. They had expended huge sums enlarging train tunnels in the Sierra Nevada, on larger cranes, and on new intermodal transfer facilities, all in order to handle larger container ships as efficiently as competing ports. In August, they announced a new disposal plan for the first 440,000 cubic yards of dredged sediment, using a Sacramento River delta site where local reclamation officials were eager for diking material. Barging the material there would be 50 percent more expensive than taking it to B1B. It also would require an environmental impact report (EIR), as required by state law.

The 400-page EIR was completed in February 1989; it included test data indicating that the project would not significantly lower water quality or adversely affect the environment. Then the public hearing and comment process began. The Central Valley Regional Water Quality Control Board, from which a "waste discharge permit" was needed, demanded additional tests, but ultimately, on July 12, 1989, voted to accept the plan, subject to stipulated protective measures and postdisposal environmental monitoring. But the Contra Costa Water District, downstream from the disposal site, asserted that the EIR contained "incorrect dilution calculations," and expressed concern that heavy metals and salts in the sediment would run off into the delta waterways, violating water pollution discharge regulations. In early August 1989, the Contra Costa Water District and the Port of Oakland each filed a lawsuit against the other, seeking a judicial determination whether the Port's EIR was legally sufficient. Not until July 1990 did the court decide, upholding the legal sufficiency of Oakland's plan. But port officials calculated that after all regulatory conditions and monitoring requirements, it would cost $21 a cubic yard to use the delta site, as compared to $2 at Alcatraz, $4 at B1B, and about $7 for an off-the-continental-shelf site. They decided to try once again to gain access to the Alcatraz and certain other delta sites for Phase I material.

As of April 1991, no decision or resolution had emerged concerning the disposal of the Phase I 440,000 cubic yards of material, much less the additional 6.5 million cubic yards called for by the full-scale harbor-deepening project. American President Lines and Maersk Lines were still compelled to reduce loads and time their sailings to catch high tide, which boosted operating and stevedoring costs significantly. Oakland lost additional port business to Los Angeles and Seattle, slowing amortization rates for existing facilities, raising costs per ton of cargo handled, and imperiling some of the 40,000 jobs estimated to flow from port activities. . . . Yet the Corps of Engineers and EPA, only recently given appropriations for ocean studies, were only at the beginning of a two-year research project that would provide the basis for designating a permanent ocean deposit site. Already confronted with vocal political challenges to its plans to designate an ocean disposal site in southern California, EPA insisted full-scale, legally defensible environmental analysis must precede approval of even a *temporary* ocean disposal site for the Oakland project. Meanwhile, the regional water quality board adopted regulations prohibiting deposit of new dredging project spoils anywhere in the bay. Four years after Congress funded the Oakland harbor-deepening project, deadlock continues in the harbor, in the courts, in the agencies, and in the assessment of scientific evidence.

The Pathologies of Adversarial Legalism

The legal procedures that gave rise to the deadlock in the Oakland case reflect fundamental ideals of pluralistic democracy—the notions, for example, that public policy should be formulated and implemented only after full and fair deliberation; that meaningful attention should be given to the claims of the individuals and groups who are not politically powerful (such as the Half Moon Bay fishermen); that environmental protection should be given special weight in planning currently urgent development projects that might deprive future generations of irreplaceable ecological amenities; and that to vindicate those values, a variety of interest groups and agencies should be able to challenge official assumptions and judgments.

But the policymaking procedures designed to protect those pluralistic values seemed to fall into the hands of the Sorcerer's Apprentice, multiplying themselves beyond control. In consequence other important values were completely undermined, such as the public interest in a reasonable degree of procedural efficiency and in decisions that retain a sense of proportion and balance among competing substantive ends. Consider, for example, the operative characteristics of the process:

1. *Irresolution.* After four years of debate about disposal of dredged sediments, there has been no authoritative determination about where to put them and what impact they would have on the environment.
2. *Legal fragmentation.* Instead of combining their concerns in one comprehensive forum, a cascading jumble of regulatory agencies, private interest groups, and courts were legally enabled or compelled to take sequential whacks at the problem.
3. *Legal complexity.* The decision process was constrained by a segmented, detailed sequence of statutorily mandated reviews, certification points, substantive specifications, and scientific standards.
4. *Legal uncertainty and inconsistency.* For all its detail and complexity, the law afforded no certainty. Three compendious, expensive environmental impact reports, scrutinized through the lenses of adversarial legalism, were stripped of legitimacy; they resolved nothing. When one court upheld a regulatory decision, another could be found to overturn it on a different legal argument. While one water agency approved a delta disposal plan, another blocked it in court for a year.
5. *Instability of compromises.* Negotiated agreements, when finally reached, were unstable. Any interest dissatisfied with the compromise could sue in court, relying on the uncompromising language of the law.
6. *Procedural extortion.* Simply by preventing definitive resolution of the issues, adversarial legal conflict shaped the outcome. No dredging has occurred. Repeatedly, the Port of Oakland felt compelled to accept more expensive disposal methods to avoid crippling procedural delays. The extortive pressures engendered by litigational and regulatory processes, not rational economic and environmental analysis, came to dominate the decisions on where the sediment would be dumped.
7. *Economic inefficiency.* In effect, and despite the law's intent, the social and economic benefits of more efficient transportation and trade were not weighed against but were totally subordinated to concerns about local environmental preservation. While uncertainties persisted about the ecological risks presented by disposal of the sediments in question,

adversarial legalism seemed to enable virtually any claim of potential ecological harm, no matter how minimal or remote, to take precedence over development projects, no matter how beneficial to human beings.

8. *Demoralization.* Governmental authority and conviction collapsed, as officials retreated to a position of litigation avoidance. A Corps of Engineers official recently asserted that in harbor-dredging matters, the Corps does not want to "get caught in the middle," to battle environmental activists, or "to tell state agencies what to do." The Corps wants local entities—that is, port authorities—to "carry the ball," whether that entails doing the research, making concessions, or fighting for their position in court. As if hoping the problem will go away, EPA promises no decision on a permanent ocean disposal site before 1992.

9. *Diversion of attention.* Because of the institutionally fragmented, sequential way environmental issues were treated, with each agency myopically perusing expansion plans in terms of the particular environmental problems contemplated by its particular governing statute, environmental protection may actually have been reduced. In April 1990, the Exxon Long Beach, a sister supertanker to the infamous Exxon Valdez, laden with 50 million gallons of Alaskan crude oil, ran aground on a "high spot" in Long Beach Harbor. Fortunately, no damage occurred; the ship was moving very slowly. . . . In other recent harbor accidents, the environment was not so lucky; tons of petroleum did foul the environment. The message should be clear: Shallow harbors, by threatening damage to ships, may pose a far larger danger to water quality and marine habitats than properly "capped" dredging spoils. Yet these risks received little or no attention in the regulatory and judicial processes that have held up the deepening of Oakland's harbor. For adversarial legalism tends to focus attention on only those problems implicated in the claimants' legal briefs, not on those which have no advocate in court.

The Roots of Adversarial Legalism

. . . The question is whether the policymaking and dispute-resolution process in NIMBY [Not In My Backyard] cases must be as legalistic, adversarial, protracted, costly, and insensitive to economic values as in the Oakland dredging story. The answer, I would argue, is no. Adversarial legalism is a product of a particular way of articulating and implementing public policies—one that invites, exacerbates, and extends legal conflict.

Administrative Finality Versus Administrative Fragmentation

With respect to the dreding problem, for example, one can imagine a very different decision system, characterized by what we might call *adminsitratively final, multifactor balancing*. Suppose the national legislature established a few regional port planning agencies with broad discretionary authority. Each agency's governing board could include representatives of recognized environmental groups. The agency would be expected to meet, in private, with local advocates of port expansion, environmental agencies, conservation groups, and representatives of the fishing industry. Based on those discussions, the agency would commission research it agrees is necessary, taking into account the apparent seriousness of the environmental risks and the social costs of delaying port expansion. It would attempt to build consensus around the plans and mitigation measures it deemed best, but if no consensus could be reached, the agency would be empowered to make a final decision. The planning agency's decisions, if appealed to courts or to other political bodies, would command considerable deference, unless shown to have been substantively arbitrary or the product of unfair influence. That discretionary administrative authority to make final and binding decisions probably would encourage serious efforts by participating interests to reach a negotiated accommodation. Ideally, the agency would keep a series of port expansion issues on the table at the same time, so that concessions to intensely advanced economic or environmental concerns in one case might be traded off against reciprocal concessions in another.

Now contrast the existing system. The many congressional statutes and state laws that structured the decisionmaking process in the Oakland dredging case created a *legally constrained, fragmented decisionmaking system.* . . .

In addition, each agency's authority was weakened and constrained by the prospect of *judicial review.* . . .

These legal constraints discourage informal, binding compromises on difficult scientific, technical, and political issues surrounding major harbor projects. . . .

Finally, because of the delays associated with lawsuits, rehearings, revisions of impact statements, and so on, the legal process empowered opponents of the project to block it simply by making legal claims, regardless of whether they were ultimately vindicated. . . .

. . . The legal traditions and structural features of American government create the conditions create the conditions under which adversarial legalism can flourish, not only in NIMBY situations, but in a wide array of policy domains.

Something Changed

If it's a matter of legal traditions and pluralistic political structures, why has the incidence of adversarial legalism in the United States increased in recent decades? . . .

For the most part, adversarial legalism of the type at issue in the Oakland case did not arise until late in the 1960s and the 1970s. . . . Moreover, adversarial legalism persisted—and seems to have continued to grow—through the more conservative 1980s. . . . It might be useful to consider why. . . .

Legal Culture

Conceivably, increases in adversarial legalism reflect changes in Americans' legal attitudes and capacities. . . .

Lawrence M. Friedman argues that despite some rear-guard opposition. . . . American legal culture indeed has shifted its center of gravity, manifesting a widespread expectation of "total justice." Most Americans, Friedman suggests, no longer accept injury, ill-treatment, environmental degradation, or poverty as acts of God or as the inevitable by-products of capitalism and modern technology. . . .

Popular demands for "total justice" probably help explain the demand for more law. But a puzzle remains: Expectations of fair treatment, compensation for misfortune, and environmental protection do not necessarily lead to laws and legal institutions that encourage adversarial legalism. . . . Fierce controversies sometimes erupt in Western Europe, as in the case of siting decisions for such large-scale developments as nuclear power plants or the planned third London airport. . . . But they usually are resolved in political and administrative forums, not in courts, and outcomes rarely are shaped by the manipulative use of legal procedures and standards. . . .

Aaron Wildavsky has argued that Americans (along with everyone else) tend to favor one of three political cultures that contend for influence over government and legal institutions. . . .

Wildavsky presumably would see egalitarians as the truest believers in Friedman's "total justice," and even more crucially, as active proponents of adversarial legalism, which helps them enforce their views against mistrusted adherents of competing political cultures. From this perspective, if adversarial legalism has grown in the United States, it is because American egalitarians, previously in a weaker position, have enjoyed a period of remarkable political success, penetrating the legislatures, judiciaries, law faculties, and news media more fully than have their political antagonists. In Western Europe, in contrast, believers in hierarchy presumably remain politically stronger, influencing European egalitarians toward corporatist rather than toward litigational strategies. . . .

Still, it is not quite clear why the egalitarian impulse relied so heavily on adversarial, legalistic policymaking and implementation methods. Western Europe proponents of equality were active and often successful in the same time period; but they did not demand, or at least did not achieve, similar laws and court decisions. And why did legal measures that encouraged adversarial legalism in the United States persist, even in the face of conservative presidential opposition, even after national concerns shifted to oil shortages, inflation, declining productivity, trade imbalances, and international competitiveness?

Political Structure

Adversarial legalism in the United States, I would argue, has been stimulated by a fundamental mismatch between a changing legal culture and an inherited political structure. Americans have attempted to articulate and implement the socially transformative policies of an activist, regulatory welfare state through the legal structures of a reactive, decentralized, nonhierarchical governmental system. In the absence of a strong, respected national bureaucracy, proponents of regulatory change and social welfare measures have advocated methods of policy implementation that emphasize citizens' rights to challenge and prod official action through litigation.

This argument is inspired by Mirjan Damaska's *The Faces of Justice and State Authority*. Damaska formulates a typology of legal processes built on two dimensions. One dimension concerns the organizational structure through which administrative and legal processes flow; the other, varying cultural visions of the proper role of government. One mode of organizing authority Damaska labels *hierarchical,* an ideal-type toward which continental European legal systems incline. It features a limited number of strong, highly professional, national bureaucracies, topped by a central ministerial authority responsible to political leaders. Fidelity to official norms and policies, along with uniformity and predictability of case-by-case decisionmaking, are the reigning ideals. Officials are relatively insulated from the potentially corrupting influence of local politicians and citizens.

American legal and administrative processes, on the other hand, lean toward a *coordinate* organization of authority. Designed to limit central authority's potential for tyranny or political bias, power is fragmented among many governmental bodies, often staffed by locally elected officials. Control is exercised horizontally, through one governmental body's capacity to check another and through citizens' rights to challenge governmental decisions in court. The best decisions will emerge, the coordinate model presupposes, not from uniform imposition of (potentially flawed) official rules, but from the clash of arguments proferred by a pluralistic welter of organizations and citizen-representatives. Outcomes in the coordinate ideal thus are to be closely attuned to the particular circumstances of each dispute, responsive to evolving local notions of justice. Adversarial claiming, negotiation, and compromise, rather than uniform hierarchical norm imposition, will be common and will be favored.

With respect to the political culture dimension, Damaska poses two further polar types: one that values an *activist state,* dedicated to the aggressive management or even transformation of economy and society; and at the other pole, a political culture that prefers a *reactive state,* expected only to provide an orderly framework for private economic and social interaction, to formulate and implement policy primarily by resolving conflict among competing interests.

There are obvious affinities between an activist state and a hierarchical organization of authority, with its respected bureaucracy and judiciary, willing and able to implement official policy faithfully. Similarly, a reactive state fits nicely with a coordinate organization of authority, with its wide openings for civilian influence, its skepticism about state-enforced norms, its reliance on adversarial argument, its openness to private negotiation. . . .

In the twentieth century, however, American government, like government in other industrialized democracies, has experienced powerful political pressures to become more activist—to steer and stabilize the economy, to bring about the "total justice" Friedman describes. In the United States, those political demands had to be channeled through political structures designed for reactive government and decentralized conflict resolution, not for centralized, top-down social engineering. That meant trouble. As Damaska writes, "A state with many independent power centers and a powerful desire to transform society can be likened to a man with ardent appetites and poor instrument for their satisfaction" (p. 13). . . .

Conclusion

Adversarial legalism arises from a vicious circle. Americans want government to do more, but governmental power is fragmented and mistrusted. So Americans seek to achieve their goals by simultaneously demanding more of government and by fragmenting and regulating it still further. Legislatures and courts mandate new goals, new benefits, and new regulatory protections. Yet implementing agencies are constrained by formal legal requirements, buffeted by threats of litigation and judicial review. In this harried condition government seems doomed to fail—incapable of living up to the demanding legal duties imposed on it, bogged down in costly legal disputes or in legal defensiveness. Perceiving governmental failure, public cynicism grows and governmental authority is diminished further. Those seeking to achieve their ends or influence government feel compelled to arm themselves with lawyers, insist on strict observation of legal rules, and threaten to go to court, simply because their opponents are likely to do the same.

Increasingly, scholars are calling for alternative, less litigious ways of solving social problems, making public policy, and resolving disputes. Their solutions call for a reversal of the antiauthority spiral—to get less adversarial legalism, we must somehow reconstitute governmental authority.

Legal scholars, for example, call for an administrative process based more on informal discussion and debate, a search for shared values, a spirit of compromise and cooperation. They criticize a body of administrative law that squeezes policymaking through a court-like litigational mold. Instead, they call for decisionmaking methods that foster "public deliberation". . . . They call for informal negotiation of regulatory rules among contending interests. . . .

In social benefit programs, some scholars suggest, the adversarial assertion of due process rights should give way to mechanisms designed to support a "dialogic community" between administrators and beneficiaries. . . . Administrative law, Mashaw argues, should focus less on judicial review and more on building and supporting administrative competence.

Cross-national studies of administrative rulemaking and implementation point in the same direction. Western European regulatory agencies, Badaracco demonstrates, avoid adversarial legalism because they have the final say. The laws give them broad discretion. Their decisions, absent major misfeasance, generally are not reversible by courts. They meet informally, privately, and repeatedly with a relatively small network of interest-group representatives who, to retain influence, must develop a reputation for integrity and reasonableness. The participants, lacking any escape route to the courts or to individual legislative allies, and knowing the agency will decide if they can't agree, are compelled to bargain seriously, to reach compromises on scientific issues and on how regulatory values should be balanced against concerns about compliance costs. The "dialogic community" arises because the law fosters, rather than undermines, what I earlier called "administratively final, multifactor balancing."

Corporatist policymaking structures have their own deficiences, of course. They lack some of the valuable features of the American system—contestability of expert opinion and official plans, openness to a wide array of opinions and interests, sensitivity to individual rights. But in the United States, merely to discuss corporatist models stimulates great suspicion. If administrative discretion and behind-closed-doors negotiation supplant legal constraint and review, Americans ask, how can we be sure that discretion will not be abused, that the politically weak will not be overwhelmed by the politically or economically powerful, that the Corps of Engineers will not revert to environmental insensitivity, that regulators will not be captured by the regulated? In short, the key to diminution of adversary legalism seems to be a bit of magic—in a disbelieving age, to restore faith in the competence and public-spirited nature of governmental authority. . . .

. . . Deadlock sometimes generates institutional changes, designed to make progress on particular problems. Learning from the Port of Oakland's experience, Port of Los Angeles officials work to build a multiagency, multicity political forum in which to negotiate port expansion plans. To avoid the delays of litigation, regulatory agencies constantly try new ways of forging consensus on particular regulatory standards and methods of implementation.

NOTES AND QUESTIONS

1. The administrative law commentators' propensity to proliferate models can be rather confusing, although sometimes enlightening. Compare the ones discussed by Mashaw (transmission belt, expertise, participatory), Richard Stewart (transmission belt, expertise, interest representation), Christopher Edley (politics, science, adjudicatory fairness), and Gerald Frug (formalist, expertise, judicial review, market/pluralist). Stewart's taxonomy is developed in his article excerpted in Chapter VI, section 5. Edley's is developed in his book, *Administrative Law: Rethinking Judicial Control of Bureaucracy* (New Haven: Yale University Press, 1990). Frug presents and criticizes his models in "The Ideology of Bureaucracy in American Law," 97 *Harvard Law Review* 1276 (1984). These models seem designed more to improve conceptual clarity by calling attention to distinctive features characterized at high levels of generality than to generate implications that can be empirically tested, as social science models often are.

Mashaw's participatory model and Stewart's interest representation model bear a strong family resemblence to one another, yet there are subtle differences in emphasis. Both speak to the configuration of interests represented before the agency, but Mashaw seems more attentive to the dignitary aspects associated with different participatory forms, while Stewart seems more concerned about the effect of participation on the agency decisionmakers as they weigh and balance the competing claims. Edley's "trichotomy" refers to different paradigms of reasoning or problem-solving, and to that extent are models of bureaucratic legitimation analogous to those of Mashaw, Stewart, and Frug. In an approach characteristic of the critical legal studies scholars, Frug is at pains to demonstrate that each of his four models is incoherent in the sense that its assumptions are meaningless, contradictory, or false. Do Mashaw, Stewart, Edley, and Frug draw the same positive and normative conclusions from their analyses of the models' deficiencies?

2. Mashaw, at the conclusion of the excerpt, notes that the courts increasingly insist that agency decisions be substantively rational; in this, he says, they look to "the contemporary incarnations of rationality, cost-benefit and risk-benefit analysis." Is such an orientation inevitable, given the standards of the Administrative Procedure Act and constitutional review by courts? Elsewhere, Mashaw has expressed considerable skepticism about the coherence and workability of cost-benefit analysis as applied to assess administrative procedures in the absence of more empirical data and a determinate normative theory of procedure. See Mashaw, "The Supreme Court's Due Process Calculus for Administrative Adjudication in Mathews v. Eldridge: Three Factors in Search of a Theory of Value," 44 *University of Chicago Law Review* 28 (1976).

3. Thibaut and Walker attempted to test their theory of procedure experimentally. The results strongly establish the experimental subjects' preference for adversarial procedures regardless of whether they played the roles of parties, nonparties affected by the outcome, or unaffected members of the public. The parties, though not the other groups, also linked their satisfaction with process to their satisfaction with outcomes. The research is reported in Walker, E. Allen Lind, and Thibaut, "The Relation Between Procedural and Distributive Justice," 65 *Virginia Law Review* 1401 (1979). For discussion of this and related research, see John Monahan and Walker, *Social Science in Law: Cases and Materials* (Westbury, N.Y.: Foundation Press, 2d ed., 1990), chapter 6.

Assuming that these findings are correct, how much weight should be given to them in the design of administrative procedures? Is a respondent's subjective satisfaction with a process or outcome a valid guide to public policy or even to the determination of what is "fair"? Are there factors distinctive to the administrative context that bear on this question? For example, should the costs associated with different procedural modes be viewed as more significant in this context than in, say, private disputes, or as less so? Within the realm of public administration, are there contexts in which these findings should carry special weight? Note that although the theory was not specifically tested on administrative proceedings, the authors, at the beginning of the excerpt, claim that it applies universally.

What do you make of the authors' sharp distinciton between pursuing a truth goal, where "mere" cognitive conflict is the problem, and pursuing a justice goal, where conflict of interest is the problem? Into which category does the legal dispute over the toxic effects of Agent Orange fall? The dispute over whether post–traumatic stress disorder causes violent behavior?

4. Reich does not imagine that all individual interests in government largess should enjoy equal procedural protection, and he seems to recognize some countervailing efficiency and other considerations that might limit the nature and scope of

those protections. Does he provide any determinate criteria for making the necessary distinctions in the procedural entitlements of, say, excludable versus deportable aliens? Of persons whose benefits are being terminated versus initial applicants?

Setting aside the clear abuses of power denounced by Reich and the question raised by Rabin about Reich's relevance limitation on substantive power, do Reich's prescriptions take adequate account of the constraints that make administration of mass benefit programs so problematic? Do they take adequate account of the virtues of low-level discretion described by William Simon in Chapter VI, section 4, or of the possible superiority of other kinds of controls discussed in that chapter? Would the formal procedural rights that he advocates eliminate the discretion that he fears, or would they disguise the discretion or push it into other corners of the system? In this connection, consider Mashaw's "law of conservation of administrative discretion" in the next chapter.

In a footnote to Rabin's article, he calls attention to Simon's criticism of Reich's article on the ground that its individualistic, rights-oriented approach actually had conservative, antiredistributive tendencies. How might this argument run? See Simon, "Rights and Redistribution in the Welfare State," 38 *Stanford Law Review* 1431 (1986).

A symposium of articles (including another by Reich) marking the twenty-fifth anniversary of Reich's "The New Property" was published in 24 *University of San Francisco Law Review* (1990).

5. Consider Rabin's observation, directed at Reich's "new property" article, that a participatory model of administrative law "has no substantive content." What might Rabin mean by this? Isn't it likely that additional procedural rights will improve the right holder's bargaining position with the agency and enhance her ability to obtain favorable outcomes? And aren't there some procedural features, such as the locus of the burden of proof, that will often be decisive on questions of great factual uncertainty?

6. My article on regulatory equity contends that policymaking through the granting of exceptions to existing rules is in part an administrative response to the constraints on informal ruling that grew more powerful during the 1970s and 1980s. These constraints were imposed not only by the courts, as described in several articles in earlier chapters, but also by Congress and by presidential executive orders. See, for example, the Regulatory Flexibility Act of 1980, 5 U.S.C. Sec. 601 et seq.; Executive Order 12291, 46 Fed. Reg., February 17, 1981. In their detailed study of the National Highway Traffic and Safety Administration, *The Struggle for Auto Safety* (Cambridge, Mass.: Harvard University Press, 1990), Jerry Mashaw and David Harfst show how rulemaking's growing political and other costs to the agency led it to pursue its policy goals through the relatively ineffective regulatory technique of recalls rather than through issuance of new safety standards. John Mendeloff's study of Occupational Safety and Health Administration standard setting demonstrates other perverse consequences of these high rulemaking costs, including underregulation. Mendeloff, *The Dilemma of Toxic Substances Regulation: How Overregulation Causes Underregulation* (Cambridge, Mass.: MIT Press, 1988). Richard Pierce has reached similar conclusions in his study of the Federal Energy Regulatory Commission (FERC). Pierce, "The Unintended Effects of Judicial Review of Agency Rules: How Federal Courts Have Contributed to the Electricity Crisis of the 1990s," 43 *Administrative Law Review* 7 (1991).

7. If, as Kagan argues, the social consequences of "adversarial legalism" are dire, alternative models are already working in other liberal democracies, and "learning" from such cases as the Oakland port project is possible, why hasn't the system been

reformed in the United States? For discussions of these questions, see, for example, James Q. Wilson, *The Politics of Regulation* (New York: Basic Books, 1980); a review essay on that book by Peter Schuck, 90 *Yale Law Journal* 702 (1981); Eugene Bardach and Robert Kagan, *Going By the Book: The Problem of Regulatory Unreasonableness* (Philadelphia: Temple University Press, 1982); R. Shep Melnick, *Regulation and the Courts: The Case of the Clean Air Act* (Washington, D.C.: Brookings Institution, 1983); John M. Mendeloff, *The Dilemma of Toxic Substance Regulation: How Overregulation Causes Underregulation* (Cambridge, Mass.: MIT Press, 1988). A number of regulatory reform proposals, most notably negotiated rulemaking, which is discussed in Chapter VI, are designed to ameliorate this set of problems.

VI

Controlling Administrative Discretion

If legislation is the skeleton of the administrative state, discretion—the official's freedom, within the limits of her power, to make a choice among possible courses of action or inaction—is its musculature. Discretion vitalizes agencies, infusing them with energy, direction, mobility, and the capacity for change. It bespeaks both our recognition of our limited ability (for reasons analyzed in the last article) to control the future through rules and our optimism about officials' ability to exercise sound judgment when they confront new problems. It helps to reconcile the values of law and equity, of justice in the aggregate and justice in the individual case.

But discretion has its decidedly dark side as well. While it is not the only form of bureaucratic power, it is surely the most troubling. Discretion enables and even invites officials to overreach, to discriminate invidiously, to subordinate public interests to private ones, to conceal bureaucratic reasons and purposes, and to tyrannize over the citizenry in countless large and small ways. It is also the aspect of bureaucratic power that is most difficult for any legal regime to control. Concern about the abuse of that power, especially in its discretionary form, is probably universal. In the United States, however, this fear is so chronic and pervasive that it constitutes a defining element of our political and legal cultures, accounting for much that is admirable—and much that is pathological—about American government.

This chapter explores this tension by considering a range of techniques— legal, political, managerial, processual, and programmatic—for controlling discretion. Since the nature and consequences of these control techniques vary significantly from one another, the chapter is quite long. Accordingly, it is divided into six sections. The first discusses the problem of administrative discretion at a general level, then each of the remaining five sections considers a particular form of control: review by courts and specialized tribunals; presidential and congressional review; managerial and professional norms; public participation; and market-oriented discipline.

1. The Problem of Administrative Discretion

Kenneth Davis, a leading authority on administrative law, has written the classic modern exposition of the subject in his 1969 book, *Discretionary Justice*. There, he mounts a sustained argument in favor of greater use of administrative rulemaking and other legal techniques in order to "confine" discretion (i.e., keep it within designated boundaries), "structure" it (i.e., control how it is exercised within the boundaries), and "check" it (i.e., have discretionary decisions reviewed by other officials). In contrast, Theodore Lowi takes a decidedly baleful view of administrative discretion. Locating its source in lawless pluralistic bargaining, he proposes a number of reforms designed to reduce it, including the resuscitation of the nondelegation doctrine. Jerry Mashaw, while sharing Davis's enthusiasm for broad legislative delegations of discretion to agencies, offers quite different arguments in favor of it. Discretion's virtue, in this view, is not simply that it is necessary in order to make policy decisions more individualized, contextual, and technically competent. Even more important, discretion increases their political accountability and legitimacy by reducing the effect of certain obstacles to rational public choice.

2. Review by Courts and Specialized Tribunals

The traditional form of legal control on administrative discretion has been review by more or less independent tribunals of various kinds, principally generalist article III courts and specialized administrative bodies. This section is primarily concerned with judicial review, although one should always bear in mind that the number of disputes heard by administrative review tribunals, such as the Social Security Appeals Council or the Board of Immigration Appeals in the Department of Justice, utterly dwarfs the number heard by courts.

The structure of judicial review of agency action for constitutionality and for conformance to statutory authority was firmly established by *Marbury v. Madison* (there, review of inaction by the secretary of state). Moreover, analogous modes of judicial review through the prerogative writ system had been recognized in England since at least the early eighteenth century. Even for discretionary decisions, judicial review for abuse of discretion and for procedural violations has long been a staple of administrative law.

Richard Stewart and Cass Sunstein sketch a remedial structure for the modern administrative state. They distinguish among a variety of legal forms by which agency action can be subjected to judicial review. Some of these forms—notably the "private right of action" against alleged violators of a regulatory statute, the "private right of initiation" against an agency for failing to enforce its statute, and the "new-property hearing right"—are relatively novel and controversial remedies for administrative illegality or neglect. Stewart and Sunstein argue that they should be legitimated and extended on behalf of the often diffuse beneficiaries of regulatory and redistributive programs.

Cynthia Farina's article on statutory interpretation uses the Supreme Court's controversial decision in Chevron U.S.A., Inc. v. Natural Resources Defense Council, Inc., 467 U.S. 837 (1984), to explore the new constitution of the administrative state—the structural relations among court, agency, and legislature. *Chevron,* which promised to be among the most important administrative law decisions of the modern era, directed the lower courts to defer to any "reasonable" agency interpretation of regulatory statutes unless Congress had unambiguously and "directly spoken to the precise question at issue." Farina's criticisms of *Chevron* typify what is now a large literature on the decision. Some of this literature suggests that even the Supreme Court has had second thoughts about according the level of agency power over statutory meaning that the Court seemed to demand in *Chevron.*

Most of the legal academic writing on judicial review evinces a rather low opinion of agencies' performance; it is only a slight exaggeration to say that law professors tend to view bureaucratic rationality and lawfulness as oxymorons. On the other hand, they tend to have a decidedly benign view of judicial review of administration. To overgeneralize a bit about what is an immense, often rich literature, legal commentators generally assume that searching judicial review of agency action is both desirable in principle and efficacious in practice in improving the rationality, fairness, and legality of the administrative process. Their real concern is doctrinal: how should a reviewing court construe the governing legal norms in order to vindicate these values?

It is all the more remarkable, then, that virtually all of the social scientists (and some lawyers) who have studied judicial review *empirically* in order to determine whether this optimism about its actual effects on the administrative (and legislative) process is warranted have come to strikingly different conclusions. R. Shep Melnick, a political scientist specializing in public law and administration, recently summarized these studies. In doing so, he has drawn sobering implications for the ways that judges, teachers, and students think about administrative law. Supposing that this social science literature is unfamiliar to most readers, I have departed from my practice of seeking to eliminate footnotes, retaining those that provide useful references.

In evaluating the consequences of judicial review of agencies, it is well to remember that as a practical matter, the remand, in which the reviewing court sends the case back to the agency for a new decision, is probably the court's single most important tool for influencing the agency. The courts, after all, affirm the agency in the vast majority of cases; in a study of direct circuit court review of federal agency decisions that Don Elliott and I conducted (mentioned in the Melnick excerpt), the affirmance rate exceeded 80 percent. When the courts cannot affirm, they either reverse the agency outright or, more commonly, remand so that the agency can apply its expertise to the issue again, under a correct legal standard. The remand's central significance as a judicial control technique in theory raises a number of far-reaching empirical questions about the remand process and its practical effects. In our study, we tried to answer some of them through telephone interviews with the lawyers on both sides of approximately 180 cases remanded to agencies in 1984 to 1985. An excerpt from this study presents our findings.

As noted earlier, most review of agency adjudications occurs not in article III courts but in specialized administrative tribunals established by statute or agency regulation. In an important sense, these tribunals compete with article III courts and agency adjudicators for the adjudicative "business" of the administrative state. The final article in this section, by law professor Harold Bruff, reviews many of the institutional design considerations that are relevant to evaluating and reforming this system.

3. Presidential and Congressional Review

Conflict among Congress, the Executive Office of the President, the agencies, the courts, "public interest" groups, and more traditional private groups to control the policy process is an endemic feature of the administrative state. With the unprecedented expansion of regulatory authority since 1970, the prolonged division between Democratic congresses and Republican presidents, and the aggressive campaign by presidents since Gerald Ford to discredit some of the New Deal and Great Society legacies, this conflict has intensified. The weapons in this struggle, moreover, assume a bewildering variety of forms, some of which raise constitutional issues that go directly to the character and legitimacy of the administrative state itself.

Peter Strauss's article presents a theory that seeks to resovle many of these questions by offering a coherent account of the place of administrative agencies in the constitutional structure of government. His theory touches upon many current constitutional controversies such as those involving claims of executive privilege, the status of independent agencies, and the legality of presidential efforts to shape regulatory decisions outside the framework of the Administrative Procedure Act's formalities.

Presidential control of the bureaucracy is far more problematic in the United States than in other advanced democratic systems. The fragmentation of political authority encourages agency officials to seek political support wherever they can find it, often in Congress and contrary to the president's political and policy interests. As a result, the president must view the bureaucracy as a potential renegade and make extraordinary efforts to assert control. There are numerous ways in which presidents attempt to influence agency decisions, including personnel policies, budget allocations, coordination, and administrative guidance. Perhaps the most bitterly contested control device in recent years is the system of Office of Management and Budget (OMB) review of agency proposals. Two articles, one by "public interest" litigator Alan Morrison and the other by two former OMB officials, Christopher DeMuth and Douglas Ginsburg, present sharply differing views about the wisdom and consequences of this approach.

Congress influences agency decisions in many ways. Although most of these modes of influence do not raise serious legal issues, some do. Louis Fischer, an analyst with the Congressional Research Service and a prolific commentator on separation of powers issues, reviews the responses of Congress and the executive branch to the Supreme Court's decision in Immigration and Naturalization Service v. Chadha, 462 U.S. 919 (1983), invalidating the legislative veto, which had been a ubiquitous technique for congressional review of agency actions. He also discusses the item veto, a perennial presidential proposal, arguing that such a reform would have consequences that its advocates do not anticipate.

4. Internal Controls: Management, Culture, and Professional Norms

All of the techniques for controlling agency discretion that have been discussed in this chapter share a common feature: each is deployed either by an independent branch of government or by another executive branch institution, such as OMB, that is not only bureaucratically separate from the agency but often has rather different political interests. These institutional and political separations (perhaps "chasms" would be a more appropriate word) generate inevitable conflicts between the agency and its would-be masters. Under the best of circumstances, it is difficult to ensure that an agent's behavior conforms to the principal's wishes. Even if the agent's duty is well-defined, the principal must bear either the cost of monitoring the agent's performance or the cost of her noncompliance. Both costs, of course, increase to the extent that the agent enjoys discretion and her performance is hard to measure; they are even higher when the agent is also embedded in a separate institution with different incentives, perceptions, and values. This "agency cost" problem is so ubiquitous that a subspecialty within the economics field systematically studies ways to minimize it, as mentioned in the first note after chapter I.

This section considers two strategies for dealing with this problem. Because each strategy depends upon controls that are internal to the agency, each promises to reduce the costs entailed by efforts to transcend institutional boundaries and barriers. By seeking to conform official conduct to common organizational or professional norms, these controls also promise to be more successful in achieving agency management's goals.

Jerry Mashaw's article on the "management side of due process" presents an alternative to the conventional model of judicial review of agency action in which courts, which are poorly equipped to obtain reliable information on the relevant policy variables, apply vague "due process" norms to a complex administrative program. His proposal bears a family resemblance to Kenneth Davis's call for more agency rulemaking and could, if effective, have similar effects—reducing the need for judicial review but also focusing whatever judicial review still occurred.

James Q. Wilson considers two important tools of, and constraints on, bureaucratic management of low-level decisionmakers. The first is the organization's culture, which he (following Philip Selznick) analogizes to character in individuals, and the second is the agency's task structure, which conditions the incentives that motivate its workers. In bureaucracies in which managers cannot closely monitor or control what low-level employees do—especially in what Wilson calls procedural, craft, and coping agencies—strong professional norms may be essential to achieving their missions. These norms, however, may conflict with other bureaucratic values, such as hierarchical control. William Simon's article examines this tension in the Massachusetts welfare agency, which conducted what he views as a deprofessionalization or "proletarianization" of the Aid to Families with Dependent Children caseworkers by adopting, among other things, the kinds of quality of control measures advocated by Mashaw.

5. Public Participation

In a highly decentralized policy process, public participation is more than a legitimating norm and a source of citizen satisfaction with procedures and outcomes. It can also be a form of control on administrative discretion, shaping the factual record and legal and policy arguments that are supposed to form the basis for agency action.

Richard Stewart, writing at the inception of the regulatory reform movement of the mid-1970s, developed what is perhaps the classic elaboration of the "interest representation" model of administrative law, premised on the inadequacy of existing modes of public participation in administrative processes and the need to develop new, more robust ones. This excerpt from a very long article is an admirably balanced analysis of the competing arguments about interest representation through participation.

The excerpt from Mashaw's book on the Social Security disability program applies some of these arguments for enhanced participation in an adjudication context, where he proposes to augment the hearing process with face-to-face interviews and representation by lay advocates.

The Harter article describes and defends the technique of negotiated rulemaking, which has achieved some prominence in recent years. Harter is particularly concerned to specify the role of courts in reviewing regulations developed through this process.

6. Market-Oriented Controls

One front in the campaign for regulatory reform—perhaps its most consequential element in the long run—is the effort to shape and limit agency discretion by infusing the bureaucratic environment with incentives either created by the market or contrived to mimic its supposed efficiencies. These incentives may take a variety of forms ranging from complete deregulation of an activity to the use of market-like incentives to augment a regulatory program that remains essentially bureaucratic. Ronald Cass's article maps this landscape and considers its legal and political implications.

1

The Problem of Administrative Discretion

Discretionary Justice
KENNETH DAVIS

Paradoxically today's excessive discretionary power is largely attributable to the zeal of those who a generation or two ago were especially striving to protect against excessive discretionary power. . . . They tended to oppose all discretionary power; they should have opposed only unnecessary discretionary power.

The two governmental or legal doctrines that are largely responsible for the damage are (1) an extravagant version of the rule of law or supremacy of law and (2) the unconstitutionality of legislative delegation of power unless accompanied by meaningful standards. These two doctrines were developed by conscientious people, including legal philosophers and judges, whose worthy purpose was protection against governmental excesses. But both doctrines grossly overshot, and both have been decisively defeated. The worst of it is that milder and sounder opposition to excessive discretionary power became identified with the extravagant version of the rule of law and with the nondelegation doctrine and was largely pulled down in the defeat of those two doctrines. And our legal system has not yet recovered its balance. . . .

What I am calling the extravagant version of the rule of law, asserted by a good many writers, declares that legal rights may be finally determined only by regularly constituted courts or that legal rights may be finally determined only through application of previously established rules. What this version of the rule of law especially opposes is discretionary power exercised outside courts and not fully subject to judicial control. . . .

The Franks Committee-Dickinson-Dicey-Hayek versions of the rule of law express an emotion, an aspiration, an ideal, but none is based upon a down-to-earth analysis of the practical problems with which modern governments are confronted. . . .

The extravagant version of the rule of law is incompatible with any regulatory program. Indeed, it is at variance with the fundamentals of any modern government. The very identifying badge of the American administrative agency is power, without previously existing rules, to determine the legal rights of individual parties. For instance, the Attorney General's Committee on Administrative Procedure said: "The Committee has regarded as the distinguishing feature of an 'administrative' agency the power to determine, either by rule or by decision, private rights and obligations." The Congress of the United States did not respond to that statement by saying it violates the rule of law. Instead, it accepted the statement as a foundation for the Administrative Procedure Act. . . .

The Nondelegation Doctrine

. . . The Supreme Court at one time paid lip service to a nondelegation doctrine embodying something like the extravagant version of the rule of law: "That Congress cannot delegate legislative power to the President is a principle universally recognized as vital to the integrity and maintenance of the system of government ordained by the Constitution." If that were true, one would have to say that as of the 1960s we no longer have the system of government ordained by the Constitution. But of course it never was true. The court gradually changed its doctrine to a requirement of standards: "Congress cannot delegate any part of its legislative power except under the limitation of a prescribed standard." But the court has upheld many delegations without meaningful standards and even many without any standards. When a district court took literally what the Supreme Court said about the requirement of standards, the Supreme Court reversed, upholding the delegation despite absence of standards. The Supreme Court was yielding to realism when it acknowledged: "Delegation by Congress has long been recognized as necessary in order that the exertion of legislative power does not become a futility.

Why has the court about given up in its one-time efforts to require that delegations be accompanied by meaningful standards? Perhaps the most important reason is that Congress has gone right on delegating, with vague standards or with no standards, and the Supreme Court has gradually bowed to the will of the people's representatives. The main reason that Congress resisted the nondelegation doctrine is stated above in our discussion of the major questions of policy decided by the Civil Aeronautics Board without significant guidance from the statute: Congress deemed itself both unequipped and unwilling to answer the major questions of policy; it decided that the job could better be done by the board, within the framework established by the statute. Anyone who will study the list of major questions the board has decided, as well as the kind of work that has gone into those decisions, is likely to begin to appreciate the wisdom in this arrangement. . . .

Legislative clarification of objectives may sometimes be undesirable, even though it is always desirable when the legislative body knows what it wants. When the society is sharply divided, when the problems are new and opinions have had insufficient time to crystallize, when biting off one concrete problem at a time is clearly preferable to trying to legislate in gross, or when sustained staffwork may contribute significantly to policy choices, a legislative body may wisely keep the policy objectives largely open. Vague or meaningless standards may then be preferable to precise and meaningful ones. This is so not merely from the standpoint of legislators' normal desires to escape the disadvantage of turning the losers into political opponents but also from the standpoint of facilitating the process of ultimately arriving at policies which are both sound and acceptable. When both the understanding of the merits and the weighing of political pressures are about in even balance, or when one of these factors points one way and the other the other way, or when these factors and a host of additional factors impel hesitation, a period of pulling and hauling, of vagueness, of inconsistencies, and of ad hoc decisions without standards may often be better than an early clarification, depending, of course, on an appraisal of the losses that may stem from the uncertainty.

Of course, some statutory delegations are clearly deficient in the extent to which they fail to clarify objectives. Some questions can be better answered by the legislative process than by the administrative process, such as broad questions on which major political parties oppose each other. Occasionally inadequate draftsmanship or haste in the closing days of a legislative session yields products that deserve to be struck down through use of the nondelegation doctrine. Perhaps the most frequent deficiency of legislative bodies has to do with failures to follow through in the development of policy after the delegation has been made. Even when vagueness or absence of standards has been necessary or desirable in the initial delegation, meaningful standards may become feasible after a few years of experience with a new program. The legislators then should either clarify the standards through amending the statute or they should prod the administrators to clarify through regulations or otherwise.

By and large, however, the emphasis should not be on legislative clarification of standards but on administrative clarification, *because that is where the hope lies*. Yet the objective is seldom complete clarification; the objective should be clarification to the extent that the subject matter and the available understanding permit, but not more than is consistent with needed individualizing. . . .

Both the rule of law and the nondelegation doctrine could have been turned into effective and useful instruments. The goal of the rule of law could have been to distinguish between necessary discretionary power and unnecessary discretionary power, the one welcomed and encouraged and the other forbidden or discouraged. *The courts could have been enlisted to help determine what discretionary power is necessary and what is unnecessary; if that had been their assignment, I think a good deal of today's excessive discretionary power could have been avoided.* Similarly, the nondelegation doctrine should never have aimed at always requiring meaningful standards; the courts could have succeeded if they had limited their objectives to designing legal devices for proper confinement and control of discretionary power.

The End of Liberalism

THEODORE LOWI

Liberal jurisprudence is a contradiction in terms. Liberalism is hostile to law. No matter that it is motivated by the highest social sentiments; no matter that it favors the positive state only because the positive state is the presumptive instrument for achieving social good. The new public philosophy is hostile to law. . . .

Interest-group liberalism has little place for law because laws interfere with the political process. The political process is stymied by abrupt changes in the rules of the game. The political process is not perfectly self-correcting if it is not allowed to correct itself. Laws change the rules of the game. Laws make government an institution apart; a government of laws is not a simple expression of the political process. A good clear statute puts the government on one side as opposed to other sides, it redistributes advantages and disadvantages, it slants and redefines the terms of bargaining. It can even eliminate bargaining, as this term is currently defined. Laws set priorities. Laws deliberately set some goals and values above others.

In brief, law, in the liberal view, is too authoritative a use of authority. Authority has to be tentative and accessible to be acceptable. If authority is to be accommodated to the liberal myth that it is not power at all, it must emerge out of individual bargains.

The legal expression of the new liberal ideology can be summed up in a single, conventional legal term: *delegation of power.* . . .

. . . Delegation of power enjoys strong standing in the courts. . . .

The doctrine of delegation of power also meets with strong support among academic political scientists and historians. Too much law would obviously be intolerable to scientific pluralist theory. In a vitally important sense, *value-free political science is logically committed to the norm of delegation of power because delegation of power is a self-fulfilling mechanism of prediction in modern political science.* Clear statutes that reduce pluralistic bargaining also reduce drastically the possibility of scientific treatment of government as simply part of the bundle of bargaining processes and multiple power structures. A good law eliminates the political process at certain points. A law made at the center of government focuses politics there and reduces interest elsewhere. The center means Congress, the president, and the courts. To make law at a central point is to centralize the political process. If this is too authoritative for interest-group liberalism it is too formal for modern political science.

Hostility to law, expressed in the principle of broad and unguided delegation of power, is the weakest timber in the shaky structure of the new public philosophy. This, more than any other single feature of interest-group liberalism, has wrapped public policies in shrouds of illegitimacy and ineffectiveness. This, more than any other feature, has turned liberal vitality into governmental and social pathology.

It is of course impossible to imagine a modern state in which central authorities do not delegate functions, responsibilities, and powers to administrators. Thus the practice of delegation itself can hardly be criticized. The practice becomes pathological, and criticizable, at the point where it comes to be considered a good thing in itself, flowing to administrators without guides, checks, safeguards. Historically, delegation had a rather technical meaning that emerged as the price to be paid in order to reap the advantages of administration. Delegation today represents a bastardization of earlier realities, and it is the bastardization that is at issue.

Evolution of Public Controls in the United States: The Delegation of Powers and Its Fallacies

Delegation of power did not become a widespread practice or a constitutional problem until government began to take on regulatory functions. The first century was one of government dominated by Congress and virtually self-executing laws. Congressional government, as Woodrow Wilson could view it in the 1880s, was possible for two reasons: Either its activity was insignificant, or it sought only to husband private action. Between 1795 and 1887, the key federal policies were tariffs, internal improvements, land sales and land grants, development of a merchant fleet and coastal shipping, the post offices, patents and copyrights, and research on how the private sector was doing. Thus, after a short Hamiltonian period when the Economic Constitution was written—including assumption and funding of debts, the taxation system, currency and banking structure, establishing the power to subsidize—the federal government literally spent one century in the business of subsidization. It was due to this quite special and restricted use of government that Congress could both pass laws and see to their execution. . . .

Delegation Defined

. . . In the Interstate Commerce Act, there was delegation of the "full ambit of authority"—executive, legislative, and judicial—in a single administrative body. It was given the power to be flexible, but it was relatively well shackled by clear standards of public policy, as stated in the statute and as understood in common law. From the beginning, the whole notion of vesting great authority in administrative tribunals was never separated from the expectation that standards of law would accompany the delegation. Much fun has been made of the myth the courts tried to create, that agencies were merely "filling in the details" of acts of Congress. But neither the myth nor the mirth can hide the fact that when a delegation was broad, positive rules of law attached themselves to it.

Many have argued that restriction on delegation was never intended. They reason that prior to the Panama [Refining Co. v. Ryan, 293 U.S. 388,] and [A.L.A.] Schechter [Poultry Corp. v. U.S., 295 U.S. 495,] cases in 1935 the Supreme Court had not declared delegations of power unconstitutional, that after 1936 no further delegations were invalidated, and therefore is it not possible that "Schechter is only of historical significance." However, the question is badly put. The question of proper standards can be posed, and was indeed posed, without necessarily involving constitutionality. Obeying its own rule of restraint on constitutional issues, the Court merely filled in the congressional and presidential void with specifications of its own—or so construed the statutes as to render the agency powerless. Jaffe himself provides the best example. The Federal Trade Commission Act of 1914 so poorly defined the key term, "unfair method of competition," that the Supreme Court invalidated order after order issued by the FTC through most of the first twenty years of its life. The question of accompanying delegations with proper standards of law did not become a dead issue—in the courts or in congressional debate—until the 1930s and 1940s.

Delegation to the Administrative Process: A Developmental Analysis

Throughout the formative period of federal control of economic life, delegation was considered a problem. It had to be encountered because an administrative component had to be added to government, and delegation was the only way to accomplish it. Delegation was nonetheless the central problem, and it was faced with various partial solutions. However, a curious thing happened in the history of public control after 1887: As public control extended to wider and more novel realms, delegation became a virtue rather than a problem. *The question of standards disappeared as the need for them increased.* To pursue this proposition it is necessary first to pursue the actual developments in the practice of delegation itself. . . .

Summary: The Rise of Delegation, The Decline of Law

Empirically, the stages of development in the administrative process can be summarized as follows:

1. There is an expansion of the scope of federal control in the sense of the number and types of objects touched by directly coercive federal specification of conduct. This is what has been meant traditionally by the expansion of government.
2. There has also been an expansion of the scope of power in the more philosophical sense of expansion to the whole universe of objects or qualities of objects in a predefined category. This expansion proceeds as categories grow larger, eventually to include certain characteristics

(such as trade) which any and all persons in the country might at one time or another possess.

3. Implied in (2) and necessary to it is the development from the concrete to the abstract. This development was observed at two levels. First, the jurisdictional categories became more and more abstract. Second, the standards by which actual qualities are designated moved rather quickly from the specific to the general.

4. The development also involved changes in sanctions. The movement seemed to be from the negative to the positive, the proscriptive to the prescriptive. The latter did not replace the former but only supplemented them. . . .

Obviously *modern law has become a series of instructions to administrators rather than a series of commands to citizens.* If at the same time (1) public control has become more positive, issuing imperatives along with setting limits, and if at the same time (2) application of laws and sanctions has become more discretionary, by virtue of having become more indirect as well as more abstract, why should we assume we are talking about the same governmental phenomena in 1968, and 1978 as in 1938, or 1918, 1908? The citizen has become an *administré,* and the question now is how to be certain he remains a citizen.

As has been said already, and as will be reiterated thematically, delegation has been elevated to the highest of virtues, and standards have been relegated to the wastebasket of history because that is the logic of interest-group liberalism. Bargaining—or, as Schlesinger might call it, participation in the "interior processes of policymaking"—must be preferred over authority at every level and phase of government. The idea that the universal application of bargaining solves the problem of power was just appealing enough to win out over alternative doctrines that appeared so conservative in light of the desperation of the 1930s. . . .

By 1968, perhaps even by 1958 or 1948, none of this reserve was left. There was only untarnished exuberance for a system built upon unregulated regulation. By 1978 these attitudes were fully congealed into the intellectual components of a Second Republic.

Law Versus Liberalism

. . . Many express concern lest the revival of law destroy bargaining. Bargaining, they say, is one of the great virtues of democracy. It maintains flexibility, and this is supposedly the way to avoid turning citizen into *administré.* But this concern only reveals the extent to which an interest-group-liberal view disorders meanings and narrows vision. An attack on delegation is an attack not on bargaining but on one type of bargaining—logrolling. The attack is on a confusion of the two meanings of bargaining and the consequent misapplication of the whole idea. First, there can be bargaining over the decision on a particular case. This type of bargaining over the stakes is logrolling. It has to do with whether certain facts are to be defined as identical to some earlier set of facts. It has to do with whether the case will be prosecuted at all, and with how much vigor. It has to do with whether monetary sanctions apply; if so, how much; whether contrition plus compliance in the future are sufficient. Second, and worlds apart, there can be bargaining on the rule or rules applicable to the decision. This can take the form of simple insistence that the authority state some rule. It can be a quibble over the definition of a concept or a profound analysis of what Congress could possibly have intended. It could be a process of defining the agency's jurisdiction.

Once this distinction between two kinds of bargaining is stated it is impossible to imagine how they could possibly become fused or confused. The differences in their consequences ought to be clear. In any libertarian, unmobilized society, some logrolling is likely to occur on every decision at every stage in the governmental process. However, if by broad and undefined delegation you build your system in order to insure the logrolling type of bargain on the decision, you are very likely never to reach bargaining on the rule at all. If, on the other hand, you build the system by stricter delegation to insure bargaining on the rule, *you will inevitably get logrolling on the decision as well.*

These propositions are not true merely because the morality of the general rule is weaker than the stakes of the individual case. Bargaining on the rule is especially perishable because broad delegation simply puts at two great disadvantages any client who wishes to bargain for a general rule rather than merely to logroll his case. Here is what he faces: First, the broad delegation enables the agency to co-opt the client—that is, to make him a little less unhappy the louder he complains. On top of that, the broad delegation reverses the burden of initiative and creativity, the burden of proof that a rule is needed. If the client insists on making a federal case out of his minor scrape, he must be prepared to provide the counsel and the energy to start a rulemaking process himself. This means that the individual must stop his private endeavors and for a while become a creative political actor. Most behavioral research agrees that this is an unlikely exchange of roles. . . .

Regulation in the 1970s: The Sky Is No Longer the Limit

Were there ever to be a change in the direction of historical development, 1969 would have been the time for it. The Nixon administration was unusually partisan; and partisan Republicans felt they differed most from partisan Democrats in precisely such matters as government intervention and state's rights. Republicans . . . could embrace the agriculture subsidies and the varieties of indulgences to all economic interests. But the regulatory arena is quite another matter. Controlling the conduct of corporations and the private affairs of individuals hit at the very core of the Republican concept of liberty.

However, the Republicans did not pick up their option after 1969. The celebration of federal government activism in the 1960s seemed virtually to become a binge in the 1970s. . . .

The second objection to the argument that Republicans merely continued the Democratic patterns is really more important to deal with because it is quite undeniably true that a strongly Democratic Congress often imposed its will on the Republican administration. Few of the regulatory laws identified in Table VI.1 were distinctive parts of President Nixon's program; and the most important enactment—wage and price controls—was passed despite President Nixon's grumbling that such authority was not needed and that he would not use it if it were handed over to him. Nonetheless, the Republicans did more than merely accept the domination of the Democrats during this period. There is ample evidence to support the argument that the continued expansion of the regulatory authority of the federal government is the direct consequence of a modern liberal state that had come to be accepted as much by Republicans as by Democrats. Democrats may stress the social benefits of regulation while Republicans may worry more about the costs. Criticisms from Democrats may focus upon improved efficiency of regulation while criticisms from Republicans may stress a shift in the burden of regulation. But as far as can be determined by their behavior, Republicans have become just as much as the Democrats committed to the contemporary liberal state and to the principles of liberal jurisprudence.

Universal Protection of Workers and Consumers

[The Occupational Safety and Health Act] OSHA and the Consumer Products Safety Commission (CPSC) are prime examples of the continuation and reinforcement of 1960s liberalism applied to public policy. OSHA was the first national effort to deal with all industry. And it was quite patently an effort by the federal government to take this vital function almost entirely away from the states. CPSC was a consolidation of several existing consumer-protection activities, but went far beyond the scope of preexisting federal law by attempting to cover virtually all sources of consumer risk from the purchase and use of commodities. In scope, jurisdiction, and purpose, these two pieces of legislation are open-ended, abstract, and universal. . . . The legislation creating OSHA covers every employer in the country (except public agencies) and seeks to provide a safe environment for employment by obliging the new agency to set standards the observance of which would produce the desired state of affairs. CPSC legislation seeks also to provide the safe environment for all consumers by having that agency set standards whose observance will create such an environment.

In neither piece of public policy did Congress attempt by law to identify a single specific evil that the regulatory agency was to seek to minimize or eliminate. In neither case did the statute attempt to identify a single cause of action against which aggrieved employees or consumers would have an easier day in court. Quite to the contrary, OSHA and CPSC laws provided only an

Table VI.1 Federal Regulatory Laws and Programs Enacted since 1970

Year Enacted	Title of Statute
1969–70	Child Protection and Toy Safety Act
	Clear Air Amendments
	Egg Products Inspection Act
	Economic Stabilization Act
	Fair Credit Reporting Act
	Occupational Safety and Health Act
	Poison Prevention Packaging Act
	Securities Investor Protection Act
1971	Economic Stabilization Act Amendments
	Federal Boat Safety Act
	Lead-Based Paint Elimination Act
	Wholesome Fish and Fisheries Act
1972	Consumer Product Safety Act
	Equal Employment Opportunity Act
	Federal Election Campaign Act
	Federal Environmental Pesticide Control Act
	Federal Water Pollution Control Act Amendments
	Motor Vehicle Information and Cost Savings Act
	Noise Control Act
	Ports and Waterways Safety act
1973	Agriculture and Consumer Protection Act
	Economic Stabilization Act Amendments
	Emergency Petroleum Allocation Act
	Flood Disaster Protection Act
1974	Atomic Energy Act
	Commodity Futures Trading Commission Act
	Consumer Product Warranties/FTL Improvement Act
	Council on Wage and Price Stability Act
	Employee Retirement Income Security Act
	Federal Energy Administration Act
	Hazardous Materials Transportation Act
	Housing and Community Development Act
	Pension Reform Act
	Privacy Act
	Safe Drinking Water Act
1975	Energy Policy and Conservation Act
	Equal Credit Opportunity Act
1976	Consumer Leasing Act
	Medical Device Safety Act
	Toxic Substances Control Act

expression of sentiments for the desired end result. The OSHA legislation took as its purpose "to assure so far as is possible every working man and woman in the nation safe and healthful working conditions and to preserve human resources" (Section 2B). The CPSC Act is, if possible, even more vague and universal, in that it ordains the commission to reduce unreasonable risk of injury from use of household products but does not identify a single risk, does not suggest what might be a reasonable undertaking of that risk, and definitely does not suggest how to make such risks reasonable. Instead, in both of these instances, Congress provided that the agency could set standards of behavior, respectively, for employers and for the producers of consumer products. Congress provided no standards whatsoever for employers or producers, nor did Congress provide any standards for the conduct of these two regulatory agencies. All Congress did was to assume that the agencies in their own wisdom would be able to provide such standards.

Where, indeed, were these standards to come from? The law is fairly explicit in both instances, and both instances provide almost pristine examples of interest-group-liberal resolutions of the problem of balancing power and interest against policy choice. Under OSHA, a standard is defined as any practice or method thought by the secretary of labor or the agency to be appropriate to provide safe and healthful employment. There are two routes to such standards. The first is a standard for a given industry or sector of the economy which is developed within the department itself on the basis of work with an advisory committee appointed by the secretary consisting of fifteen members, including one or more drawn from HEW and the rest from experienced and qualified representatives of employers and employees, as well as one or more representatives of health and safety agencies of the states. The second route to such industry standards, and the route most frequently employed, is the adoption of what the law calls a "national consensus standard" which is defined in the law to mean any occupational safety standard that has already been adopted and promulgated by a "nationally recognized standards-producing organization under procedures where it can be determined by the Secretary that persons interested and affected by the scope or provisions of the standard have reached substantial agreement on its adoption" (Section 3[9]). The result basically has been that OSHA adopts the criteria of industrial safety prevailing in a given trade association at the time. These consensus standards are so closely tied to trade association standards that some manufacturing concerns have found themselves in a bind, because trade association practices change faster than the OSHA bureaucracy can keep up with. These so-called consensus standards would appear to be indistinguishable from the practices under the old NRA codes that were declared unconstitutional in the 1930s. The difference is that in the 1970s they are not merely constitutional; there is so little suspicion as to their constitutionality that there is no particular urge to take these issues to court.

Regulation under CPSC is an example of a different, though related, extreme. Rather than look to industry to find existing industry standards to embrace, CPSC keeps its own counsel, not by adopting its own standards and regulations that can be expected to apply to a broad class of consumer items for a certain period of time, but by not making any advance rulings at all. It takes each case on its merits and moves toward consumer items and toward regulation of those items with a vigor that is matched only by its unpredictability. It can ban a product altogether. It can issue standards that regulate the design of a product. It can respond to consumer complaints about individual products, or it can take initiatives and pursue one consumer item or a broad category of consumer items; and it can do so by specific identification of an item as a risk or by a more general standard of the degree of riskiness of the product or products. It may, or it may not, take into account alternative ways of dealing with the riskiness of a product. In brief, the agency began with an almost unlimited mandate from Congress, and it can go on avoiding limits on itself by the simple expedient of not issuing rules of general legislative content. It can stick to specifications. This agency is an example of pure administrative power.

Congress and the administrators seem to be perfectly well aware of the basic illegitimacy of what they are doing with CPSC. Having created a monster of unadulterated administrative power, Congress then turned around and granted the agency broad discretion to subsidize persons who would come in and speak for consumers in opposition to agency decisions. These people were called offerors and were expected to be professional standard developers from the private sector. In 1976 Congress added still another provision to the original act authorizing the courts to award to plaintiffs the costs of the suit, including attorney fees and the fees of expert witnesses, in order "to enable interested persons who have rights under the CPSC to indicate those rights." Since that was also not enough assurance that CPSC would have any claim to legitimacy, a move was already afloat at that time to create a Consumer Protection Agency that would perhaps more effectively speak for aggrieved persons. But no amount of after-the-fact tinkering with administrative or clientele arrangements will counterbalance the unhappy tendencies set in train by any piece of regulation as badly drafted as CPSC.

Between the OSHA method of embracing current industrial standards and the CPSC method of arbitrarily pouncing without any administrative guidelines at all, it would be extremely difficult to choose a good method of administrative regulation. But why choose either? How many abuses of contemporary government do we need before we look beyond the practices to the basis of the liberal state itself? And the need for this is all the more pressing if Republican administrations embrace that same liberal state.

Environmental Protection: Intimations of Nixon

Although President Nixon did cooperate with Congress in the drafting and the acceptance of the new and vast OSHA and CPSC regulatory programs, his own approaches and preferences can be seen more directly through the Environmental Protection Agency actions and through wage-price controls. EPA was established in 1970 by executive orders issued by President Nixon. Through these actions President Nixon sought to consolidate several major programs in a single agency. EPA was then to do its job in a manner already established by OSHA and CPSC—by "setting standards consistent with national environmental goals." In other words, "EPA would be charged with protecting the environment by abating pollution." Nixon's concept of the agency was typical of the liberal approach. EPA would "monitor the condition of the environment," "establish quantitative environmental base lines," and then "set and enforce standards for air and water quality and for individual pollutants." To do this, EPA would take guidance from itself but could also encourage broader participation "through periodic meetings with those organizations and individuals interested in that regulatory package."

The whole universe is covered by the EPA's jurisdiction. Since pollution can come from anywhere, we must naturally equip our agency with power to cover anything and everything. How can anyone be against clean air or water? And let us, indeed, have it by 1976; and if not, then by 1986; if not by then, at least let there be satisfaction that authority was exercised on behalf of the people. It is as though there were a trade-off between pollution and the number of regulations concerning pollution. Congress knew nothing in the beginning and admitted it by mandating clean air and water to administrators to pursue entirely as they saw fit. And neither Congress nor the president has reviewed the substance of the thousands of standards and regulations emanating from EPA in order to determine whether there is any relation at all among these regulations or between them and the original legislative enactments. It seems that Nixon's main preoccupation was to be sure that he had a hand in environmental protection regulation, not that he wanted to restrain or to regularize it, or to make it more explicit, or to impose upon it the sense of justice that a strict constructionist might have had. Our experience with the Democrats of the 1960s, including the Democratic-dominated Congress, would have led us to expect little more from them. But the important point here is that the Republican administration did nothing different. . . .

Liberal Jurisprudence in the 1970s: The End of the Rule of Law

These case histories involving the presence or absence of standards in legislative delegations and rules in administrative practice overwhelmingly support the simple proposition that *law begets law*. Agencies that begin in a context of statutes that associate guidance with power are agencies that begin with legal integrity and have histories of greater legal integrity. Agencies that begin with little or no legal integrity are very unlikely to develop any along the way.

An attack on the practice of delegation turns out actually to be a hopeful view. If the rise of delegation and the decline of law were the mere result of technical complexity it would be an irreversible process because technical complexity is the law of modernity. However, this review suggests that the real problem is one of abstraction rather than one of complexity. It is from abstraction that uncontrolled discretion flows. Abstraction is reversible, with the increase of knowledge, *but only if the leadership desires to reverse it*. This is why a change of public values is so essential.

Liberal jurisprudence persists as the ideological support structure for interest-group liberal legislation because it still holds so many of the top legal minds in its thrall. The most important of these is Kenneth Culp Davis, whose writings through three decades have been carefully studied and widely cited.

Davis's argument can be paraphrased as follows: The reason why Congress makes policy-without-law and commits full discretion to administrative agencies is simply that we had to have each program and no one was willing or able to make a clear decision or set any guidelines. Since we must have these programs, and since no one is able to say what the rule of law should be, we must throw ourselves upon the mercy of those given authority; then we must do whatever we can to be sure these officials are of good quality and that there are procedural safeguards against arbitrary use of government power. No proof is offered, only an assertion that the "objective of requiring every delegation to be accompanied by meaningful statutory standards had to fail, should have failed, and did fail." This offers a jurisprudential carte blanche for poor legislative drafting and at the same time sweeps away all concern for the consequences.

The decline of Congress, the decline of independence among regulatory agencies, the general decline of law as an instrument of control are all due far more than anything else to changes in the philosophy of law and the prevailing attitude toward laws. Admittedly the complexity of modern life forces Congress into vagueness and generality in drafting its statutes. Admittedly the political pressure of social unrest forces Congress and the president into premature formulations that make delegation of power inevitable. But to take these causes and effects as natural and good, and then to build the system around them, is to doom the system to its present slide toward its lowest common accomplishment. A government of statutes without standards may produce pluralism, but it is a pluralism of privilege and tight access, evidence of which is seen throughout this volume. A government of policies built on legal integrity will not destroy pluralism but will lift pluralism, and political discourse, to the highest levels of political responsibility where decisions on rules are supposed to be made.

From this formalist or constitutionalist standpoint, a bad program is worse than no program at all. If, from time to time, political pressure or the force of circumstance makes a standardless regulatory program necessary, it will always be better to admit the necessity than to pronounce it a good thing. If the empty enactment is seen as a necessary evil, there might then be something of an urge to give it, after a few years' experience, the standards and guidelines the legislators had been unable to give it the first round.

This does not represent so extreme or global a change in point of view, yet it could contribute substantially toward revitalization of Congress, toward the opening up, if not the complete severing, of agency-clientele relations, and toward some modest elevation of the level of political discourse. The effort to regulate, especially when it comes as late as in the United States federal system, immediately attaches a morality to government. When that morality is a criterion for the regulatory act, it is bound to have a characteristic influence of some sort on the political process. When the regulation is done permissively and in the deliberate absence of the morality implied in the very use of coercive language, the result can only be expected to be demoralizing, to the clientele, the agency, and the law itself. The group process is dynamic and cumulative when groups have an institutional structure against which to compete. Without that formal structure the group process is not truly pluralistic at all. It is merely co-optive. And it is ineffective. Worse, it converts mere ineffectiveness into illegitimacy.

Liberal sentiments remain, and, indeed, they may be the best sentiments. But the interest-group-liberal method is as inappropriate for our time and for those sentiments as was the laissez-faire liberal method for 1929. The interest-group method was an ideal means of achieving a bit of equity. Its day is done, for equity is no longer enough. It has proven itself unequal to the tasks of planning and achieving justice. A grant of broad powers to administration is not a grant of power at all. It is an imposition of impotence.

Cases drawn from the 1970s show mainly that neither Congress nor the executive branch learned anything significant from the 1960s. Liberal jurisprudence—belief in the goodness of policy-without-law—seems to be as strong as ever, despite the growth and expansion of criticism of the way our national government is conducted. Unfortunately, that criticism has not turned fully enough to the real source of the problem. We are fast being done in by our national government, but not by its inefficiency or its corruption, its size or the irresponsibility of its senior officials. We are being done in by the very efficiency and zealousness with which officials pursue the public interest. It is not the officials who are corrupt, but their jurisprudence, which seems to render them incapable of making right decisions.

Prodelegation: Why Administrators Should Make Political Decisions

JERRY MASHAW

The Critics Criticized

The critical literature is thus of two general types: one predicts that vague delegations will delegitimize representative governance; the other suggests that statutory vagueness leads to an overall reduction in public welfare. Either effect would be serious; taken together they are perhaps calamitous.

Interestingly, none of the literature surveyed provides more than anecdotal support for the hypotheses offered. Indeed, none of the literature is sufficiently precise about the difference between vague and specific legislation to permit confident assignment of legislation to one or another category. But before worrying about empirical testing and better specification of the behavioral hypotheses, we should consider the plausibility of the normative claims that are being made.

Vagueness and Legitimacy

The delegitimation critics might be divided further into two camps. Lowi seems to make an argument from authoritativeness; Ely and Justice Rehnquist seem to be making arguments about the need for accountability. . . .

Yet, while focusing on the rule of law and its undeniable importance in maintaining liberty, we should not forget the apparent equal importance of a contradictory demand: the demand for justice in individual cases. Moreover, the demand for justice seems inextricably linked to the flexibility and generality of legal norms, that is, to the use of vague principles (reasonableness, fairness, fault, and the like) rather than precise rules. . . .

The Ely-Rehnquist demand for legislative decisionmaking as a prerequisite to accountability is similarly incomplete, and even more perplexing.

The sort of specific issue accountability that Ely seems to applaud, indeed to advocate, is hardly transparently desirable. Do we really want to choose our representatives (or hold them accountable) on the basis of specific votes concerning specific legislation which, but for constitutional necessity (a nondelegation doctrine with bite), they would have cast in more general terms? How exactly does it help us in choosing legislators to judge them on the basis of preference expressions that are not the expressions they would give but for the constitutional necessity of being specific?

Even if we were to imagine that statutory precision is informative, it is hard to envisage how rational voter calculation is improved appreciably. When one votes for Congressman X, presumably one votes on the basis of a prediction about what X will do in the next time period in the legislature. How much better off are voters likely to be in making that prediction—that is, in determining how well Congressman X is likely to represent them over a range of presently unspecified issues—by knowing that he or she voted yes or no on the specific language in certain specific bills in some preceding legislature? After all, the voter will also know that X could not have controlled all or even a substantial portion of the language of those bills. Votes must have been cast "all things considered." Therefore, when making a general appraisal of X's likely behavior in the future, it is surely much more important that voters know the general ideological tendencies that inform those votes (prolabor, probusiness, prodisarmament, prodefense) than that X votes for or against the particular language of particular bill. I know of no one who argues that statutory vagueness prevents the electorate from becoming informed on the general proclivities of their representatives. . . .

Vagueness and the General Welfare

The Aranson et al. thesis fills in the behavioral gaps in Ely's account and goes on to claim that negative welfare effects flow from vague delegations.

What are the bases for this pessimistic view of legislation? The first proposition is that legislators are oriented primarily to their own reelection. Thus they will consider legislative actions from the perspective of their ability to claim credit with relevant constituencies. Second, because legislators represent different constituencies with different interests, all of them will find it useful to make trades with each other (logrolling) so that the demands of a wide variety of constituencies can be satisfied. Finally, even where the costs of legislation affect all constituencies equally, bare majority coalitions of the beneficiaries of legislation remain possible, provided that the benefits of legislation exceed one half its costs. Given this possibility, all legislation (or, sometimes, most legislation) can be presumed to be the satisfaction of some coalition of private groups at a public cost that exceeds the legislation's benefits to the coalition. Notwithstanding these routinely negative results, voters will not chuck the legislators out because of something called "high voter perceptual thresholds." Voters can see the direct benefits to them but are relatively impervious to the high indirect costs to them of benefits to others.

The empirical gaps and logical leaps in this model are quite wonderful. First, all politics is interest-group politics. All legislation and all legislator-constituency relationships are "pure pork barrel." Legislation with respect to abortion, prayer in the public schools, environmental protection, and river and harbor improvements can all be modeled in precisely the same way. More critically, a possibility theorem (it can happen) is transformed into a behavioral prediction (it will happen). Because bare majority coalitions *could* pass disbeneficial legislation, they do. . . .

Quite apart from the doubts engendered by modest reflection on the empirical bases of this attempt to extract welfare consequences from (sometimes axiomatically diverse) public choice theorems, the analysis presented appears quite incoherent when placed in the context of a choice between vague or specific legislation. Presumably, vague delegations foster the march toward pareto pessimal results by reducing legislative decision costs. But there is a problem. Logrolling is one of the major driving forces behind the proposed worst-case scenario. Vague delegations, however, would seem to inhibit logrolling, for delegations of power transfer policy decisions to the jurisdiction of administrative agencies who have grave legal difficulties, and little apparent incentive, to trade values across programs. Any delegation thus restricts the policy space across which logrolling can be orchestrated and thereby limits the number of deals available to legislators in the next time period. . . .

In short, while Aranson's general theory of legislation may capture the dynamics and welfare consequences of certain classes of legislation—appropriations bills for defense installations or for river and harbor improvements—it is a theory which seems to explain specific, not vague, legislation. And to the extent that we believe that such "Christmas tree bills" are indeed instances of private interest legislation that reduce general welfare, we perhaps should favor statutory vagueness as a potential correction.

The Legitimation Value of Broad Delegations

Strangely enough it may make sense to imagine the delegation of political authority to administrators as a device for improving the responsiveness of government to the desires of the electorate. This argument can be made even if we accept many of the insights of the political and economic literature that premises its predictions of congressional and voter behavior on a direct linkage between benefits transferred to constituents and the election or reelection of representatives. All we need do is not forget there are also presidential elections and that, as the Supreme Court reminds us in *Chevron,* presidents are heads of administrations.

Assume then that voters view the election of representatives to Congress through the lens of the most cynical interpretation of the modern political science literature on congressional behavior. In short, the voter chooses a representative for that representative's effectiveness in supplying governmental goods and services to the local district, including the voter. The representative is a good representative or a bad representative based upon his or her ability to provide the district with at least its fair share of governmental largesse. In this view, the congressperson's position on various issues is of modest, if any, importance.

The voter's vision of presidential electoral politics is arguably quite different. The president has no particular constituency to which he or she has special responsibility to deliver benefits. Presidents are hardly cut off from pork-barrel politics. Yet issues of national scope and the candidates' positions on those issues are the essence of presidential politics. Citizens vote for a president based almost wholly on a perception of the difference that one or another candidate might make to general governmental policies.

If this description of voting in national elections is reasonably plausible, then the utilization of vague delegations to administrative agencies takes on significance as a device for facilitating responsiveness to voter preferences expressed in presidential elections. The high transactions costs of legislating specifically suggests that legislative activity directed to the modification of administration mandates will be infrequent. Agencies will thus persist with their statutory empowering provisions relatively intact over substantial periods of time. Voter preferences on the direction and intensity of governmental activities however, are not likely to be so stable. Indeed, one can reasonably expect that a president will be able to affect policy in a four-year term only because being elected president entails acquiring the power to exercise, direct, or influence policy discretion. The group of executive officers we commonly call "the Administration" matters only because of the relative malleability of the directives that administrators have in their charge. If congressional statutes were truly specific with respect to the actions that administrators were to take, presidential politics would be a mere beauty contest. For, in the absence of a parliamentary system or a system of strict party loyalty, specific statutes would mean that presidents and administrations could respond to voter preferences only if they were able to convince the legislature to make specific changes in the existing set of specific statutes. Arguments for specific statutory provisions constraining administrative discretion may reflect therefore a desire merely for conservative, not responsive, governance.

Of course, the vision of a president or an administration having to negotiate with the Congress for changes in policy is not one that is without its own attractiveness. Surely, we desire some limits on the degree to which a president can view a national election as a referendum approving all the president's (or the president's colleagues') pet projects, whether disclosed or undisclosed during the campaign. Those who abhor the policies of the Reagan administration, for example, might surely be attracted to a system that would have required that particular president to act almost exclusively through proposals for legislative change. Yet it seems likely that the flexibility that is currently built into the processes of administrative governance by relatively broad delegations of statutory authority permits a more appropriate degree of administrative, or administration, responsiveness to the voter's will than would a strict nondelegation doctrine. For, if we were to be serious about restricting the discretion of administrators, we would have to go much beyond what most nondelegation theorists seem to presume would represent clear congressional choices.

This last point is so neglected in the nondelegation literature that it is worth spelling out in some detail. While most discussions of the nondelegation doctrine focus on the question of substantive criteria for decision, establishing criteria is but one aspect of the exercise of policy discretion. In the formation of regulatory policy, for example, at least the following general types of questions have to be answered: What subjects are to be on the regulatory agenda? What are their priorities? By what criteria are regulations to be formulated? Within what period of time are they to be adopted? What are the priorities for the utilization of enforcement machinery with respect to adopted policies? What are the rules and procedures by which the relevant facts about the application of legal rules will be found? What are the rules by which facts and law will be combined to yield legal conclusions, that is, "findings" that there have or have not been violations of the regulations? What exceptions or justifications are relevant with respect to noncompliance? If violations are found, what corrective action or remedies will be prescribed?

Each of these questions can, of course, be broken down into a multitude of others and the answer to each question is a policy choice. Virtually any issue that can be specifically controlled by legislative answers to one of these questions can be reopened and redetermined when considering another. Thus, for example, even if Congress had adopted as legislation every specific and detailed rule that subsequently has been adopted by the Occupational Safety and Health Administration, the influence of the OSHA statute might still be vastly different (indeed has been vastly different) in the Carter and Reagan administrations. In the broadest term, the statute has had a different meaning in the years 1976 to 1980 than in the years 1980 to 1984.

Were the Congress to attempt to make statutory meaning uniform over time, it would have to specify the most extraordinarily elaborate criteria for exercising enforcement initiative, for finding facts, for engaging in contextual interpretation, and for determing remedial action. Indeed, to insure uniformity it would have to specify some objective criteria for all these judgments and some algorithm by which they were unified into decisions. Squeezing discretion out of a statutory-administrative system is indeed so difficult that one is tempted to posit a "Law of Conservation of Administrative Discretion." According to that law the amount of discretion in an administrative system is always constant. Elimination of discretion at one choice point merely causes the discretion that had been exercised there to migrate elsewhere in the system. If Congress were able to adopt specific regulatory criteria in some particular instance, say the OSHA statute, it would only have begun the process of making the real regulatory choices itself.

Nor will it do to suggest that activities beyond the setting of substantive criteria really do not raise the broad policy issues that concern nondelegation theorists. How the facts will be found often determines who wins and who loses. What cases are important enough to pursue entails policy discretion of the broadest sort. When to withhold remedial sanctions or alternatively to make an example of some offender raises issues of basic moral and political values. In short, as Aranson, Gellhorn, and Robinson recognize, making legislation specific and congressional choice determinative means addressing all these issues in great detail.

Such a strategy would, of course, result in wonderfully wooden administrative behavior and on that ground alone be highly objectionable. More important for present purposes, were Congress forced to repeal the Law of Conservation of Administrative Discretion in order to comply with a reinvigorated nondelegation doctrine, it would thereby eliminate executive responsiveness to shifts in voter preferences. For in this scenario, the high transaction costs of specific legislation will give an enormous advantage to the status quo, and the status quo will be susceptible to change only by a statute of the same kind.

Responsiveness to diversity in voter preferences is not limited to changes through time. It is surely plausible to imagine that, with a large land area and a heterogeneous citizenry, governmental responsiveness also entails situational variance at any one time. If our laws were truly specific, this would also be impossible. . . . Broad delegations recognize that tight accountability linkages at one point in the governmental system may reduce the responsiveness of the system as a whole.

There is one additional reason to believe that broad delegations to administrators might improve responsiveness. As voting theorists seldom tire of telling us, whenever three or more alternative policies exist there is the ever-present possibility of a voting cycle which can be broken only by resort to some form of "dictatorship" result. Legislators must, therefore, often delegate decisive authority somewhere in order to decide. There are any number of ways to deal with this problem—rules committees, forced deadlines, random selection, allocations of vetoes, or the like. Lumping alternatives together in a broad or vague statutory pronouncement and delegating choice to administrators is but another way of avoiding voting cycles through the establishment of dictators.

. . . [I]f the choice is among rules committees, random selection, providing vetoes to particular parties, or delegations to administrators, the legislature may sensibly conclude that the latter is likely to be the more responsive mechanism. Administrators at least operate within a set of legal rules (administrative law) that keep them within their jurisdiction, require them to operate with a modicum of explanation and participation of the affected interests, police them for consistency, and protect them from the importuning of congressmen and others who would like to carry logrolling into the administrative process. In short, if Arrow's theorem makes us uncertain about the responsiveness of majority rule voting procedures to citizens' or even legislators' desires, perhaps vague delegations to administrators can be a technique for avoiding the more disheartening aspects of breaking out of voting cycles.

Moreover, delegations to administrators may become particularly attractive where the alternative preference orderings that would produce collective intransitivities are interpreted as conditional on alternative perceptions of states of the world. For in this situation it is possible that administrative research, fact-finding, or "natural" experimentation with alternative policies will produce a unified view or construction of reality that ultimately yields a rational or transitive collective ordering of preferences. Interpreted in this way, delegation to experts becomes a form of consensus building that, far from taking decisions out of politics, seeks to give political choice a form in which potential collective agreement can be discovered and its benefits realized.

NOTES AND QUESTIONS

1. Does Davis address the implications for his approach of the increasingly costly, formalized, protracted character of rulemaking proceedings (a problem discussed in note 6 of the previous chapter's notes and questions)? If an agency confined its discretion by issuing rules, would it increase its exposure to tort claims alleging that it had violated its own rules and was therefore liable per se? Given these incentives, why would any agency mindful of institutional interests willingly surrender its discretion?

2. Davis writes that the goal of rulemaking "should be clarification to the extent that the subject matter and the available understanding permit, but not more than is consistent with needed individualizing." Can one say anything in general about how this norm is to be applied? Are the considerations too numerous to be combined into an abstract but determinate theory of rulemaking? For one such effort, applied to several specific policy areas, see Colin Diver, "The Optimal Precision of Administrative Rules," 93 *Yale Law Journal* 65 (1983). On the competing virtues of agency adjudication and rulemaking, see, for example, Richard Pierce, "The Choice Between Adjudicating and Rulemaking for Formulating and Implementing Energy Policy," 31 *Hastings Law Journal* 1 (1979); Glen Robinson, "The Making of Administrative Pol-

icy: Another Look at Rulemaking and Adjudication and Administrative Procedure Reform," 118 *University of Pennsylvania Law Review* 485 (1970); and the Schuck article excerpted in Chapter V.

3. In part, discretion is a problem because it can lead to erroneous and illegal decisions, not only unwise or unfair ones. If agency illegality originates in different aspects of agency life and if each source implies a different set of remedies, it may be useful to analyze the distinct kinds of agency error or illegality. At least four such types can be distinguished: comprehension-based, capacity-based, motivation-based, and negligence-based. For an analysis, see Peter H. Schuck, *Suing Government: Citizen Remedies for Official Wrongs* (New Haven: Yale University Press, 1983), pp. 3–12.

4. Lowi might be said to subscribe to what Davis derides as the "extravagant version of the rule of law." Indeed, Lowi's passionate indictment of "liberal jurisprudence" for having abandoned it in favor of "interest-group liberalism" is probably the most sweeping, and among the most frequently cited, analyses in the legal and political science literatures. A revealing feature of the book is the disjunction between the severity and comprehensiveness of his diagnosis and the nature of his prescriptions. In a final chapter entitled "Toward Juridical Democracy," Lowi proposes that (1) the Supreme Court revive the nondelegation doctrine of *Schechter Poultry,* (2) the president veto laws that are too standardless, (3) the agencies issue more discretion-confining rules (here he endorses Davis's approach), and (4) Congress review and codify or repeal what the agencies have done under broad delegations and also adopt sunset (or "tenure-of-statues") laws. If the present regime, which Lowi calls the "Second Republic," is as pervasively "rotten" as he says, can we not assume that the sources of pathology are deeply structural and systemic—so much so that his remedies, which are neither structural nor systemic, are probably unrealistic, ineffective, or both? Does Lowi advance a sophisticated causal theory, or is his analysis largely concerned with the *symptoms* of a policymaking system that is constitutionally fragmented, localistic, and politicized, that reflects deep popular suspicion of and hostility to centralized public power, and that lacks an elite civil service establishment that could regulate the outcomes of pluralistic bargaining? For other explorations of how these characteristics affect American government, see, for example, the Moe article in Chapter IV, the Kagan article in Chapter V, and James A. Morone, *The Democratic Wish: Popular Participation and the Limits of American Government* (New York: Basic Books, 1990). The extent of American exceptionalism in this regard is discussed in Chapter VII.

Do you understand Lowi's distinction between bargaining on the rule and bargaining on the decision? Does Mashaw's "Law of Conservation of Administrative Discretion" suggest that this distinction is not as clear or momentous as Lowi claims? Would unorganized citizens not be at least as disadvantaged in lobbying Congress for specific standards as they are in coping with broad delegations to agencies?

Is Lowi's description of the nature of public law in the Second Republic, at least in its post-1960s phase, accurate? Compare, for example, Lowi's discussion of standardless environmental legislation with what Bruce Ackerman and William Hassler, referring paradigmatically to the Clean Air Act of 1977, call post–New Deal, agency-forcing statutes. Ackerman and Hassler, *Clean Coal, Dirty Air: Or How the Clean Air Act Became a Multibillion-Dollar Bail-Out for High-Sulfur Coal Producers and What Should Be Done About It* (New Haven: Yale University Press, 1981). See also, R. Shep Melnick, *Regulation and the Courts: The Case of the Clean Air Act* (Washington, D.C.: Brookings Institution, 1983). In light of such statutes, is it true that "neither Congress nor the Executive Branch learned anything significant from the 1960s."? Can one make a case that they in fact learned some of the wrong lessons?

5. Ironically (in view of his advocacy of a participatory model of bureaucratic legitimation), Mashaw's argument for broad delegations of authority to administrators fails to exploit the participatory advantages of administrative rulemaking over legislation. Consider, for example, that it is only in the relatively late stages of the policy development process—the stages at which an agency formulates proposed rules—that the actual consequences of a proposed policy for the less organized groups will become sufficiently transparent that they can analyze those effects and act politically upon their analyses. Even legislation specific enough to pass muster under a revitalized nondelegation doctrine would often still be too general to make this kind of focused participation practicable. Are there other aspects of participation that are affected by the breadth of legislative delegations to agencies? A recent call for vigorous judicial control of delegation is David Schoenbrod, *Power Without Responsibility: How Congress Abuses the People Through Delegation* (New Haven: Yale University Press, 1993).

6. Is it true, as Mashaw's "Law of Conservation of Administrative Discretion" states, that discretion constrained at one point in the policy process will "merely" reappear at others? Does it follow from the law's logic that all policies are equally discretionary in the end so long as one defines the policy process broadly enough (which Mashaw quite properly is at pains to do) to include not simply the formulation of rules but all other policy-relevant decisions as well? Is Mashaw going so far as to suggest that insofar as ultimate administrative outcomes are concerned it does not matter whether, at what point, or in what form the substantive rules constrain discretion?

7. Government procurement policy constitutes a particularly striking example of the extent to which public concerns about bureaucratic discretion and the corruption opportunities that it affords have produced what some analysts regard as a pathological system of control. See Steven Kelman, *Procurement and Public Management: The Fear of Discretion and the Quality of Government Performance* (Washington, D.C.: American Enterprise Institute Press, 1990).

2

Review by Courts and Specialized Tribunals

Public Programs and Private Rights

RICHARD STEWART AND CASS SUNSTEIN

The Development of Remedies for Administrative Beneficiaries

The Traditional Model of Administrative Law: Private Rights of Defense

Under the traditional model of administrative law, courts police the boundary between two realms. The first is the realm of private law, in which citizens enjoy liberty and property defined by common law entitlements, and contract with one another within a juridically defined framework. The second is the realm of government, in which elected representatives and other government officials make decisions under procedures laid down by the Constitution and by statutes. The boundary between the two realms is crossed when government decisions are implemented against private persons through the coercive exercise of official power. An invasion of common law rights is legitimate only if it has been authorized by the elected legislature—the representative mechanism through which members of society collectively consent to invasions of that sort. A common law action by a citizen against an officer becomes the occasion for a judicial determination whether the legislature has authorized what would otherwise be a common law wrong.

By creating private rights of defense, the traditional model of administrative law curbs official bias or arbitrariness in the enforcement process and thus promotes impartial treatment. At the same time, the system limits the power of government, maintains a well-ordered sphere of private liberty, and preserves the system of market exchange. These are the classic functions of the rule of law.

Limitations of the Traditional Model

The traditional model contains no ready basis on which courts could entertain claims by the beneficiaries of a regulatory scheme that an agency had improperly failed to take enforcement action against third parties. If enforcement was declined, the beneficiary was denied an advantageous opportunity to which he had no common law right.

An alternative basis for judicial review—the prerogative writs, such as mandamus—was deprived of its practical utility by encrusted restrictions on its availability. Traditionally, courts also refused to endow many governmental benefits, such as employment, with the protections of due process. Beneath the technical learning of these decisions lies the conviction that private rights to regulatory protection or to other benefits would be inconsistent with needed administrative flexibility, and the associated fear that recognition of such rights would thrust judges into a supervisory role for which they lack ability and authority.

The Modern Development of Beneficiary Remedies

Contemporary government furnishes two types of advantageous opportunities that concern us here. The first consists of regulatory benefits obtained through government control of the conduct of regualted firms or persons. These benefits, exemplified by the cleaner air resulting from pollution controls, are typically collective in character, in that they cannot be afforded to one person without simultaneously being provided to many others. Courts have sought to protect collective interests in regulatory benefits through initiation rights. The second type of advantageous opportunity consists of individual benefits such as social assistance payments and individual rights such as the right to be free from racial discrimination. The private right of action and the new-property hearing right have increasingly been limited to the protection of such individual benefits and rights. . . .

Initiation rights. Statutes designed to protect regulatory beneficiaries would be undone if agency implementation and enforcement were inadequate or nonexistent. Responding to this possibility, courts have relaxed traditional principles of standing, ripeness, and prosecutorial discretion in order to permit review of agency inaction or of action that is assertedly inadequate. These developments amount to the judicial creation of a private right of initiation.

Occasionally, a court reads a statute to require an agency to take enforcement action whenever a statutory violation is shown. The duty to enforce is held to override agency claims of inadequate resources, competing responsibilities, the need for discretion, and the desirability of further study or negotiation. Such decisions create a strongly responsive right of initiation.

More commonly, right of initiation decisions adopt a deferential standard of review. Courts require agencies to show that they have considered the evidence and claims submitted by beneficiaries and to explain allegedly unlawful inaction; nevertheless, courts acknowledge the importance of discretion, competing interests, and limited budgets. Courts using this approach will not compel enforcement simply because a violation has been shown; the claims of the agency or of third parties must be demonstrably outweighed by the interests asserted by the beneficiary. Such cases recognize a weakly responsive right of initiation.

Whether strongly or weakly responsive, private initiation rights may raise serious problems for regulatory administation. Successful suits could squander agency resources on isolated, minor controversies, thereby diverting energy from larger patterns of misconduct.

Private rights of action. At common law, courts created private rights of action either by relying upon statutes to give specific content to the open-textured "reasonable man" standard of negligence, or by creating an action in damages for statutory wrongs. In creating private rights of action, federal courts have relied on the notion of federal common law, bolstered at times by reference to legislative intent and the need to effectuate the goals of regulatory statutes. . . .

Judicial creation of private rights of action raises greater difficulties when the legislature has entrusted enforcement of a statutory scheme to a specialized administrative agency that is empowered to issue rules or to adjudicate controversies under the statute. In this context, private rights of action may usurp the agency's responsibility for regulatory implementation, decrease legislative control over the nature and amount of enforcement activity, and force courts to determine in the first instance the meaning of a regulatory statute. *New-property hearing rights. Goldberg v. Kelly* and later decisions established that agencies may not withdraw or reduce certain individual statutory benefits, such as welfare payments, without providing procedures ranging from a statement of reasons to a trial-type hearing. A statute creates a constitutionally protected "property" entitlement if it limits the discretion of administrative officials so as to mandate the provision of a benefit to those meeting specified terms. Court-ordered procedural safeguards are designed to promote accuracy in the agency's resolution of the entitlement claim; the extent of protection required is determined by considering the interest of the beneficiary, the probable value of additional safeguards, and the burden on the government of providing those safeguards. . . .

Remedies for Regulatory Beneficiaries: Four Systems

Assume that a regulatory program provides for public enforcement. Given the two private enforcement remedies—rights of action and initiation—four different systems of regulatory enforcement are possible. Table VI.2 illustrates the possibilities. Although courts have generally not framed their decisions as choices among alternative enforcement systems, analysis of the alternatives clarifies the considerations at stake in judicial decisions about private remedies. *Exclusive public enforcement: Neither right of initiation nor right of action.* The exclusive public enforcement approach rejects any role for private enforcement in a system of administrative regulation. Courts that follow this approach conclude that the enforcement of a statute cannot be divorced from the agency's delegated task of giving life to that statute.

This approach emphasizes three defects in private enforcement remedies.

Table VI.2

| | | Private Right of Action | |
		Not available	Available
Private Right of Initiation	Not available	Exclusive public enforcement	Supplementary private enforcement
	Available	Judicial supervision	Maximum enforcement

First, courts lack the self-starting investigatory and analytical capacities needed to deal with complex social and economic problems. This point assumes particular force when Congress, in writing vague or general laws, relies on agency specialization to particularize and narrow statutory norms. Second, courts are unable to ensure centralized and coordinated enforcement. A passive and largely decentralized judiciary, dependent on the vicissitudes of private initiative, is likely to create ad hoc and inconsistent regulatory policies. Finally, courts are largely isolated from the processes of political oversight that serve to legitimate regulatory programs. Congress can control an agency's enforcement processes through appropriations and other forms of review; the president can also assert a measure of supervision over agency enforcement policies. Congress, the president, or agency officials may well conclude, on the basis of administative experience or changing social norms, that full enforcement is not warranted. . . .

Courts have refused to recognize private rights of initiation for similar reasons. An agency's decision not to undertake a regulatory initiative may be based not only on the legality of private conduct, but also on a wide variety of other managerial, political, and substantive considerations. Agencies are thought specially competent and experienced at weighing such considerations, and the weighing process may be difficult to recreate on judicial review. Furthermore, decisions not to act are ordinarily made informally, without elaborate records. If such decisions were subject to judicial review, some would later have to be explained in detail; agencies might have to use more formal procedures to make enforcement decisions; and regulatory programs might be severely disrupted. Public funds would also be expended in defending against initiation requests; resources needed for the protection of other beneficiaries might be depleted. And it would often be difficult for courts either to judge the merits of agency allocations of enforcement resources or to contribute to improved agency performance.

Judicial supervision: Private right of initiation, No private right of action. The judicial supervision approach is based on doubt that the political process can be trusted to ensure proper performance of administrative responsibilities. Courts that follow this approach acknowledge the value of an agency's specialized experience, political accountability, and capacity to plan for coordinated enforcement. For these reasons, they refuse to create private rights of action under statutes that provide for public enforcement. They believe, however, that initiation rights can improve agency performance while avoiding many of the drawbacks of private rights of action.

This belief is based on three grounds. First, the scope of judicial review in initiation cases normally permits considerable deference to agency discretion. Only rarely will a relevant statute unambiguously command enforcement in a particular case; hence rights of initiation are generally weakly responsive. By contrast, if a plaintiff has a private right of action, the court must give de novo consideration to every asserted regulatory violation.

Second, recognition of a private right of action is much more likely to undermine coordinated administration and orderly legal development than is recognition of a right of initiation. Dangers of ad hoc and inconsistent judgments are minimal under a weak right of initiation. Even if a strong initiation right is recognized, agencies can attempt to coordinate initiation orders with their overall enforcement schemes.

Third, initiation rights are less likely than rights of action to subvert legislative oversight and fiscal control of the nature and amount of enforcement. In initiation cases, courts will ordinarily respect budget constraints aimed at limiting implementation and enforcement activity. Rights of action, by contrast, circumvent budgetary limits by enlisting private enforcement resources.

In *Dunlop v. Bachowski,* the Supreme Court endorsed an application of the judicial supervision approach. The plaintiff sought to compel the secretary of labor to bring suit to set aside a union election allegedly conducted in violation of the Labor-Management Reporting and Disclosure Act of 1959 (LMRDA). Distinguishing cases holding the exercise of prosecutorial discretion unreviewable, the Third Circuit had noted that the LMRDA "demonstrate[d] a deep concern with the interest of individual union members," and that judicial review was appropriate to ensure that the members were not "left without a remedy." The court had also noted the absence of a private right of action. The Supreme Court expressly endorsed the Third Circuit's ruling that enforcement failure was judicially reviewable.

Supplementary private enforcement: Private right of action, no private right of initiation. The supplementary private enforcement approach regards the creation of administrative agencies as a useful but by no means exclusvie method for implementing statutory requirements. Public regulation may be needed because of the inadequacies of the common law system in coping with industrial conditions. Public enforcement is, however, frequently inadequate because of budget constraints; private actions can be a useful supplementary remedy by providing additional enforcement resources.

J.I. Case Co. v. Borak reflects these considerations. The Supreme Court created a private right of action under Section 27 of the Securities and Exchange Act of 1934 on behalf of shareholders challenging management's proxy statements as deceptive, notwithstanding the power of the SEC to bring suit and the failure of Congress explicitly to authorize private enforcement. The Court emphasized the act's "broad remedial purposes" as well as the apparent inability of the SEC to effectuate those purposes adequately. The Court did not discuss the possibility that private enforcement might subvert political control over enforcement. Instead, the Court asserted that the statutory goal of " 'protection of investors' . . . implies the availability of judicial relief where necessary to achieve that result."

Private rights of action, unlike rights of initiation, do not divert limited agency resources from other violations that may be more important. The supplementary private enforcement approach rejects private rights of initiation, both because of this problem of diversion of resources and because of the drawbacks discussed earlier in this section.

Maximum enforcement: Both right of action and right of initiation. Under a maximum enforcement approach, the values of political control, specialization, and centralization are considered less important than maximizing private participation in the enforcement process. A powerful concern with the possibility of agency abdication of regulatory responsibilities is characteristic of the maximum enforcement approach.

In *Medical Committee for Human Rights v. SEC,* the District of Columbia Circuit followed this approach. The court held reviewable a decision by the SEC not to initiate proceedings against a corporation that had refused to include a shareholder proposal in its proxy statement, even though the shareholders had a private right of action against the corporation. The shareholders were permitted to forgo a private suit and proceed against the SEC for two reasons—the desirability of augmenting private enforcement with public resources and experience, and the "independent public interest in having the controversy decided" with agency participation.

The Evolution of Beneficiary Remedies

Forty years ago, the exclusive public enforcement approach was dominant in the federal courts. The supplementary private enforcement approach prevailed during the past several decades, as courts reacted to perceived inadequacies in agency enforcement efforts by creating private rights of action. Today, judicial supervision appears to be the most widely accepted approach. The Supreme Court's narrow construction of the exception to reviewability for those decisions "committed to agency discretion," as well as the Court's willingness in *Bachowski* to supervise enforcement discretion, confirms the tendency of lower courts to permit regulatory beneficiaries to challenge agency failure to enforce regulatory programs adequately. It seems increasingly clear, however, that a deferential standard of judicial review will ordinarily be followed in initiation cases. At the same time, the Court's restrictive approach to private rights of action makes it far less likely that such rights will be recognized.

This is not to say that the cases follow a consistent pattern, for courts sometimes depart from the judicial supervision approach. When litigants seek to protect important personal liberties, such as freedom from discrimination, courts often create both initiation rights and rights of action. By contrast, when an agency has been granted broad discretion in an area such as regulation of rates and competitive practices, courts are likely to deny both private remedies. . . .

While recognizing the limitations of courts and of the forms of action that they create, we conclude that courts have authority to create rights of action and rights of initiation. These remedies should, we believe, stand with rights of defense and new-property hearing rights as legitimate and useful correctives for deficient administrative performance.

The presumption in favor of review of administrative decisions should be extended to beneficiaries as well as to regulated entities, an extension that is achieved by countering rights of defense with rights of initiation. Unless direct private enforcement is clearly a preferable remedy, courts should generally recognize weakly responsive private rights of initiation. On occasion, however, courts may be justified in viewing a relevant public value—such as the government's responsibility to eliminate racial discrimination in federally supported activities—as an overriding norm that compels enforcement action through a strongly responsive initiation right.

Judicial creation of private rights of action is an alternative method of protecting statutory interests. When the regulatory statute creates a right–duty relation similar to those established by the common law, courts may ordinarily provide private remedies without exceeding their competence, subverting legislative control of regulatory policy, or contradicting any of the reasons for the creation of an administrative enforcement system. The common law analogy will not, however, support private rights of action when a statutory norm is vague or ambiguous, delegates broad managerial authority, or presents a context in which coordination and consistency in regulatory policy is important. A private right of initiation is more appropriate in such circumstances, which will be common in the modern regulatory setting.

We have explained that, without background understandings of institutional purpose, it is impossible to give content to legislative silence on remedies or to determine in which circumstances particular remedies are appropriate. The three conceptions of institutional purpose that we have identified are a means of liberating courts from the formalist thesis. Such conceptions justify the judicial creation of remedies for defective administrative performance and, at the same time, guide the process of judicial lawmaking by demarcating the larger purposes that these remedies can serve.

But the conceptions are a form of bondage as well. Grounded in eighteenth-century principles of public and private law, they generate remedial regimes that are narrow and often dysfunctional in an industrially developed polyarchy administered by bureaucracies. The traditional notions of private and public law on which the three conceptions are bottomed are not fully compatible with the administrative process. The four paradigm remedies are thus partial and perhaps anachronistic approaches to inadequate agency performance. In this sense, the four remedies suffer from shortcomings similar to those presented by the common law forms of action in the nineteenth century.

Statutory Interpretation and the Balance of Power in the Administrative State

CYNTHIA FARINA

Judicial attempts, over the years, to cut through . . . conceptual complexity have produced a large number of statutory interpretation opinions that defy easy reconciliation. To the extent the cases can be categorized, they seem to cluster around two very different models of the proper functions of court and agency in interpreting regulatory statutes. In one model, the interpretive authority rests principally with the court. Using traditional techniques of statutory construction, the court exercises its own judgment to determine de novo what the statute means. In this "independent judgment model," the agency's function can be analogized to that of an expert witness: its view of the proper meaning becomes a factor in the court's analysis, to be given whatever persuasive effect it appears to merit in the circumstances. In the second model, the principal interpretive responsibility rests with the agency. The court must accept any reasonable construction offered by the agency, so long as the statutory language or, possibly, the legislative history is not patently inconsistent. In this "deferential model," the agency's function is to give meaning to the statute: the court determines only whether the interpretation the agency has chosen is a "rational" reading, not whether it is the "right" reading. . . .

Late in its 1983 Term the Supreme Court, in *Chevron U.S.A., Inc. v. Natural Resources Defense Council, Inc.*, announced the end of judicial vacillation between these two interpretive models. All six paticipating justices joined an opinion which endorsed deference in emphatic terms that quickly caught the attention of lower courts and commentators. Two years later, *Young v. Community Nutrition Institute* reiterated and strengthened *Chevron*'s message in an opinion in which the remaining three justices joined. Subsequently, the appointment of Justice Scalia added an even more determined proponent of *Chevron* and the deferential model. . . .

Independent Judgment as Judicial Usurpation?

Chevron's conclusion that deference is necessary to prevent the judiciary from arrogating power Congress wished to place in administrative hands requires acceptance of two propositions: First, Congress intends, as a factual matter, that agencies take principal responsibility for determining what regulatory statutes mean; and second, Congress is unconstrained, as a matter of constitutional doctrine, in choosing to entrust such responsibility to agencies. While both these propositions are critical to the "judicial usurpation" justification for deference, neither is self-evident—and neither is critically examined by the *Chevron* Court. In fact, the first proves to be extremely problematic, while the second can be accepted only by repudiating fundamental aspects of existing theory.

Chasing the Will-o'-the-Wisp of "Interpretive Intent"

If Congress, when it enacts regulatory legislation, intends that the implementing agency will resolve a particular question of statutory meaning, then a court would indeed be contravening the legislative will to treat the issue as a matter for independent judicial determination. On these facts, the validity of the "judicial usurpation" justification for deference would turn solely on whether any constitutional principle prevents effectuation of the legislature's "interpretive intent." However, this set of facts—in which Congress *both* perceives the need for future interpretation *and* formulates an intent that it be accomplished by the agency—is only one of four logically possible scenarios.

Alternatively, Congress may recognize that its words will require interpretation and assume that the *court* will resolve questions of statutory meaning in the normal course of appropriate litigation. In this scenario, the legislature has an interpretive intent, but its principal object is the judiciary, not the agency. A third possibility is that Congress means to provide the substantive answer itself, but instead unwittingly creates an interpretive question by failing to express its answer clearly. In such a case, the legislature's intention—unsuccessfully implemented—is not to delegate to any other entity the power to determine meaning. Finally, Congress may have no intent whatsoever about interpretive responsibility or its allocation. This could occur because the substantive question is unknowable or, if theoretically knowable, is unrecognized at the time the legislature acts, or because the question is lost or deliberately bypassed in the proccss of hammering out the regulatory package.

Whenever one of these latter three scenarios accurately describes the state of Congress's "intent" on a given issue of statutory meaning, it is simply inaccurate to suggest that the independent judgment approach would constitute judicial usurpation of power legislatively designated for another. Of course, discerning, for each statutory provision requiring interpretation, which scenario correctly describes the enactors' intent is likely to be a formidable, if not an impossible, task. *Chevron* sidesteps this difficulty by announcing a presumption. If Congress has not itself *unambiguously* resolved the *precise* substantive issue, then we are to assume that it either "explicitly" or "implicitly" delegated to the agency the interpretive task. Apparently, this presumption—that ambiguity equals legislative intent to empower the agency— is irrebuttable. . . .

If we can find no certain evidence of Congress's "typical" interpretive intent either in the form and content of regulatory statutes themselves or in the APA [Administrative Procedure Act] and its legislative background, we might turn, finally, to larger theoretical models of the legislative process. Landes and Posner's economic analysis of the relationship between the legislature and the courts posits that Congress relies upon an independent judiciary to enforce the original terms of statutes because the certainty of such enforcement, over time, will enhance the value of any given legislative deal to the interest groups who are bidding for it.

. . . [Landes and Posner's] analysis implies that at least with respect to important questions of statutory meaning—those terms of the legislative "deal" that would have mattered to the original bargainers—Congress typically intends the court rather than the agency to control interpretation.

The accuracy and significance of Landes and Posner's model can, of course, be vigorously debated, but this is precisely the problem shared by all of the "evidence" on Congress's interpretive intent. Despite *Chevron*'s casual equation of ambiguity with a deliberate delegation of power to the agency, we have little basis for divining with any acceptable degree of confidence the legislature's "typical" expectation regarding the role of courts and agencies in determining statutory meaning. . . .

The dilemma is a serious one *if* the constitutional touchstone is congressional intent. The proposition that Congress is unqualifiedly free to give agencies the power to declare statutory meaning—the second critical assumption of the "judicial usurpation" argument—proves, on close examination, to be as troublesome as the assumption that Congress in fact generally intends to do so.

Constraints on the Allocation of Interpretive Power

It is surely a far more remarkable step than *Chevron* acknowledged to number among Congress's constitutional prerogatives the power to compel courts to accept and enforce another entity's view of legal meaning whenever the law is ambiguous. . . . Some commentators have argued, however, that the interpretation of statutes, particularly regulatory statutes, stands on a different footing. Professor Monaghan has stated the position most forcefully: "[O]nce the delegation of lawmaking competence to administrative agencies is recognized as permissible, judicial deference to agency interpretation of law is simply one way of recognizing such a delegation."

At the threshold, this nonchalant classification of law interpretation as simply a species of lawmaking is troubling. . . . Our mainstream political thought has always included the belief that there is *and should be* a real distinction between making law and interpreting law. . . . Liberating agencies from the precept that statutory interpretation is not a guise for flights into policymaking seems a peculiar way to vindicate legislative supremacy.

A defense of deference that would abandon any formal distinction between making and interpreting regulatory law thus has disturbing implications. But the problem with recognizing an unconstrained legislative prerogative to allocate interpretive authority run deeper still. To conceptualize Congress's ability to delegate power to agencies in absolute, "on-off" terms—"*if* Congress may give lawmaking authority to agencies, *then* it may give law-interpreting authority to agencies"—deceptively oversimplifies the doctrinal course that sanctioned the growth of administrative agencies. . . .

. . . [T]he way in which mature nondelegation analysis reconciled agencies to the constitutional scheme was faithful to separation of powers theory and furthered the purposes which that structural principle was intended to serve. When *Chevron* asserts that agencies may be given the principal responsibility for defining the statutes under which they function, it not only repudiates a major strand of administrative law doctrine, but also undermines the constitutional reconciliation of regulatory power which that doctrine represented. . . .

. . . By funneling enormous power into agencies, the regulatory statutes enacted in this century have radically reconfigured the pattern of government authority. As the Court's nondelegation analysis ultimately apprehended, this development can be reconciled with separation of powers principles if, but only if, the new concentration of power is offset by correlative checks. And, as Judge Leventhal recognized, the interpretive authority is one critical dimension along which such a check can be interposed. Given the many factors that cause legislatures to fail to speak clearly and precisely in regulatory statutes, the quantum of power represented by the responsibility to resolve statutory ambiguity is not insignificant. The choice between deference and independent judgment is, at bottom, a choice between whether this authority will further swell agency power or will instead help to counterbalance it. Whether or not there existed some original constitutional mandate that the judiciary hold the interpretive authority, separation of powers may require that, in the twentieth century administrative state, the choice be made in favor of counterbalance.

. . . If Congress chooses to delegate regulatory authority to agencies, part of the price of delegation may be that the court, not the agency, must hold the power to say what the statute means. In other words, Cass Sunstein's pithy criticism of deference according to *Chevron*—"foxes shouldn't guard henhouses"—is a counsel not merely of prudence, but of constitutional necessity.

It might be objected, however, that one can accept the proposition that the flow of power to agencies must be offset by some correlative counterweight without concluding that the *judiciary* be the branch providing the necessary external control. And this objection might be supported (although the Court did not attempt any such connection) by invoking *Chevron*'s second constitutionally based justification for choosing deference. The "interference with legitimate political control" argument asserts that courts, as the branch least accountable to the people, are the branch least suitable to make the policy choices inherent in interpreting regulatory statutes. It argues that the deferential model is constitutionally preferred because it permits agencies to make those choices under the supervision and control of the political branches. The assumptions underlying this second justification for deference must next be examined.

Independent Judgment as Interfering with Legitimate Political-Control?

. . . Appealing as it seems, this "interference with legitimate political control" justification for deference can nonetheless be challenged on many levels. In the first place, treating the interpretation of regulatory statutes as simply an exercise in policymaking has troubling implications, some of which have already been noted. In particular, when this view of interpretation is superimposed on *Chevron*'s rule that a statute is "ambiguous" (and hence fair game for "interpretation") whenever it does not provide unequivocal direction on the precise point at issue, great potential is created for overlooking substantive directions and limitations that the enactors had intended, and attempted, to incorporate in the statute. Even in the unlikely event that the "meaning" selected by the agency accurately reflects the wishes of the people's current elected representatives, the prospect that regulatory statutes will routinely be amended or even repealed by "interpretation" should at least give us pause. In the second place, the Court's reasoning reflects a singularly narrow conception of the judicial function. *Chevron* implies that the shaping of public policy is so foreign to the judiciary's proper task that courts must avoid responsibility for resolving policy questions whenever possible. The primacy of the political branches in establishing domestic and foreign policy is beyond dispute; surely, however, it overstates theory and distorts reality to suggest that the judicial role in defining public policy is an extraordinary one, legitimately undertaken only as a last resort. Far-reaching policy resolutions are a regular part of both common law and constitutional adjudication. Even if common law examples are dismissed as irrelevant to the federal courts and constitutional adjudication is distinguished as sui generis, the federal judiciary has long administered sweeping yet open-textured statutes in such areas as antitrust, labor, and civil rights. Whatever one's views of the wisdom of this, it simply cannot be considered aberrant for federal courts to resolve policy questions as part of deciding cases properly before them.

Even if we were to accept that interpretation is merely policymaking and that public policy can legitimately be shaped only by the political branches, the most fundamental problems with *Chevron*'s second justification for deference remain. The proposition that deference is constitutionally compelled in order to legitimate regulatory policymaking rests on two assumptions. First, if courts follow the deferential model, agencies will exercise the interpretive power subject to the direction of the political branches; second, the direction agencies receive will indeed respond to the constitutional concerns raised by the delegation of regulatory power. Once again, neither of these critical assumptions is self-evidently correct, and neither is scrutinized by the Court. In fact, the first assumption may be wholly unfounded and certainly is misleading to the extent it implies that *Congress* will regularly play an active directory role in agencies' interpretive decisions. And since the president is the "political branch" most likely to achieve control over such decisions, the second assumption is also inaccurate—at least in view of how principles of legitimacy and separation of powers traditionally have been framed.

. . . If we remain committed to separation of powers as a central structural credo, then choices such as the one made in *Chevron* are steps in the wrong direction. At stake in *Chevron* was the fate of one relatively small but not insignificant slice of the regulatory power pie: the authority to interpret the statutes that define the policymaking universe. The Court's resolution deliberately moves that power squarely into the president's domain. By relinquishing the authority to determine statutory "meaning" to agencies whenever Congress has failed to speak clearly and precisely, *Chevron* enlarges the quantum of administrative discretion potentially amenable to direction from the White House. It then goes even further and exhorts agencies to exercise this discretion, *not* by attempting to intuit and realize the objectives of the statute's enactors, but by pursuing the regulatory agenda of the current chief executive.

Had the Court instead chosen the independent judgment model, this slice of power would have been placed beyond the reach of the executive in the hands of another branch. The negative act of withholding authority from a power center that tends to have too much could have yielded positive constitutional benefit. Concededly, independent judgment is not a constitutional panacea. To the extent that giving meaning to regulatory statutes involves law*making* as well as law *finding,* judicial control over the interpretive process does not satisfy the legitimacy ideal any better than does presidential control. However, as the ideal has traditionally been defined, it can be satisfied by little short of some mechanism for remanding important policy questions raised by interpretation to the legislative process for clarification. Several commentators have called for such a remand mechanism, and this would, at least in theory, be the most direct response not only to legitimacy but also to separation of powers concerns. While searching for the optimal solution, though, we ought to be alert to take what we can get. Allocating principal responsibility for interpreting regulatory statutes to the judiciary would significantly further separation of powers by placing power where it will counterbalance, rather than contribute to, the concentration of regulatory authority in the executive. In the interminable struggle to make peace between the Constitution and the administrative state, that would be no small victory.

Conclusion

Honoring the original commitment to separation of powers in the twentieth-century administrative state requires a heightened sensitivity to attaining equilibrium among the power centers of government. Contriving the new balance does not mean we must forfeit all the advantages presidential direction can bring to the administrative process. Even were it possible to preclude presidential involvement in regulatory policymaking—which, as a practical matter, it surely is not—the result would simply be an imbalance of a different sort. What it does mean is that we must recognize the constant tendency of regulatory power to flow, centripetally, toward the head of the executive branch and think deliberately and carefully about where to find counterbalance for this tendency.

I have argued here that the Court in *Chevron* failed to do these things and, as a result, made a choice that skewed the balance even further. But the shortcomings of the Court's analysis in that case is only one manifestation of a larger problem. Just as statutory interpretation is but one aspect of the task of defining federal regulatory policy, so the choice of who holds the interpretive authority is but one dimension of the question of who controls the power that Congress has delegated. In the last few years, the Court has confronted several other dimensions of that question. It has considered who may appoint agency heads and who may remove them. It has considered whether agency actions may be disapproved by formal legislative action short of the full lawmaking process. It has considered whether the reasons for agency enforcement strategy may be subjected to judicial scrutiny. In each case, it has decided in favor of the president and the agency, and against the possibility of an external check by Congress or the judiciary.

The outcome in some or even all of those cases may be correct, but the Court's analysis surely has been inadequate. It continues to invoke the Framers' fear of legislative dominance, and disregards the fact that much of the power which bred that fear has passed out of Congress's grasp. It recalls the Framers' concern that the president be a strong and independent player in the power struggle and overlooks the superiority that structure and custom have given him in a world of delegated power. It emphasizes the safeguards that the full legislative process was intended to provide against ill-considered, unrepresentative policy choices and ignores the reality that much of our most significant policymaking now occurs in an administrative process where substitutes for the original safeguards must somehow be provided. It casts agencies in the role of law *executors,* worries about shielding them from undue legislative or judicial management, and blinks at their far more important role as law*makers,* whose vulnerability to White House influence ought to raise equally worrisome questions of undue control from the chief executive.

. . . The assessment of whether power has shifted to the point that counterbalance is required is one that cannot be conducted under any certain, or universally agreed upon, standard. The question of when an assumption of power by one branch *becomes* constitutionally necessary because of the flow of power to another branch is a question that even (or, perhaps, especially) the judiciary will be hardpressed to answer in a principled fashion.

Yet one can acknowledge the difficulty of the task and concede the undesirability of giving it to judges and still insist, for one simple reason, that they must attempt to do it: whenever the Court decides one of these cases, it alters the balance of power. Refusing to recognize such cases as opportunities for adjusting the balance will not change the fact that the consequence of deciding them *inevitably* will be either to augment, or to stem, the flow of power toward agencies and the president.

Administrative Law and Bureaucratic Reality

R. SHEP MELNICK

From the perspective of a political scientist, what is both puzzling and exasperating about the literature on administrative law is how little it has to say about administrators and administration. Administrative law in the United States is almost entirely about *courts*. Articles abound on judicial doctrines relating to scope of review and rulemaking procedures. But no one pays much attention to the tasks performed by administrators, the agency's sense of mission, the conflicting pressures placed on it, or even what happens after judges hand down their decisions.

Others have noticed this lacuna. In a review contribution to [the *Administrative Law*] *Review,* Peter Schuck and Donald Elliott wrote,

> Although the study of administrative law started in earnest more than fifty years ago, we still know little about what is perhaps the central question in that field: how does judicial review *actually* affect agency decisionmaking? This question goes to the fundamental nature and quality of the modern administrative state, yet academic specialists have largely neglected it; the subject remains a matter for uninformed speculation. Despite (or perhaps because of) the lack of data, strong opinions on this question are common.

One such strongly held (but unsubstantiated) opinion is that without "searching and thorough" judicial review, bureaucrats would be running wild, issuing arbitrary and capricious orders with utter abandon. Another view is that without such judicial review agencies would be in the hip pocket of "special interests," captured, we are always told, "by the very interests they are supposed to control." It is bad enough that these two views are incompatible. What is worse is that usually they are both wrong.

Judges' images of the administrative process are of the utmost importance because the central issue in administrative law is the extent to which judges should defer to administrative judgment and expertise. Judges are unlikely to defer to administrators they do not trust—regardless of what the Supreme Court said in *Chevron v. NRDC.* It is time administrative law spent less time dissecting the words of judges and more time observing the activities of bureaucrats.

In recent years a small group of scholars have taken a closer look at how court decisions have reshaped administrative agencies and their policies. The purpose of this article is to acquaint students of administrative law with some

Reprinted by permission from 44 *Administrative Law Review* 245 (1992). © 1992 American Bar Association. All rights reserved.

of these findings.[1] As will soon become clear, most of these studies show that judicial intervention has had an unfortunate effect on policymaking. Judicial review has subjected agencies to debilitating delay and uncertainty. Courts have heaped new tasks on agencies while decreasing their ability to perform any of them. They have forced agencies to substitute trivial pursuits for important ones. And they have discouraged administrators from taking responsibility for their actions and for educating the public.

To those who suspect that these findings reflect either the biases of the author or the peculiarities of the cases chosen for close analysis, I offer the following challenge: provide us with detailed studies of cases in which the courts have improved policymaking. There surely must be some.[2] Unsupported assertions about the malevolence of bureaucrats and the good will of judges do not constitute a convincing reply to these empirical studies.

Rulemaking

Let me start with the most obvious problem, rulemaking delay. In the late 1970s William H. Rodgers, Jr., summed up the conventional wisdom about aggressive judicial review of agency rulemaking by saying, "The hard look doctrine plays no favorites; it is advanced as enthusiastically by industry as it is by environmentalists. Its acceptance is deep." Ten years later many people are having second thoughts.

In retrospect, what happened is quite clear. The courts said, consider all the "relevant" evidence, respond to all "significant" comments, and weigh all "reasonable" alternatives. Who could object to that? The problem was that judges failed to explain what they meant by "relevant," "significant," and "reasonable." Like pornography, they knew it only after they saw it. Since agencies do not like losing big court cases, they reacted defensively, accumulating more and more information, responding to all comments, and covering all their bets. The rulemaking record grew enormously, far beyond any judge's

---

[1] I will focus primarily on the following books: Robert A. Katzmann, *Institutional Disability: The Saga of Transportation Policy for the Disabled* (1986); Jerry L. Mashaw & David L. Harfst, *The Struggle for Auto Safety* (1990); R. Shep Melnick, *Regulation and the Courts: The Case of the Clear Air Act* (1983); Jeremy Rabkin, *Judicial Compulsions: How Public Law Distorts Public Policy* (1989); and Martin M. Shapiro, *Who Guards the Guardians? Judicial Control of Administration* (1988). A recent article by Robert Kagan combines a brief case study with an insightful analysis of what he calls "adversarial legalism." Robert A. Kagan, "Adversarial Legalism and American Government," 10 *Journal of Policy Analysis & Management* 369 (1991).

[2] Some studies of the consequences of *constitutional* rulings have shown that federal courts have at times improved the performance of *state* bureaucracies. See, for example, the discussion of mental health facilities in Phillip J. Cooper, *Hard Judicial Choices: Federal District Court Judges and State and Local Officials* (1988), and *Remedial Law: When Courts Become Administrators* (Robert Wood, ed., 1990). Both these books, though, show that in many other policy areas judges' lack of understanding of administrative behavior has produced unintended consequences. See also *Courts, Corrections, and the Constitution: The Impact of Judicial Intervention on Prisons and Jails* (John Dilulio, ed., 1990) and Donald Horowitz, *The Courts and Social Policy* (1977).

ability to review it. As a result, it took longer and longer to complete the rulemaking process. Richard Pierce reports that "the time required to make policy through rulemaking has been stretched to nearly a decade." In some instances the final rules appeared just as the underlying problem had disappeared or changed fundamentally.

Thus began a vicious cycle: the more effort agencies put into rulemaking, the more they feared losing, and the more defensive rulemaking became. Even then, they lost a significant number of cases—it was just too hard to predict what a randomly selected panel of judges would do. Draw two Reagan appointees on a three-judge panel and you might well lose because you regulated too much; draw two Carter appointees, you might lose because you regulated too little. Little wonder that many agencies looked for ways to avoid the rulemaking quagmire. Some agencies decided it would be easier to use adjudication to establish agency policy. Others set policy through interpretive rulings or internal enforcement policies. So instead of more rules, we have more discretion; instead of uniformity, we have particularism. Just as water runs downhill, agencies run away from uncertainty, which is what the judicial review often represented.

Jerry Mashaw and David Harfst's recent book presents a particularly graphic example of how a federal agency has responded to "hard look" judicial review by virtually abandoning rulemaking. After the courts rejected its first set of safety rules, the National Highway Traffic Safety Administration replaced rulemaking with a recall strategy. Decisions of other federal courts made this strategy easy to purrsue. The only problem was that recalls do little to improve auto safety. In a similar vein, Richard Pierce has shown that the courts made it impossible for the Federal Energy Regulatory Agency to use rulemaking to revise seriously deficient policies on regulation of natural gas.

This does not mean that agencies never engage in rulemaking or win in court. What do we know about the effects of judicial review for the substance of the rules that are issued? In the area of health and safety standards both my research and the work of John Mendeloff indicate that judicial review has decreased the *number* of standards but increased their *stringency*. Why? Because at the same time that industry is challenging an agency's evidence, public interest groups are charging that the agency has not sufficiently protected the public health.

Judges on key courts have often insisted that EPA [Environmental Protection Agency] and some other agencies take a "health only" approach to standard-setting, especially when a carcinogen is involved. In *Lead Industries Ass'n. v. EPA,* for example, Judge Skelly Wright announced that the legislative history of the Clean Air Act "shows [that] the Administrator may not consider economic or technological feasibility in setting air quality standards." In another decision involving standards for airborne lead, Judge Wright hearkened back to Judge Bazelon's famous claim about the "special judicial interest in favor of protection of the health and welfare of people."

The courts have hardly been consistent on this. The Supreme Court's rulings on OSHA's [Occupational Safety and Health Administration] statutory mandate are far from doctrinaire—in fact they are incoherent. For many years the Supreme Court's inability to speak clearly on such matters left the lower courts—especialy the D.C. Circuit—in control. The D.C. Circuit now appears to be pulling back from its previous absolutist position, but in the 1970s and 1980s the message EPA and other agencies often got was this: If you want a standard to survive, collect lots of information, make the rule very stringent, and then use enforcement discretion to avoid the politically dangerous consequences of this stringency. With so few standards being set, environmentalists insisted that the ones that survived were really tough. Seeing these stringent regulations coming down the pike, business fought them tooth and nail, adding to delay.

Mendeloff convincingly argues that this peculiar combination of underregulation and overregulation leads us to get far less safety bang for the buck than European nations, which set more standards, but make them individually more lenient.[3] The courts do not bear sole responsibility for this costly adversarial system, but they surely add to the problem. How is it possible to arrange a grand compromise—more standards for less stringency—if any trade association or environmental group can challenge it in court and have a shot at winning? How is it possible to encourage open horse-trading when the courts are insisting upon "rational" decisionmaking based on the accumulation of enormous amounts of technical information?

Action-Forcing Litigation

Many of the statutes passed by Congress in the 1970s and 1980s were filled with so-called nondiscretionary duties: deadlines, "hammers," and the like. A 1985 study co-authored by the Environmental Law Institute and the Environmental and Energy Study Institute reports that EPA alone was subject to 328 statutory deadlines. Among the other findings of this study were the following:

1. "Very few statutory deadlines (14 percent) have been met."
2. "Congress imposes more deadlines on EPA than it can possibly meet."
3. "Court-ordered deadlines almost always command top management attention at EPA. Top management takes the threat of contempt quite seriously and personally, even though the threat is not real."
4. "The multiplicity of deadlines reduces EPA's ability to assign priority to anything not subject to a deadline."

What this means is that scores of deadlines and other statutory requirements are lying around, usually unused, but still potential weapons in lawsuits—or threatened lawsuits. Do nondiscretionary duties and court en-

---

[3] Mendeloff, *The Dilemma of Toxic Substance Regulation: How Overregulation Causes Underregulation* (1988). Additional support for this argument comes from Ronald Brickman et al., *Controlling Chemicals: The Politics of Regulation in Europe and the United States* (1985), and David Vogel, *National Styles of Regulation: Environmental Policy in Great Britain and the United States* (1986).

forcement produce "the rule of law"? Of course not; they produce the rule of those who decide which lawsuits to bring. Or, to put it another way, it transfers responsibility for setting agency priority from top administrators to interest groups, some of which use attorneys' fees won in these relatively easy cases to cross-subsidize other activities. It is hard to imagine a worse way to apportion agency resources.

The study cited above asked former EPA administrators what they thought of deadlines and judicial enforcement of them. One said, "From a real world point of view, they are necessary." But another was much more negative, saying, "Deadlines reinforce the sense that we [EPA] are not getting anywhere, to the detriment of public sense of confidence in government." Surprisingly enough, the negative view came from William Ruckelshaus, one of the most outstanding public administrators we have seen in some time, and the positive view came from Anne Gorsuch Burford, who is probably among the worst.

One example will help explain why Ruckelshaus found the deadline-litigation syndrome so detrimental to environmental protection. The Clean Air Act gives the EPA one year to publish regulations setting standards for pollutants it determines are "hazardous." During the Carter administration, EPA determined that radionuclides are carcinogenic, and therefore listed them as hazardous air pollutants. It failed to take any further action. About one year into the Reagan administration, the Sierra Club sued in the northern district of California. Several years later the matter finally came to a head when Judge Orrick commanded Administrator Ruckelshaus to issue a standard for radionuclides.

At first Ruckelshaus surprised everyone by refusing to obey the court. Not only was the scientific evidence unclear—it always is—but EPA estimated that the health risks were tiny and the potential cost large. Estimated risk from some facilities was one cancer death every fifty years, from other facilities, one cancer death every thirteen to seventeen years. Ruckelshaus was in the middle of a major effort to educate the American public about the nature of environmental risks—to let them know that there is no such thing as a completely safe environment, and to force the government and the public to be honest about the level of safety they are willing to pay for. He told the court, "Given the inevitable burdens that regulation imposes just by its existence, and the shortage of resources to deal with real health risk both in EPA and the society at large, these risks did not appear to me to be large enough to warrant regulation." The judge responded by calling Ruckelshaus a "scofflaw," describing his actions as "outrageous," and expressing shock that such a "responsible person" (and a lawyer to boot) would so cavalierly disregard an order of the U.S. District Court. After being found in contempt of court, Ruckelshaus published regulations that would have virtually no effect on radionuclide levels. EPA estimated that it spent $7.6 million in contract funds and 150 staff-work years on this regulation.

One might conclude from this story that the bureaucracy is filled with scofflaws like William Ruckelshaus who need to be reined in by federal judges. Let me suggest instead that the problem is that too *many* administrators are more than willing to hide behind the argument "don't blame me for stupid policy, the courts made me do it." The law reviews are filled with theoretical discussions of a dialogue between courts and agencies, about using judicial review to stimulate "deliberation." In the real world what happens is that judicial review short-circuits real deliberation about what constitutes good policy. Agencies say, "The courts made me do it"; courts say, "Congress made me do it"; and Congress says, "Protect public health, but don't put anyone out of work, and don't question our right to condemn the agency no matter what it does." The political science literature calls this "blame-avoidance." Most people call it passing the buck. Only a law professor could call it "dialogue."

Separation of Powers

The legal literature is reticent to admit what everyone knows: that a large number of administrative law cases are the outgrowth of divided government. For the first 180 years of the Republic, divided government—control of the presidency by one party and control of Congress by the other—was a rare event. In 1969, however, it became virtually a permanent condition of the American polity. The Democratic party has had a majority in at least one house of Congress in each of the twenty-two years since 1969, and in both houses for sixteen of those twenty-two years. Barring some electoral miracle, by 1996 . . . Democrats will have controlled the House for forty-two consecutive years.

Those who read the newspaper have noticed that Democrats and Republicans do not always agree on domestic policy or even on foreign affairs. Agencies often find themselves caught between congressional subcommittees (and their persistent staff) that demand more spending or more regulatory activity and the White House (or, more likely, the Office of Management and Budget) that demands less. With relaxed rules on standing, including, at times, even standing for members of Congress, many of these disputes end in court.

It is no coincidence that what Richard Stewart has aptly described as the "reformation of administrative law" began to take shape just as Richard Nixon was assuming the presidency. Nixon had a way of giving a bad odor to everything he touched—including the entire executive branch. Certainly Watergate contributed to judges' insistence that administrators be kept on a short leash. But competition between the presidency and Congress for the control of administrative agencies persisted long after the helicopter lifted Tricky Dick off the White House lawn.

In this decades-old battle between the branches, court intervention has generally given the advantage to the legislative branch. In the process it has added to the already considerable power of congressional subcommittees. The duty of the judiciary, Judge Wright announced in an important administrative law case, is "to see that important legislative purposes, heralded in the halls of Congress, are not lost or misdirected in the vast hallways of the federal bureaucracy." More recently Judge Patricia Wald has described the federal judges as a "trustee for the ghosts of Congresses past." Of course, given unprecedented reelection rates, the ghosts of Congresses past are still in Congress today and are likely to still be there in the year 2000, quite ready to defend their own interest. At least judges have been frank about their "tilt" toward the legislative branch. Congress has paid tribute to its alliance with the courts by relaxing standing and jurisdictional requirements, authorizing attorneys' fees, and recognizing private rights of action.

When courts come to the aid of Congress they usually end up strengthening its subcommittees. Judges intent upon determining the meaning of statutory phrases ambiguous on their face almost always turn to documents produced by congressional committees or other program advocates. Justice Scalia may have exaggerated somewhat when he claimed that "routine deference to the detail of committee reports . . . [is] converting a system of judicial construction into a system of committee-staff prescription," but he was not far from the mark. There are plenty of examples, ranging from the post-hoc legislative "history" produced by subcommittee staff to force HEW to write regulations under Section 504, to the stray sentences inserted in a Senate report used to justify the enormous Prevention of Significant Deterioration program.[4]

Many years ago, Justice Jackson warned that heavy reliance on legislative history may lead the courts to impose policies the president would never have accepted. More recently, Solicitor General Starr has pointed out that the use of legislative history "minimizes or ignores the role of the Executive." Moreover, by putting so much emphasis on *technical* justifications for agency policy, the courts have made *political* decisions seem somehow illegitimate. This view of policymaking weakens the position of the Executive Office of the President and of political executives within the agencies whose job it is to insist that whenever possible, agency policy be "in accord with the program of the president."[5]

Given the importance of the presidency in our constitutional system, it is remarkable how seldom the president is mentioned in the legal literature on administrative law. It is hard to know whether this is the result of cynical partisanship (if we ignore Republican presidents, maybe they will go away) or prudish distaste for electoral politics. In many cases, federal judges (not to mention the authors of law review articles) assume the pose of Louis in the movie *Casablanca:* they are shocked, shocked!, to find politics going on here. (It is appropriate that "*casablanca*" means "White House" in Spanish.) This

---

[4] Katzmann, supra note 1, at 49–58; Melnick, supra note 1, at chap. 4, 340–42, 373–79. For other discussions of the role of subcommittees and staff in the creation of legislative history, see Michael J. Malbin, *Unelected Representatives: Congressional Staff and the Future of Representative Government* (1980), and Roger H. Davidson, "What Judges Ought to Know about Lawmaking in Congress," in *Judges and Legislators: Toward Institutional Comity* (Robert A. Katzmann, ed., 1988).

[5] See Martin M. Shapiro, "The Presidency and the Federal Courts" in *Politics and the Oval Office* 141 (Arnold Meltsner, ed., 1981) and Shapiro, supra note 1 at chaps. 4–7. The weakening of the presidency is also a major theme of Rabkin, supra note 1.

was particularly evident in the famous air bag case. Given the vagueness of the underlying statute and the inability of previous administrations to decide whether to mandate airbags, one must ask what was wrong with allowing the outcome of a presidential election—a landslide, at that—to determine the outcome of the controversy? If Congress wants to require airbags, it can do so by writing a law to that effect. The courts' solicitude for Congress and its committees would be more justifiable if legislators had few other ways to influence agency activities. The fact is, however, that those members of Congress with a particular interest in a program usually have enormous influence. They have the appropriations process, investigative powers, reauthorization powers, confirmation hearings, and extensive links to the media—not to mention the power of constant harassment.

In his study of six federal bureau chiefs, Herbert Kaufman of the Brookings Institution reports that "[t]he chiefs were constantly looking over their shoulders . . . at the elements of the legislative establishment relevant to their agencies. . . ."[6] He adds, "Not that cues and signals from Capitol Hill had to be ferreted out; the denizens of the Hill were not shy about issuing suggestions, requests, demands, directives, and pronouncements." Members of Congress, he concludes, have an "awesome arsenal" to employ against administrators who ignore their wishes. Anyone who doubts this can read the huge political science literature on the power of Congress and its committees,[7] or one can save a lot of time and simply ask the FSLIC officials who tried to shut down the Lincoln Savings and Loan.

At the very least, it is time for administrative law to be more honest about the role of the Executive Office of the President and political appointees. We should remember that the Constitution makes the president part of the legislative process and gives the president—not judges—the responsibility to "faithfully execute the laws." The Supreme Court's decision in *Chevron v. NRDC*, coupled with its recent tendency to avoid reliance on legislative history, may end this tilt toward Congress. As Judge Wald has noted, compared with previous practices these judicial doctrines are "inherently executive enhancing." The Supreme Court, though, has hardly been consistent on these points. And we are only beginning to learn about the lower courts' application of Supreme Court doctrines.[8]

---

[6] Herbert Kaufman, *The Administrative Behavior of Federal Bureau Chiefs* 47 (1981).

[7] See e.g., Joel D. Aberbach, *Keeping a Watchful Eye: The Politics of Congressional Oversight* (1990): R. Douglas Arnold, *Congress and the Bureaucracy* (1979): *Congress: Structure and Policy* (Michael McCubbins & T. Sullivan, eds., 1987); Morris P. Fiorina, *Congress: Keystone of the Washington Establishment* (2d ed. 1989): Christopher H. Foreman, Jr., *Signals from the Hill: Congressional Oversight and the Challenge of Social Regulation* (1988): and Arthur Maass, *Congress and the Common Good* (1983).

[8] Schuck and Elliott explore many of the subtleties of the lower court response to *Chevron* in "To the *Chevron* Station: An Empirical Study of Federal Administrative Law," 1990 *Duke Law Journal* 984. They note, for example, that the rate of affirmances increased right after *Chevron*, but then began to drop. They also point out that the D.C. Circuit—which for years has reversed or remanded a greater percentage of agency decisions than any other circuit—actually decreased its affirmance rate after *Chevron*.

National Uniformity

In some policy areas judicial review has substantially reduced the uniformity of federal law. There is no little irony in this. Ever since the days of John Marshall, federal courts have tried (and often succeeded) to strengthen the Union by reining in the states. Now that we have created a large number of federal agencies charged with carrying out nationally uniform programs, extensive court intervention all too often creates different rules for different circuits. As Peter Strauss has pointed out, "[T]he infrequency of Supreme Court review combines with the formal independence of each circuit's law from that of the other circuits to permit a gradual balkanization of federal law."[9] He notes that the legal profession has "yet to come to grips with the problem."

Although the problem has arisen in a variety of areas—tax law, labor law, education of the handicapped, and welfare, to name a few[10]—it became most apparent in the heated confrontation between the courts and the executive branch over disability benefits. The agency charged with carrying out disability reviews, the Social Security Administration (SSA), is the very embodiment of the New Deal's commitment to national uniformity. The SSA's willingness to engage in "nonacquiescence" in order to preserve this uniformity predated the Reagan administration's attempt to purge the disability rolls in the early 1980s.[11] The heavy-handedness of the Reagan administration's policies should not blind us to the larger issue. In her recent book on the SSA, Martha Derthick has this to say about the agency's much-maligned nonacquiescence policy:

> The underlying premise—that policy must be nationally uniform—was not in the least contrived for the occasion. It was contained in law, Congress having stipulated in 1980 that administration of the disability insurance program be "uniform . . . throughout the United States." Just as important, it had long been at the core of the agency's operating code. . . . [T]his code had been reinforced in the SSA's case by the particular historical circumstance that its programs and administrative style were a reaction against the features of American federalism. For the SSA's programs to develop regional differences would be more than an monumental inconvenience to the

---

[9] Peter L. Strauss, "One Hundred Fifty Cases Per Year: Some Implications of the Supreme Court's Limited Resources for Judicial Review of Agency Action," 87 *Columbia Law Journal* 1093, 1105 (1987).

[10] For Strauss' discussion of labor and tax law, see id. For discussion of the Supreme Court's lack of control over special education law, see Mark C. Weber, "The Transformation of the Education of the Handicapped Act: A Study in the Interpretation of Radical Statutes," 24 *University of California, Davis Law Review* 349 (1990). For a discussion of AFDC, see R. Shep Melnick, *The Politics of Statutory Rights: The Courts and Congress in the Welfare State* (forthcoming).

[11] See Edward D. Berkowitz, *Disabled Policy: America's Programs for the Handicapped*, chaps. 3, 4 (1987); Susan G. Mezey, *No Longer Disabled: The Federal Courts and the Politics of the Social Security Disability* (1988), and Samuel Estreicher & Richard Revesz, "Nonacquiescence by Federal Administrative Agencies," 98 *Yale Law Journal* 679 (1989).

agency; it would constitute a humiliating retrogression to the time when state governments dominated domestic functions and citizens were treated differently depending on where they happened to live. The SSA's very existence rested on the belief that such differences were unfair.[12]

Compared with agencies like the SSA, the IRS, or even EPA, the federal judiciary is a highly decentralized institution. Constant intervention by the decentralized judiciary can lead to confusion and unfairness. It can also have a corrosive effect on the sense of mission of an agency like the Social Security Administration—a commitment to national uniformity and to prompt determinations, which has served us well since 1935.

Power Within the Agency

Aggressive judicial review creates winners and losers within administrative agencies. Clearly the biggest losers are political executives, who are less able to set priorities or to resist demands from congressional committees and interest groups. The biggest winners are lawyers with the agency's Office of General Counsel. Many studies of courts and agencies have come to this conclusion.[13] OGC attorneys are the ones who explain what the courts are likely to accept and reject. Frequently—especially on remand—they end up writing substantial portions of the regulations. At the risk of offending my readers, let me suggest two difficulties with this transfer of power to their "brother" and "sister" attorneys.

First, many of these lawyers workin the agency for a relatively short period of time; they are bright, aggresive young men and women who yearn to "cast a shadow" before moving on to a more lucrative career elsewhere. In that sense they are no different from congressional staffers, law clerks, or many lower-echelon people in the White House. More cocky neophytes with short-time horizons is hardly what Washington needs.

Second, few of these lawyers have ever run anything. They seldom have a sense of how hard it is to manage a program at the regional or state level. As Martin Shapiro has put it, the judicial demand for rationality

> tends to shift power within the agencies from those who are really concerned about making policies that work to those concerned with defending them in court . . . from real administrators responsible for the actual operations of programs to lawyers. And it will often lead to a choice of the alternative that

---

[12] Martha Derthick, *Agency Under Stress: The Social Security Administration in American Government* 141 (1990) (quoting Social Security Act of 1980, Pub. L. No. 96-265, 94 Stat. 454 [codified as 42 U.S.C. § 421 (a)–(c),(g)]).

[13] See id. at 183–84; Katzmann, supra note 2, at 113–20; Jerry Mashaw, *The Struggle for Auto Safety* (1990); Melnick, supra note 2, at 379–83; Rosemary O'Leary, "The Impact of Federal Court Decisions on the Policies and Administration of the U.S. Environmental Protection Agency," 41 *Administrative Law Review* 549, 566 (1989); and Christopher J. Bosso, *Pesticides and Politics: The Life Cycle of a Public Issue,* 183–84 (1987).

can most easily be made to appear synoptic rather than the one that seems best.

This power shift has contributed to some memorable policy disasters. For example, in the early 1970s the D.C. Circuit ordered EPA to promulgate transportation control plans adequate to meet ambient air quality standards, even in cities like Los Angeles. Top administrators considered this politically suicidal. Regional and technical personnel warned that it would be hard to formulate reasonable plans and virtually impossible to implement them. Hard chargers in the Office of General Counsel, though, saw this as their chance to remake the American system of transportation, to fight highways and automobiles, and promote mass transit. The plans, of course, were a political, technical, and management nightmare. The hardest charger in the Office of General Counsel then left the agency—to teach administrative law. As the old saying goes, if you can't do it, teach it.

Get Real, Administrative Law

This article has suggested both that courts often make a mess of policy because they have a poor understanding of administrative agencies and that administrative law has done little to correct judicial misperceptions. What understanding of bureaucracy might be more accurate? Let me propose the following generalization about public bureaucracies in the United States: They are almost always given huge, even utopian, goals and are then saddled with a large number of constraints that prevent them from achieving these goals efficiently—or even at all. We tell EPA, for example, to protect the public health with an adequate margin of safety, but advise it not to spend too much money or put anyone out of work. We tell them to use the best scientific evidence, but refuse to let them pay enough to recruit top-flight scientists, and then we tell them, "By the way, do it within 90 days." We expect bureaucrats to account for every penny of public money, to record every conversation with a member of an interest group, to show that they have treated everyone equally, and to consider all the relevant information and alternatives—but to stop producing all that red tape and being so damn slow.

Courts are particularly likely to make these conflicting demands because they are so decentralized, their exposure to policymaking is so episodic, and the opportunities for forum-shopping are so apparent to interest groups. Courts are good at responding to complaints, and people have lots of complaints about bureaucracy. The problem is that these complaints often require incompatible responses. Today the D.C. Circuit hears a case brought by environmental groups complaining about unconscionable delay. In two years the Sixth Circuit will hear industry complain about shoddy evidence and exorbitant costs. And two years after that a district court judge in Cleveland will be asked to balance the "equities" in order to keep a particular local factory in operation. Who considers the connection among all these decisions? The Supreme Court? Simply to ask this question is to answer it—no one does.

What is to be done? Most obviously, judges should remember a key part of the Hippocratic oath: First, do no harm. In administrative law that translates into the command, defer! defer! Just as importantly, administrative law needs to become less self-absorbed and more concerned about how public bureaucracies work. The best place to start is with James Q. Wilson's recent book, *Bureaucracy: What Government Agencies Do and Why They Do It*[14] which offers an unequalled discussion of the nature and the varieties of public bureaucracies. In addition to the works previously mentioned, there are a number of other case studies which will acquaint students of the law with the perspective of public administrators.[15]

It is especially important that law students become familiar with some of this nonlegal literature. Like most Americans, law students are contemptuous of bureaucrats. Moreover, in class after class they are asked to adopt the perspective of the judge. If law students fail to learn about public administration in administrative law courses, they never will. Adding studies of agencies to the curriculum will have the further advantage of forcing law professors to descend the Olympian heights of appellate review to wallow in the details of the day-to-day life of government bureaucrats.

My hunch is that this second suggestion will buttress the first. The more that students, law professors, and judges learn about the travails and dilemmas of public administration, the less likely they will be to second-guess agencies or to impose rigid rules and procedures. To purloin Schuck and Elliott's wonderful pun, a better understanding of bureaucratic reality will lead administrative law "to the *Chevron* station."

[14] James Q. Wilson, *Bureaucracy: What Government Agencies Do and Why They Do It* (1989).

[15] See John J. DiIulio, Jr., *Governing Prisons: A Comparative Study of Correctional Management* (1987); Robert A. Kagan, *Regulatory Justice: Implementing a Wage-Price Freeze* (1978); Herbert Kaufman, *The Forest Ranger: A Study in Administrative Behavior* (1960).

To the *Chevron* Station: An Empirical Study of Federal Administrative Law

PETER SCHUCK AND E. DONALD ELLIOTT

What Happens after Remand?

Judicial opinions are stories—albeit elaborate, highly stylized ones—and only the most unimaginative reader of opinions could finish them without wondering how the stories actually end. Courts in administrative law cases are especially likely to pique our curiosity when, forswearing their power to resolve the dispute themselves, they instead remand it to the agency to write the story's conclusion. In such cases we naturally want to know about the denouement. Specifically, we ask: What did the agency do procedurally once it got the case back? What was the outcome, and did the court's intervention really matter? How long did the proceedings take after remand? And were there intervening events that affected the outcome? Taken together, these questions pose what is probably the least studied but most important issue surrounding administrative law: What difference does judicial review of agency action make?

Unfortunately, our study cannot resolve this issue definitively. Our data on post-remand events, which relate to the 1984–85 period, consist entirely of the responses of the parties' lawyers to a questionnaire and follow-up telephone interviews. These data are, therefore, inevitably impressionistic. In addition, the lawyers may have tended to put a somewhat better (from their clients' perspective) face on the outcomes than more detached observers would. Nevertheless, the data do permit us to shed some light on some of the empirical questions noted above.

Post-Remand Procedures

We began with some general hypotheses concerning what happens procedurally once a case is returned to an agency. First, we expected that many remands would lead to informal settlements or withdrawals. They delay incident to any remand, together with the further agency proceedings that a remand might necessitate, could be so costly that the parties would seek a cost-minimizing alternative to continued litigation. More generally, remand decisions would almost certainly alter the parties' prospects, bargaining positions, and incentives to litigate. Second, the reasons displacement effect that we found should support the hypothesis that agencies to which cases were re-

Peter H. Schuck and E. Donald Elliott, "To the *Chevron* Station: An Empirical Study of Federal Administrative Law," 1990 *Duke Law Journal* 984. Reprinted by permission of the authors and Duke Law Journal.

manded in the post-*Chevron* [*Chevron USA, Inc. v. Natural Resources Defense Council, Inc.*] period would hold additional hearings more frequently or otherwise supplement the record. The agencies, we supposed, would need to institute such procedures in order to cure the factual or explanatory defects identified by the reviewing courts.

The data confirmed our first hypothesis. The lawyers reported that in about 40 percent of the remanded cases in 1985, no further proceedings occurred. The number of remands that "wash out" in this way increased by about 10 percent after *Chevron*. Among the agency clusters, the NLRB, DOL, and "other regulatory" group were the ones least likely to undertake further post-remand proceedings. This tends to confirm our suspicion that adjudictions, which accounted for virtually all of the NLRB and DOL cases, are easier to settle or drop than other kinds of proceedings.

Our second hypothesis—that remands after *Chevron* would more frequently cause the agencies to hold additional hearings or revise the record—was not borne out. Indeed, the opposite was true. Although proceedings of that kind were instituted in roughly 30 percent of the remanded cases during the pre-*Chevron* period, that figure declined to about 18 percent in the post-*Chevron* period. This mystery only deepened when we disaggregated the remands in order to focus upon the fact-based and/or rationale-based cases, which were more common after *Chevron* and were expected to necessitate record-revising procedures after remand. Contrary to expectations, we found that such procedures were more likely to be employed before *Chevron* than after. Although the post-*Chevron* increase in cases that "wash out" probably accounts for some of this decline, we are unable to explain the rest of it.

More generally, the data on post-remand procedures reveal certain features of the structure of post-remand activity. At a gross level, we can see what happens procedurally. As already mentioned, pre-*Chevron* about 40 percent of the remanded cases prompted no further action while only 30 percent of the remands (18 percent post-*Chevron*) actually led to procedures in which the administrative record would be revised. The data also indicate that in another 15 to 20 percent of the remands, the agency instead issued a new opinion—which usually sought to justify its earlier position by adopting a new legal theory or interpretation—rather than simply supplying additional explanation. The remainder of the cases were coded "other."

When we disaggregate the post-remand activity data by circuit court, the most interesting finding is that remands in the D.C. Circuit were less likely to "wash out" than those in all other circuits combined. It is not at all clear why this should occur unless the greater significance of D.C. Circuit cases makes them more difficult to settle informally.

Post-Remand Outcomes

We began our study with a strong belief in what might be called the "agency gets the last word" hypothesis. That is, we expected that an agency to which a reviewing court remands a case usually will end up reaffirming the position that it originally took.

This hypothesis was grounded in the same kind of Realist considerations that had led us (erroneously, as it turned out) to doubt that the Supreme Court's *Chevron* decision would have much effect on how the courts of appeals disposed of agency cases. Agencies generally make policy decisions for what they think are compelling political, institutional, and programmatic reasons. A court does not eliminate those goals when it remands to the agency; it simply strengthens the agency's incentives to find other ways to achieve them. Whether the agency gets the last word, however, depends not only upon the agency's motivation, but also upon the degree of constraint created by the control techniques available to a remanding court. Courts generally employ these techniques in ways that are designed to allow the agency considerable discretion and flexibility in how to respond to the remand—this flexibility permits agencies to exploit the technical and policy expertise for which they were established in the first place.

In order for a reviewing court to have the last word, it must either write remand opinions prescribing quite specifically what the agency may and may not do with the remanded case, or engage in repeated remands until the agency complies with the court's wishes. Unless the court is prepared to adopt one of these strategies, however, it may not be able as a practical matter to prevent the agency from adhering to the agency's original position. Yet both these strategies are problematic; each risks undermining the principle of judicial respect for the agency's expert judgment and programmatic responsibilities. We therefore expected that courts usually would acquiesce, perhaps grudgingly, in allowing agencies the last word.

These considerations led us to formulate another, more specific hypothesis about the interaction between the nature of controls exercised by reviewing courts and the specific reasons for the remand. We supposed that agencies are most likely to reaffirm their original decisions after rationale-based remands and are least likely to do so after law-based remands. In particular, the agency should be least constrained when the reviewing court does not reject what the agency has done, but simply demands a fuller explanation of, and justification for, that action. It should be most constrained when the court reverses the agency for having applied the wrong legal standard; in that event, application of the correct legal standard may well preclude the agency from doing what it would like to do, regardless of which form of words and justifications the agency uses to defend its result.

The data provide only weak support for our "agency gets the last word" hypothesis. Our most striking findings are that the agencies reaffirmed their original decisions in only 20 to 25 percent of the remanded cases; *Chevron* did not really alter this percentage. However, in a substantially larger share of cases—about 40 percent of the remands during the pre-*Chevron* period and about 37 percent of those after *Chevron*—the post-remand process actually produced what the lawyers agreed could be characterized as "major changes." . . .

This "major changes" finding is especially impressive for two reasons. First, lawyers for both the petitioners and the agencies, who might have been expected to diverge greatly on this point, in fact came up with roughly similar estimates of the occurrence of "major changes." Second, the agencies were much more likely to adopt these changes by relying upon the "old facts" (i.e., facts that were already in the pre-remand record) than by generating and relying upon "new" ones. This suggests that the mere occurrence of a remand, without more, frequently causes an agency to alter its original position in important ways. . . .

It would be interesting to know why, and under what circumstances, agencies change their original position after remand. When we broke down the data by agency cluster, the only pattern that we found was that the health, safety, and environmental agencies and the INS [Immigration and Naturalization Service] were least likely to make major changes after remand. The circuit-by-circuit breakdown failed to reveal any strong pattern: Agencies were less likely to make major changes in cases remanded from the D.C. Circuit than from other circuits generally, but the differences were not great.

More surprising was the fact that breaking down the data according to remand type failed to provide any real support for our hypothesis, noted above, that agencies subjected to law-based remands would be the least likely simply to reaffirm. The agency's impetus to adopt changes apparently is fairly independent of the degree of freedom allowed to the agency by the particular type of remand. As a minimum, this would seem to demonstrate that agencies are more open-minded than most commentators have believed. A more speculative proposition is that even when courts make rationale-based or fact-based remands—that is, remands that leave the agency most free to reaffirm its original position—they often communicate doubts about the merits of the underlying agency policy, doubts that agencies apparently take seriously.

These inferences are strengthened further by the lawyers' responses to two other questions. Approximately 40 percent responded affirmatively to the query, "Did the court's remand affect the ultimate result reached by the agency?" Here again, we were struck not only by how large this response was but also by how closely the responses by the lawyers for both agencies and opposing parties converged on this point. And when we asked the nonagency lawyers to describe "the practical effect on your client of the ultimate resolution by the agency after the court's remand," approximately 40 percent selected the response "much more favorable to client," while only about 17 percent selected the response "about the same effect on the client as before remand." . . .

The Duration of Post-Remand Proceedings We initially hypothesized that the elapsed time between the date of the remand and the date of the final agency action after remand would vary according to the type of administrative proceeding, the type of remand, and the type of agency. In particular, we expected that these proceedings would take longer in the following situations: in rulemaking proceedings (as opposed to adjudications); in proceedings involving fact-based remands (as opposed to remands involving law-based or rationale-based remands); and in health, safety, and environmental agencies (as opposed to other types of agencies). Our reasons for these expectations were discussed earlier. . . .

We obtained responses on the duration of the proceedings for 127 cases, or about 90 percent of the cases that were remanded in the 1984–85 period. The data reveal that the post-remand proceedings at the agency level alone took about seventeen months to complete on average. Almost two-thirds of the remands were completed within a year, but one in ten was still pending almost *five years* after the court remanded to the agency. . . .

The data on the duration of different types of proceedings seemed to contradict our hypothesis that rulemakings would take longer than adjudications. But because the total number of rulemakings in our dataset was small (only ten, or about 8 percent of the total during the entire 1984–85 period), and the standard deviations around all of the duration averages were large, we cannot say that the hypothesis has been refuted. Moreover, the differences in duration between the two proceeding types were generally small and the evidence pointed in different directions. Ratemakings took the longest time to complete (more than thirty months on average), but there were relatively few of them (six during the entire period).

When we analyzed the duration data according to the type of remand ordered by the court, we failed to confirm our hypothesis that fact-based remands would take longer to complete than law-based or rationale-based ones. Law-based remands took the least time and rational-based remands took the longest time. The differences, however, were small (generally less than two months) and the standard deviations were large.

Similarly, our analysis of the data broken down according to particular agency cluster failed to confirm our hypothesis that health, safety, and environmental agencies would take longer to complete. Indeed, the only pattern that we could discern was that the remands to the INS took longest to complete, but even here the number of cases was too small to justify much confidence in this observation.

Significant Post-Remand Events   In administrative law, the adage "justice delayed is justice denied" has a special meaning. It is not simply that the victim of agency illegality, in the absence of a stay, may continue to suffer while the agency's action remains in effect. The passage of time also affects the substance of administrative decisions more directly than it does decisions made in most other legal contexts.

This difference reflects two distinctive features of administrative law. One is that the period of time elapsing between the original agency decision and the agency's post-remand response is often so protracted that new developments— for example, changes in the applicable law, relevant facts, or the political environment—are almost bound to occur. The other is that the norms of administrative law permit, and sometimes even require, those new developments to influence the merits of the underlying agency decision in ways that would be inappropriate in a court decision. The value placed upon the agency's technical expertise, broad policy discretion, and political responsiveness mean that agencies are expected to take certain kinds of changes into account in their policy decisions to a far greater extent than are courts. For example, the advent of a new president, changes in agency leadership, and other political factors are ordinarily deemed relevant to the merits. Similarly, the agency ordinarily must take into account changing economic conditions, statutory contexts, and court rulings. Far from seeking to insulate the agency from these influences, administrative law recognizes their affirmative value in shaping the climate, and often the substance, of decisions.

These normative considerations led us to formulate several descriptive hypotheses about the kinds of events that occur after remand that may influence the agency's subsequent decision. We expected that any delay would increase the probability that intervening events would affect the outcome of many agency decisions. We thought that the most important of these events would be changes in presidential or agency leadership, changes in the relevant law (other than that represented by the remand itself), and changes in economic or competitive conditions germane to the agency's programmatic agenda. Less important, we supposed, would be changes in which lawyers handled a particular case.

The data reveal a high degree of agreement between the agency and opposing lawyers concerning the incidence of significant intervening events. Unfortunately, the wording of our question—"Were there any significant intervening events . . . which might help to explain any change?"—was more ambiguous than it might have been. We failed to make clear that we were concerned only with remands in which the agency altered its original decision. This was only a subset (albeit a surprisingly large one, as we learned) of all remanded cases. In what follows, then, we have assumed that our respondents so understood the question.

In almost one-third of the remands leading to changes (32 percent), the lawyers responded either "no" or "don't know." This suggests that when agencies change their positions on remand, those changes often may have little to do with changes in the relevant personnel, politics, law, policy, or economic conditions. In those cases, other factors—including the appellate court's remand-rebuke to the agency—may better account for the agency's change of heart.

Given that the remands we studied occurred around the midpoint of a long presidential administration, we were not surprised to find that in fewer than 2 percent of the remands was the agency's change caused by a change of national administration or of agency heads. Perhaps for similar reasons, an even smaller percentage of the changes after remand were attributed to a change in the agency staff or lawyer handling the case. Changes in economic or competitive conditions were cited as the reason for the agency's new position in only about 2 percent of the remands.

The single most frequently cited reason for the agencies' post-remand changes was the occurrence of a significant change in the law (or in legislation, court decisions, or agency policy) other than the legal change contained in the remand itself. But the fact that this affected only 17 percent of the changed decisions on remand suggests that agencies that adopt new positions usually do so for reasons having to do with factors internal to the agency itself, rather than for "environmental" reasons.

Our residual category of intervening events ("other") was also large. Fortunately, our interviewers often made notations on the coding sheets describing which "other" events the lawyers had in mind. These other events included the following: a different administrative law judge on remand; publicity about the case; the pendency of other cases; a compromise within the affected industry; a better rationale by the agency; agency policy change; changes in circumstances that facilitated settlement; publication of a congressional report; and a change in the agency's partisan composition.

We also expected that intervening changes in leadership, law, and economic conditions would be most influential in those types of proceedings and forums that we assumed would be most oriented to policy and political considerations: rulemakings, ratemakings, and D.C. Circuit cases generally. We could not draw any strong inferences from our analysis according to proceeding type, because although roughly one-fourth of adjudications were said to be influenced by those kinds of changes, the number of rulemakings and ratemakings in the sample was too small to permit a meaningful comparison. When we analyzed the data according to circuit, we did not find the relationship we expected; in fact, the results of D.C. Circuit remands were somewhat less likely to be influenced by such changes than the remands in the other circuits, but the differences were not great.

Specialized Courts in Administrative Law
HAROLD BRUFF

Today, caseload pressures on the federal courts led to renewed interest in creating or expanding specialized courts to relieve the crush. . . .

All three constitutional branches have performed specialized adjudication, in ways that present only subtle functional distinctions. Our government includes article I legislative courts, article II executive adjudicators, and article III judges with specialized dockets. Although no very clear normative or constitutional theory has emerged to limit congressional allocations of business among these entities, courts and commentators do display concerns that guide prescriptive analysis. . . .

The Benefits and Costs of Specialized Courts

The general benefits and costs of specialized courts are well known. First, they relieve the caseload burdens of other courts, perhaps substantially. Gauging docket relief is not, however, a matter of simply counting, filings shifted

Reprinted by permission from 43 *Administrative Law Review* 329 (1991). © 1991 American Bar Association. All rights reserved.

from one court to another. What matters is the amount of time and effort spared the judges. . . .

Second, specialized judges can become expert in the substantive and procedural issues surrounding particular programs, especially highly technical ones. More accurate decisions should result. . . .

Third, division of labor promotes efficiency. Due to expertise and a limited caseload, specialized courts can produce expeditious decisions. . . .

Finally, specialized courts reduce or eliminate intercourt conflicts, promoting a uniform national body of law. . . .

A primary cost of specialization is loss of the generalist perspective. . . . A broadened perspective may be especially important in those who review the action of bureaucracies that are themselves narrowly focused.

Also, specialization may diminish the prestige of a court. It will be staffed by lower-caliber judges, those who can tolerate life on the assembly line. Loss of prestige can fundamentally impair a court's power. . . .

Specialization can produce bias problems, in two ways. First, the appointments process may be distorted as nominees are selected and confirmed for their views on specific issues. . . .

Second, specialization can distort application of the review standard. Growing expertise may lead courts to substitute their judgment for an agency, creating an overly dominant oversight body. On the other hand, such a court can become too friendly with an agency that it reviews regularly, or with interests that dominate it.

A Brief History of Specialized Courts

. . .

Lessons Learned

Some guidelines for constructing specialized courts emerge from this overview. First, to minimize jurisdictional uncertainty and litigation, subject matter should be chosen for its segregability from other claims. For example, tax issues usually do not accompany others; energy issues eligible for TECA [Temporary Emergency Court of Appeals] often do. To avoid bifurcating appeals, litigation should be shunted to the specialized court in its entirety (as occurs with the CAFC) [Court of Appeals for the Federal Circuit] or left in the generalist courts. Too many forum-shopping opportunities attend creation of a new court with jurisdiction over only a portion of an integrated subject matter, such as the old CCPA [Court of Customs and Patent Appeals, now merged into the CAFC] or the Tax Court. Why endure the general disadvantages of specialization while forfeiting the primary benefit of unification of the law?

Second, the nature of a court's docket should expose the judges to both sides of pertinent controversies, instead of a set of appeals presenting skewed arguments, as in the CCPA. The CAFC, with a wider jurisdiction, sees a full range of related problems in patent law, and must directly balance interests in invention and competition. Also, the consolidation of the cases gives the new court adequate remedial scope to adjust the body of law it administers.

Third, allocations of subject matter should avoid combining generalists and specialists in ways that erode gains from specialization. It makes little sense to have the specialist trial forum of the Tax Court reviewed in the generalist courts of appeals. Similarly, difficulties that the generalist district courts have with patents litigation cannot all be cured by the CAFC, operating under the constraints of appellate review.

Fourth, each stage of judicial review should serve a distinctive function that is best performed by the court employed. Repetitive appellate review in different courts under TECA wastes resources. . . .

Fifth, the Commerce Court and TECA have demonstrated that a court for a single industry or a single agency is in jeopardy of capture by its clientele, or at least debilitating suspicion that it has occurred.

A Comparison of Federal Adjudicators

Separation of powers analysis usually places particular officers "in" one branch of government or another according to statutory provisions controlling their appointment, responsibilities, salary, and removal, and associated doctrines concerning their amenability to supervision by other officers. Hence, a statement that an adjudicator belongs with the core judiciary of article III, the "legislative" courts of article I, or the executive officers of article II implies a set of preexistinng conclusions about the particular attributes of the office in question.

Since federal adjudicators are all appointed by the president or by other executive officers, it is the other variables that determine degrees of independence. The Constitution's focus on life tenure and salary stability as attributes of federal judges suggests that the constitutional stature of other adjudicators depends mostly on their job security. . . .

Federal Judges

Of course, all federal judges share the same tenure protections, and today no one doubts their independence. I discuss them only to note some differences between district and circuit judges that bear on the assignment to them of administrative review responsibilities.

Single-judge district courts lack the collegial mechanisms by which the courts of appeals seek correct and consistent outcomes. Multimember panels dampen the idiosyncracy or incompetence of a single judge. Thus, circuit judges have independence from other branches of government, but not decisional independence from one another. . . .

District court decentralization also hinders the formation of a relatively uniform body of law over a large territory. . . .

The experience of circuit judges may make them better suited than district judges to exercise administrative review, because appellate judges always serve as restrained reviewers of decisions by others, not initial triers of fact. A district judge, possessed of tools for original fact-finding and accustomed to their use, may be reluctant to lay them aside. . . .

The courts of appeals follow procedures designed for resolution of issues of law. Their intellectual process is abstract. Their physical location, especially in the D.C. Circuit, is often remote from the impact of administration around the nation. With their high degree of insulation from political pressures, appellate judges may be tempted to cast themselves in the role of guardians of the public interest, against a tendency of bureaucrats to yield to powerful interests. District judges, on the other hand, are factfinders, given to specifics. They are dispersed around the nation, and have closer ties to their communities than do appellate judges. They see administration at the point of impact, and experience its factual context in their courtrooms. . . .

Legislative Judges

Outside the constitutional judiciary, judges of article I legislative courts most closely approximate the formal independence of federal judges. Their statutory terms are the longest in government. For example, Claims Court judges have fifteen-year terms, and are removable by the CAFC only for cause or disability. Tax Court judges also serve for fifteen years, and are removable by the president only for cause.

Renewable terms risk executive influence on legislative courts, as judges curry favor in hopes of reappointment. For several reasons, though, this risk seems small. . . .

When we seek a truly independent adjudicator, institutional separation is nearly as important a tool as a tenure guarantee. Article I judges do reside in separate organizations from the agencies they review, but they remain less independent than are district judges. The vulnerability of specialized adjudication to perceptions of capture is partly due to the effects of a steady diet of subject matter and repeated advocacy from a single source. Thus, even formally separate institutions can come to share values, if the informal links between them are strong enough.

Administrative Adjudicators

There is a fundamental difference between agencies and article I courts. The latter are true courts, in the sense that they do nothing but adjudicate. They make policy only as all courts do, incidentally to the decision of cases. Agencies, on the other hand, use adjudication along with rulemaking and enforcement processes as tools for the articulation of policy as well as its application to particular parties. This central and distinctive characteristic of the administrative process has led to some structural and legal accommodations that affect adjudicative independence. For agencies have legitimate interests in supervising the performance of their adjudicative processes in ways that might be inappropriate for article I judges. . . .

Administrative law displays a basic ambivalence about adjudicative neutrality—the benefits of obtaining knowledgeable decisionmaking are gained at the risk of introducing unacceptable levels of bias or interest. This tension between expertise and bias has existed for centuries. Today's accommodation, which is basic to the legitimacy of administrative adjudication, is a set of statutory controls that somewhat resemble judicial structure and procedure. These are: organizational separation of investigative and adjudicative staffs below the level of the heads of an agency, the statutory guarantees of independence that administrative law judges (ALJs) enjoy, and the APA's adjudicative procedures, which are designed to balance informality and fairness. . . .

Congress has sometimes separated adjudicators entirely from the rest of an agency. This step is feasible for fact-intensive matters such as enforcement proceedings, where adjudication is not central to general policymaking. Moreover, enforcement presents especially sensitive concerns for the fairness of adjudication. . . .

Within agencies, the trial judges usually have substantial guarantees of independence. This has not always been the case. Passage of the APA in 1946 was spurred partly by complaints that agencies used hearing examiners who lacked objectivity because the agency controlled their tenure. The APA included basic "separation of functions" requirements, forbidding examiners to be supervised by investigative or prosecutorial personnel.

The examiners' modern counterparts, administrative law judges, enjoy statutory tenure protections that justify their loftier title. Agencies select new ALJs from lists of qualified persons maintained by the Office of Personnel Management. OPM, not the employing agency, determines their pay and promotion. ALJs are assigned to cases in rotation and may not perform duties inconsistent with those of an ALJ. They may be removed only for good cause, as determined by the Merit Systems Protection Board after a hearing. Not surprisingly, agencies that do not use ALJs to make initial decisions endure criticism regarding the fairness of their processes.

For many years, ALJs were in little danger of removal by disciplinary proceedings. A recent, sharp upturn in the frequency of removal attempts, though, signals ALJs that their indefinite terms of office are not the practical equivalent of tenure until retirement. Nevertheless, it seems to be generally understood that removal is not proper for reasons that threaten decisional independence of the ALJs, as opposed to misconduct, unethical behavior, or poor work habits (for example, low productivity). . . .

An illustration of the difficulty of defining the appropriate degree of independence for ALJs is provided by a controversy in the SSA's disability benefits program. Turmoil followed SSA's initiation of case disposition goals for individuals ALJs, along with a quality assurance program that identified ALJs whose decisions deviated markedly from the average reversal rate of initial benefits denials. Pressure ensued to keep production up and allowance of claims down. The controversy abated somewhat when SSA modified the program to review allowances at random. This episode gave monitoring an unnecessarily bad name because the executive oversight technique seemed designed to skew the outcome of pending adjudications regardless of their merits, thereby invading the realm of fact judgment that has usually been regarded as sacrosanct.

The SSA controversy reveals the enduring tensions that attend the incompleteness of separation of functions in the agencies. The ALJs are in, but not of, the agencies. Proposals currently circulate to form them into a separate corps, free of residual agency influence. Yet the agencies do retain a legitimate interest in ALJ performance. If an ALJ systematically diverges from the agency's view of its statute or its policies on methods of fact determination, the agency bears a responsibility to try to avert the resulting inequality or illegality. The federal courts lack the perspective to monitor and correct systemic deficiencies in programs like disability benefits. The agency has the capacity to gather the necessary information and to send signals that will register, both through general policymaking and through managerial monitoring.

An agency's final decision is often vested in its head officers. At this level, sharp differences in formal tenure appear. Members of the independent regulatory agencies serve substantial terms (shorter, though, than Article I judges) and are usually removable by the president only for cause. Presidents rarely remove these commissioners, although a whiff of scandal and a threat of removal force an occasional resignation. . . .

Cabinet officers, who sometimes finally decide adjudicative matters, serve at the pleasure of the president. Nevertheless, . . . it is generally understood that presidential supervision of execution should steer clear of interference in adjudications, no matter who performs them. Nor is presidential interest in and dislike of a particular decision likely to rise to the point of firing a subordinate whose performance is otherwise satisfactory. Indeed, cabinet officers having adjudicative responsibilities often find it wise to delegate them to less politically involved subordinates, thereby divesting themselves of control over the decisions unless and until the delegation is rescinded. Unlike ALJs, these officers may enjoy no formal tenure protections, but there is often de facto job security. . . .

Still, there is a basic difference in the vulnerability of decisions of adjudicators to reversal by higher authority. The courts of appeals review findings of article I judges as they do those of district judges, under the "clearly erroneous" standard. In administrative law, officers reviewing ALJ findings are not constrained by any deferential standard. Instead, courts review the *agency's* final decision for substantial evidence on the whole record, and grant the ALJ's contribution only "such probative force as it intrinsically commands." The substantial evidence and clearly erroneous tests are approximately equivalent; if anything, courts are supposed to defer somewhat more to agencies than to district judges. Hence, it is the *agency's* final deciders and not its ALJs whose decisions are difficult to displace.

Some differences in behavior between article I trial judges and ALJs stem from differences in their assigned roles. Although the common law trial model probably fits article I judges about as well as it does district judges, ALJs often depart from the norms of the adversary system. . . . It may be that ALJs treat the interpretive and informal policies of their employing agency with more deference than an article I judge would accord to the policy of a formally separate agency. For example, IRS interpretive regulations probably receive more independent scrutiny in the Tax Court than do similar SSA policy statements before their ALJs.

More generally, those who work within an agency are subject to a multitude of open or subtle socializing pressures that do not reach a separate institution. . . .

Notes on Constitutional Architecture

. . .

. . . The basic strategy of *Crowell v. Benson,* . . . was to allow administrative adjudication as long as vital judicial controls were present. Congress was free to allocate initial authority to tribunals that might be temporary, specialized, and informal, without staffing them with federal judges, as long as appropriate appellate jurisdiction remained in the federal courts.

Crowell's theoretical compromise still holds sway, although the Court has struggled in recent years to find the limits of congressional power to convert traditional judicial functions into public rights to be adjudicated by agencies or legislative courts. In *Commodity Futures Trading Com'n v. Schor,* the Court allowed agency adjudicators to entertain state law counterclaims in reparation proceedings, in which disgruntled customers seek redress for brokers' violations of statutes or regulations. . . .

Agencies and article I courts decide issues of constitutionality, law, and fact. Of course, these categories blur at the margin, but they aid analysis. Which of these issues must an article III court review, and with what intensity? The Supreme Court has never decided whether the due process clause allows Congress to preclude review of an administrative decision completely. To avoid reaching this troubling separation of powers issue, courts often read statutes that appear to preclude all review to permit at least constitutional inquiries.

There are two principal reasons for restricting judicial review of facts. First, "retail" review of fact determinations in particular cases can absorb vastly more judicial resources than "wholesale" review of overall policy and procedure. Yet if individual rights must suffer, we should be most reluctant to invoke this ground alone. Here we must realize that "minor" administrative cases often receive only summary review in federal court, whatever their importance to affected individuals. Thus, raw statistics on review caseloads can overstate both the actual burden on the courts and the contribution of the constitutional judiciary to review efforts.

Second, stopping fact review short of federal court may actually improve the fairness and consistency of factfinding in some administrative programs. The SSA's disability benefits adjudications provide a prominent example. Internal SSA mechanisms are better able to promote fairness and consistency of outcome than are decentralized districts courts reviewing episodically. The agency's internal Appeals Council, which reviews ALJ decisions, has the potential to monitor adjudications systematically and to translate needed correctives into binding policy.

This does not mean, however, that agency factfinding should be final. Legislative courts may provide acceptable substitutes for the constitutional courts, bearing institutional benefits for both the agencies they review and the courts they partially displace. I turn to that prospect.

A Legislative Court for Initial Review of Selected Agency Programs

. . . [T]here is an institutional structure that does appear to capture most of the benefits of specialization, while minimizing its costs. That structure employs a semispecialized legislative court sitting nationwide to review both high-volume, fact-intensive agency adjudications, and some other programs that especially need specialization. From that court appeals would flow to the CAFC, itself a semispecialized court that can provide informed review and that can articulate uniform national law.

Obviously, such a structure could not concentrate all review in a single administrative court; nor should we attempt to do so. Instead, the point is to be selective, picking programs that most need institutional change and that can most benefit from the attributes of this particular scheme. Certainly, reasonable minds could differ on the particular mix of jurisdiction to assign the new court. I offer general criteria for programs to be included, and some specific nominees. The point of beginning is a unique structure that Congress recently created for review of veterans' benefits determinations by legislative and constitutional courts. This could form the core of a more generalized institution for other federal programs as well.

The Court of Veterans Appeals

In 1988, Congress ended its longstanding preclusion of judicial review of veterans' claims decisions and created the Court of Veterans Appeals (CVA), a legislative court with exclusive initial review jurisdiction over issues of law, fact, and constitutionality arising under laws administered by the Department of Veterans' Affairs (VA). The president appoints its seven judges for fifteen-year terms, and may remove them only for cause. The court may sit nationwide. Although only veterans and not the VA may invoke the court's jurisdiction, the VA does appear through its General Counsel. Review is on the administrative record, and the court may set aside VA fact determinations that are "clearly erroneous." This test was used in place of the more familiar "substantial evidence" or "arbitrary and capricious" formulas in order to allow intensified fact review.

Either a veteran or the agency may appeal to the CAFC, which may consider issues concerning the validity of statutes, regulations, and policies, but not errors of fact. Some jurisdictional litigation is sure to flow from this restriction. Where the law-fact distinction is cloudy, the legislative history stresses the difference between decisions of general importance and those limited to the correctness of a particular award. From the CAFC, review may be sought in the Supreme Court. . . .

In its other features, the scheme correctly follows principles of comparative institutional competence. . . .

The principal disadvantage of this structure is the danger of capture by an agency or an interest group, or the perception that it has occurred. . . . In an effort to dampen bias by broadening jurisdiction, proposals have floated to form a legislative court to hear appeals from many government benefits decisions, into which veterans' claims could be folded. Let us examine that possibility.

A Benefits Court

Among federal benefits programs, social security disability has produced the most frequent proposals for specialized review. In the wake of the new veterans' legislation, expansion of the CVA/CAFC model to include SSA disability has attracted support. The issues and procedures in federal disability programs, or even in benefits programs more generally, are similar enough to invite such consolidation. . . . I conclude that although a new legislative court should contain a benefits review component, it should have an even broader jurisdiction to diffuse appointments pressures.

A Semispecialized Administrative Court

The best way to maximize the benefits of specialized review while minimizing the detriments may be to broaden the veterans' review model to include a range of subjects other than federal benefits. That is, the new court's jurisdiction should be semispecialized—broad enough to remove litmus tests in the appointments process and to avoid comparisons to the assembly line, narrow enough to allow regular and informed monitoring of the agency programs selected. The exact mix of jurisdiction chosen for the new court is not crucial if these criteria are met. . . .

Hence I propose a new article I Administrative Court, to have initial review jurisdiction over selected programs, final as to fact for some of them. Review of this court could be assigned to the article III CAFC, reinforcing the current status of that court as a leader in administrative review, like the D.C. Circuit. The Administrative Court would sit nationwide, providing single-judge review for some high-volume, fact-intensive programs such as veterans and social security benefits. Other matters could be decided by panels, and without factual finality. When the new court sat in such an appellate capacity, further review in the federal courts could be made discretionary, to avoid duplicative appellate review.

Thus, either limited circuit-riding or decentralized organization would be needed for the new court. It would be fair to concentrate all initial review of important agency decisions in Washington. The benefits programs, with small money value but high importance to affected individuals, especially demand a local forum. Some specialized courts have ridden circuit in the past. The attendant inefficiencies can be avoided if the court has enough business to justify permanent locations around the nation. To avoid geographic inconsistencies in the law, it would then be necessary to provide a mechanism for part or all of the full court to review decisions of its local components.

Enrichment of the court's diet could be accomplished by reassigning some jurisdiction that is presently vested in specialized fora. Elements could include the Claims Court's nontax jurisdiction, appeals from the International Trade Commission, enforcement cases now decided by special split-enforcement tribunals, and appeals from the Merit Systems Protection Board. Patent cases now found in district court could be transferred to the Administrative Court to give it and the CAFC more complete control over that subject matter. Also, Congress could reassign the portions of immigration jurisdiction currently vested in district court. (Deportation cases, currently most reviewed in circuit court, may involve sufficiently important interests to deserve immediate resort to a constitutional court.) Appeals from decisions of the National Labor Relations Board could be shunted from the circuit courts.

Tax cases from the Claims Court, the Tax Court, and the district court may deserve separate treatment. The complexity of tax law and the breadth of the issues it raises suggest the wisdom of a specialized court deciding only tax cases. Still, the present Tax Court could be established as a separate division of the Administrative Court, to exercise all initial review functions in tax cases. Appeals from this division would then go to the CAFC, as with the Administrative Court's other cases.

NOTES AND QUESTIONS

1. Stewart and Sunstein state that courts in initiation cases "will ordinarily respect budget constraints aimed at limiting implementation and enforcement activity." What is the basis for this statement? Are there workable criteria that courts can use to decide when budget constraints are or are not a legitimate reason to grant or deny initiation relief?

Stewart and Sunstein discuss four possibilities for remedying regulatory inaction. Is this a logically exhaustive set, or are there others? What about a statutory or judicial mandate that under certain conditions the agency must provide a written statement of reasons for its inaction that includes answers to certain questions? In a variation on this scheme, an attorney general who declines to appoint a special prosecutor after being asked to do so must notify a tribunal through a confidential memorandum of his reasons. The Ethics in Government Act, 28 U.S.C. Sec. 592. What about conditioning a private right of action on the initiator having notified the agency of his or her intention to sue and the agency having failed to act during a specified period? Title VII of the Civil Right Act of 1964 contains an analogous provision. What are the arguments for and against such remedies?

2. Much of Farina's analysis flows directly from her claim that executive agencies tend to have "too much [power]" and her conception of the nature of separation of powers. As to the first, what is the basis for this assertion? As many of the contributions in this chapter indicate, this is hardly the view of most political scientists and others who have studied the agencies empirically. These observers are more impressed by the extent to which agencies are hemmed in by Congress, the White House, the courts, special interests, the system of "adversarial legalism," and other forces. The contrast between the views of Farina and Melnick is particularly striking. Whereas Farina maintains that the courts augment the president's power, Melnick claims that they tend to favor Congress and its committees. Farina and Melnick also disagree about how *Chevron U.S.A. Inc. v. Natural Resources Defense Council, Inc.* affects these tendencies. What are the essential grounds of their disagreements? As to separation of powers, compare Farina's view of the principle and of the role and effectiveness of courts in enforcing it with those of Melnick and Peter Strauss, and Louis Fisher in section 3.

Farina endorses "some mechanism for remanding important policy questions rasied by interpretation to the legislative process for clarification." In a footnote, she refers to the invalidation of statutes on nondelegation grounds, a clear-statement approach, and a federal law review commission. If the policy questions are truly important, however, why is a remand necessary to trigger congressional review? How would such a remand to Congress, as distinguished from agencies, work—that is, what would the court be saying to Congress, where would the burden of legislative inertia lie and why, and what legal state of affairs would obtain after the court's remand and before Congress resolved the matter (if it did)? An unusual (and from the traditional American constitutional perspective, extreme) variation on this proposal appears in the Canadian Charter of Rights and Freedoms, which authorizes the federal Parliament or a provincial legislature to statutorily override for five years a provision of the charter or a judicial interpretation of the charter. Constitutional Act of 1982, Sec. 33.

3. Farina's article exemplifies a recent recrudescense of interest among legal scholars in statutory interpretation. This interest seems linked to several factors: the growing importance of statutes in public law, the influence of deconstructionist literary theory on the analysis of legal texts, and the hostile reaction by a predominantly liberal academy to the Rehnquist Court's growing deference to Congress and to agency interpretations of statutes, marked by its decision in Chevron, U.S.A., Inc. v. Natural Resources Defense Council, Inc., 467 U.S. 837 (1984). Some other examples of commentary in this vein are Cass Sunstein, "Interpreting Statutes in the Regulatory State," 103 *Harvard Law Review* 405 (1989); Sunstein, *After the Rights Revolution: Reconceiving the Regulatory State* (Cambridge, Mass.: Harvard University Press, 1990); and William Eskridge, "Dynamic Statutory Interpretation," 135 *Univeristy of Pennsylvania Law Review* 1479 (1987). For a more sympathetic response to the Court's approach by a D.C. circuit judge, later solicitor general, see Kenneth W. Starr, "Judicial Review in the Post-*Chevron* Era," 3 *Yale Journal on Regulation* 283 (1986).

4. To further complicate the analysis, two empirical approaches to the question of how judicial review affects agency behavior are presented. In Melnick's review of the

behavioral literature on agencies, he asserts that administrative law scholarship and teaching, with its obsessive preoccupation with courts, is out of touch with the realities of bureaucratic behavior, especially in response to judicial interventions. Is this true? Can case studies of the kind he discusses ever establish what the marginal effect of judicial review is (i.e., the difference between agency behavior with and without it), much less the marginal effect of a particular style of judicial review? If administrative law scholarship is as misguided as this, why has it persisted for so long?

The Schuck-Elliott study approaches the judicial review question by seeking to uncover statistically the patterns of circuit court administrative law decisions during an era of great change stretching over more than twenty years. The study's finding that about 40 percent of the remands resulted in "major" changes in agency action suggests that this form of judicial relief does not simply lead to a reprise of the original decision but can effectively alter it. On the other hand, it also finds (in a section not included in the excerpt) that petitioners succeeded in obtaining a major change in the agency's position in "only" about 12 percent of the cases—8 percent in which the court reversed the agency outright, plus 40 percent of the 9 percent of the cases in which it remanded and the agency on remand adopted a major change. Id. at 1060. Does this constitute a useful measure of the effectiveness of judicial review in agency cases? If so, does this suggest that the review is (or is not) effective? If not, can you think of a measure that is better while also being researchable?

The Schuck-Elliott study also seeks to assess the appeals courts' response to the Supreme Court's *Chevron* decision. The extent of lower court compliance with a Supreme Court directive is interesting not only in its own right but for the light it may cast on the question of agency compliance with judicial directives. Is it realistic to expect that the latter would be greater than the former? Indeed, given the difference between agencies and courts, wouldn't we instead expect precisely the opposite?

In fact, the study finds that judicial affirmances of agency decisions increased by almost 15 percent after *Chevron,* and both remands and reversals declined by roughly 40 percent; that the decision seemed to have had the expected "outcome displacement effect" in which the increased affirmances "came from" reduced reversals rather than reduced remands; that the decision was immediately followed by a large decline in substantive law remands, precisely the kind that *Chevron* sought to discourage, while the total number of remands remained constant; but that by 1988, only four years after the decision, the increased affirmance rate had declined to 75.5 percent, about halfway between the pre- and post-*Chevron* rates. Id. at 1020–43.

5. The judicial and academic commentary on *Chevron* is already quite extensive. See, in addition to the articles on statutory interpretation mentioned above, Antonin Scalia, "Judicial Deference to Administrative Interpretations of Law," 1989 *Duke Law Journal* 511; Laurence Silberman, "*Chevron*—The Intersection of Law and Policy," 58 *George Washington Law Review* 821 (1990); Cass Sunstein, "Law and Administration After *Chevron,*" 90 *Columbia Law Review* 2071 (1990); Kevin Saunders, "Interpretative Rules with Legislative Effect: An Analysis and a Proposal for Public Participation," 1986 *Duke Law Journal* 346. In an article reviewing the post-*Chevron* decisions of the Supreme Court, Thomas Merrill finds that the Court has failed in these cases to adhere consistently to its own analytical framework in *Chevron.* He sets forth an alternative, "executive precedent" model of judicial deference. Merrill, "Judicial Deference to Executive Precedent," 101 *Yale Law Journal* 969 (1992).

6. In a court and administrative system as far-flung and decentralized as ours, maintaining national uniformity in administrative law, particularly in reviewing agency interpretation of statutes, is exceedingly difficult. For a view of *Chevron* emphasizing

the Court's reduced capacity to enforce such uniformity, see Peter Strauss, "One Hundred Fifty Cases Per Year: Some Implications of the Supreme Court's Limited Resources for Judicial Review of Agency Action," 87 *Columbia Law Review* 1093, at pp. 1121–22 (1987). Another aspect of this problem is the practice of agency nonacquiescence in lower court rulings. See, for example, Samuel Estreicher and Richard Revesz, "Nonacquiescence by Federal Administrative Agencies," 98 *Yale Law Journal* 679 (1989); Steve Koh, "Nonacquiescence in Immigration Decisions of the U.S. Courts of Appeals," 9 *Yale Law & Policy Review* 430 (1991) and sources cited there at note 1.

7. Much judicial review of agency action occurs in the context of tort-like actions for damages and for injunctive relief, yet this genre of litigation is often neglected in both administrative law and torts courses. These actions are generally brought against the federal government under the Federal Tort Claims Act and against individual federal officials in a so-called *Bivens* action under the Constitution. Federal court actions against state and local agencies and officials are usually brought under 42 U.S.C. Sec. 1983, the Civil Rights Act of 1871. These remedies have spawned a very complex, specialized jurisprudence. See generally, Peter Schuck, *Suing Government: Citizen Remedies for Official Wrongs* (New Haven: Yale University Press, 1983). For a treatise on Section 1983, see Sheldon Nahmod, *Civil Rights and Civil Liberties Litigation: The Law of Section 1983* (Colorado Springs, Colo.: Shepards/McGraw-Hill, 3d ed., 1991).

8. The literature on specialized administrative law tribunals in agencies and especially in the federal court system is also growing. See, for example, Richard Revesz, "Specialized Courts and the Administrative Lawmaking System," 138 *University of Pennsylvania Law Review* 1111 (1990); Stephen Legomsky, *Specialized Justice: Courts, Administrative Tribunals, and a Cross-National Theory of Specialization* (New York: Oxford University Press, 1990); Legomsky, "Forum Choices for the Review of Agency Adjudication: A Study of the Immigration Process," 71 *Iowa Law Review* 1297 (1986); Richard Posner, *The Federal Courts: Crisis and Reform* (Cambridge, Mass.: Harvard University Press, 1985), pp. 147–60; Lawrence Baum, "Specializing the Federal Courts: Neutral Reforms or Efforts to Shape Judicial Policy?" 74 *Judicature* 217 (1991); Federal Courts Study Committee, Working Papers and Subcommittee Reports, July 1, 1990, Vol. I, pp. 153–232.

3

Presidential and Congressional Review

The Place of Agencies in Government: Separation of Powers and the Fourth Branch

PETER STRAUSS

Three differing approaches have been used in the effort to understand issues such as these. The first, "separation of powers," supposes that what government does can be characterized in terms of the kind of act performed—legislating, enforcing, and determining the particular application of law—and that for the safety of the citizenry from tyrannous government these three functions must be kept in distinct places. . . .

"Separation of functions" suggests a somewhat different idea, grounded more in considerations of individual fairness in particular proceedings than in the need for structural protection against tyrannical government generally. It admits that for agencies (as distinct from the constitutionally named heads of government) the same body often does exercise all three of the characteristic governmental powers, albeit in a web of other controls—judicial review and legislative and executive oversight. As these controls are thought to give reasonable assurance against systemic lawlessness, the separation-of-functions inquiry asks to what extent constitutional due process for the particular individual(s) who may be involved with an agency in a given proceeding requires special measures to assure the objectivity or impartiality of that proceeding. The powers are not kept separate, at least in general, but certain procedural protections—for example, the requirement of an on-the-record hearing before an "impartial" trier—may be afforded.

"Checks and balances" is the third idea, one that to a degree bridges the gap between these two domains. Like separation of powers, it seeks to protect the citizens from the emergence of tyrannical government by establishing multiple heads of authority in government, which are then pitted one against another in a continuous struggle; the intent of that struggle is to deny to any one (or two) of them the capacity ever to consolidate all governmental authority in itself, while permitting the whole effectively to carry forward the work of government. Unlike separation of powers, however, the checks-and-balances idea does not suppose a radical division of government into three parts, with particular functions neatly parceled out among them. Rather, the focus is on relationships and interconnections, on maintaining the conditions in which the intended struggle at the apex may continue. From this perspective, as from the perspective of separation of functions, it is not important how powers below the apex are treated; the important question is whether the relationship of each of the three named actors of the Constitution to the exercise of those powers is such as to promise a continuation of their effective independence and interdependence.

In the pages following I argue that, for any consideration of the structure given law-administration below the very apex of the governmental structure, the rigid separation-of-powers compartmentalization of governmental functions should be abandoned in favor of analysis in terms of separation of functions and checks and balances. . . . What would be the consequences of so viewing *all* government regulators? I believe such a shift in view would carry with it significant analytical advantages by directing our focus away from the truly insignificant structural and procedural differences between the "independent regulatory commissions" and other agencies to the relationships existing between each such agency and the three named branches. Each such agency is to some extent "independent" of each of the named branches and to some extent in relationship with each. The continued achievement of the intended balance and interaction among the three named actors at the top of government, with each continuing to have effective responsibility for its unique core function, depends on the existence of relationships between each of these actors and each agency within which that function can find voice. A shorthand way of putting the argument is that we should stop pretending that all our government (as distinct from its highest levels) can be allocated into three neat parts. The theory of separation-of-powers breaks down when attempting to locate administrative and regulatory agencies within one of the three branches; its vitality, rather, lies in the formulation and specification of the controls that Congress, the Supreme Court and the president may exercise over administration and regulation. . . .

From the perspective suggested here, the important fact is that an agency is neither Congress nor president nor Court, but an inferior part of government. Each agency is subject to control relationships with some or all of the three constitutionally named branches, and those relationships give an assurance— functionally similar to that provided by the separation-of-powers notion for the constitutionally named bodies—that they will not pass out of control. . . .

Presidential Direction of Agencies

Viewed from any perspective other than independence in policy formation, the legal regime within which agencies function is highly unified under presidential direction. . . .

Even in the arena of policy, one readily finds major respects in which agencies' work is centrally managed. . . . Most prominent may be the OMB's annual creation of a national budget expressing the president's view of the relative priorities to be accorded the various efforts of national government. While Congress has occasionally limited the discipline of the budget process by requiring agencies simultaneously to provide the appropriate congressional committees their submissions to OMB, the president's coordinative, policy-setting function is recognized in these provisions also. Overall, presidential coordination is an activity of importance, one in which the agencies generally cooperate and from which they receive benefits as well as occasional constraint.

The independent agencies are often free, at least in a formal sense, of other relationships with the White House that characterize the executive branch agencies. The president's influence reaches somewhat more deeply into the top layers of bureaucracy at an executive agency than at an independent commission. . . .

Presidential influence over the independent agencies is heightened by the special ties existing between the president and the chairmen of almost all of the independent regulatory commissions. . . .

The Character of Presidential and Congressional Oversight Relations with Agencies is Determined as Much by Politics as by Law

Perhaps the central fact of legislative-executive management of oversight relationships with the agencies is the extent to which behavior is determined by political factors rather than law. The White House's treatment of cost-benefit analysis by independent regulatory commissions in conjunction with major rulemakings is a notable example. . . .

The reasons for this acceptance of presidential input are clear. The president's effective power over the independents would counsel against excluding his concerns even if political loyalties did not command attention. . . .

The Limited Constitutional Instructions About the Place of Agencies in Government

. . . The preceding review of the existing institutions of American government and of the body of textual, contextual, and interpretational constraints bearing upon them should cast doubt on the idea that our Constitution *requires* that the organs of government be apportioned among one or another of three neat "branches," giving each a home in one and merely the possibility of relations with the others. President, Congress, and Supreme Court are undoubtedly to be distinct in form and in function; below that level the text does not speak, sharp distinctions are frequently hard to find in fact, and the Court's occasional efforts to find them in theory have repeatedly led to embarrassments.

Thus to recognize that most of administrative government lies outside the constitutionally defined structure would not defeat the purposes either of separation of powers or of the system of checks and balances. The notions of checks and balances and (as an identifier of strong claims to attenuate political controls across the board) separation of functions are more vital in understanding the place of agencies in government. So long as separation of powers is maintained at the very apex of government, a checks-and-balances inquiry into the relationship of the three named bodies to the agencies and each other seems capable in itself both of explaining fully the results of past inquiries into the permissible structures of government below its apex, and of preserving the framers' vision of a government powerful enough to be efficient, yet sufficiently distracted by internal competition to avoid the threat of tyranny. This approach reflects common political science and presidential perceptions of the way government actually works without the evident conceptual embarrassments of "the qualifying 'quasi.' "

A focus on checks and balances could tend to emphasize struggles among the three branches for positions of control, and to ratify the bureaucrat's sense that he constitutes a legitimate fourth force in government, making control that much harder. Such a focus would be consistent with article II's references to the president, rather than an executive branch, a limitation which seems to have figured so prominently in the result in *Nixon v. Fitzgerald.* . . .

"Checks and Balances" as a Limit on Congress's Authority to Create the Structure of Government

This part of the essay supposes that government agencies charged with administering public law are *not* to be regarded as having been placed in one or another branch but rather exist as subordinate bodies subject to the controls of all three. Having thus put the agencies, in some sense, "out" of the executive branch, perhaps the most pressing question that remains is what if any relationship between an agency and the president might Congress (and the courts) still be required to honor. The issue what constraints exist on Congress in giving structure to government is thus to be assessed under the somewhat elusive checks-and-balances approach partially adopted in *Buckley* [*v. Valeo*] and *Nixon v. Administrator of General Services.*

From the text, three principal constraints emerge: that the president must appoint at least the head of any agency doing the work of government; that the agency in doing that work must have a relationship with the president consonant with his obligation to see to the faithful execution of all laws; and that, in particular, the president must have the authority to demand written reports of the agency prior to its action on matters within its competence, with the strong implication that consultation if not obedience will ensue. Central to the overall judgments of the Constitution, and reflected in these textual passages, is the elementary judgment that we were to have a unitary, politically responsible head of government, possessed of sufficient independent authority to serve as an enduring counterweight to the political muscle of Congress. Arrangements destructive of such a role—whether creating a "multiple executive," defeating the possibility of presidential political responsibility for the work of government, or threatening the president's continuing capacity to resist the Congress—are for this reason suspect.

This part argues that these constraints at a minimum require that Congress observe a principle of parity in its treatment of the possibility of political control of agency action by itself and by the president. Fairness and separation-of-*functions* considerations may often support exclusion of an agency or at least certain types of agency action from the domain of politics generally. However, Congress cannot expect to reserve political oversight for itself without recognizing corresponding oversight responsibilities in the president. Yet parity *is* a minimum; the president by virtue of his office as chief executive may be able to claim relationships beyond the constraints of parity. Recognition by the courts of a constitutionally based claim of executive privilege—that is, of private communication with agencies directly responsible for law-administration—is the most obvious example of such a claim. A more controversial claim would be for a requirement that Congress recognize presidential authority to resolve or mediate at least some types of internal policy disputes—for example, those placing separate agencies in direct confrontation with each other—or requiring judgments beyond a particular agency's ordinary responsibility and expertise.

Thus, Congress's authority to create the government's structure must be constrained in a manner that will preserve essential conditions of the president's intended political responsibility for the day-to-day, law-implementing activities of government. Even the most modern notion of what constitutes executive power suggests that the president must retain substantial lines of communication and guidance. To deny the president that authority would be to deprive him and the public of that responsibility, and effectively to permit the Congress, again, to establish multiple centers of law administration primarily under its control. Similarly, the execution of not a single law but many inevitably raises questions of priority, conflict, and coordination that rarely are addressed in any of the acts concerned; particular departments, with their narrow responsibilities, may be incapable of appreciating the interplay. Attending to these conflicts seems an inevitable aspect of a chief executive's function. That a legislature creates a statute and makes its application mandatory, or perhaps dependent upon stated circumstances, does not mean that the legislature will bestow the resources necessary to achieve that end, that it ever "intended" full enforcement, or that it has carefully thought through the relationship of the new mandate to those that have preceded it and those that will follow. In addition to resolving these conflicts and setting priorities among statutes, is the practical requirement of coordinating law-administration with political program and molding both to changing circumstances. The day-to-day course of national affairs generates new issues to which a coherent response must be made and for the resolution of which the public will hold the president politically responsible.

As we saw in the preceding part, the Court has found that the inquiries suggested by the checks-and-balances notion often lead to imprecise results. Yet the difficulties at the margin of saying whether given arrangements threaten "core functions" that are themselves imprecise—whether one is speaking of the president's functions, the Court's, or the states'—are not sufficient justification for refusing the inquiry. Although the stakes may be higher, the task thus facing a court does not differ in principle from the paradigmatic judicial task in reviewing acts of administrative discretion of determining whether that discretion has been abused. . . . Undoubtedly Congress has been given enormous discretion in shaping the instruments of government within the constitutional scheme; yet maintaining the Constitution as an instrument of law does not permit us to characterize this discretion as being without ascertainable constraint. . . .

Constraints Arising Out of the Constitution

Whether or not an agency is to be equated with the president in some constitutional sense, the constitutional text directly requires three forms of relationship between the president and the head of that agency: the president is to appoint the agency head with the advice and consent of the Senate; the president may require the agency head to give an opinion, in writing, "upon any Subject relating to the Duties of [the] . . . Office[]," and, most diffusely, the relationship must otherwise be such as to permit the president effectively to "take care that the laws be faithfully executed." . . .

"Take care that the laws be faithfully executed." As Professor Corwin and others have noted, the Delphic responsibility of the president to "take Care that the Laws be faithfully executed" does not tell us whether the president has authority to direct the affairs of government beyond that which statutes confer. What is important to note, however, is that the uncertainty is not unlimited. Whether this phrase implies that the president is to be a decider or a mere overseer, or something between, it requires that he have significant, ongoing relationships with all agencies responsible for law-administration. The unitary responsibility thus expressed, and sharply intended, does not admit relationships in which the president is permitted so little capacity to engage in oversight that the public could no longer rationally believe in that responsibility. The charge to "take care" implies that congressional structuring must in some sense admit of his doing so. An effort, then, to establish an agency over which the president's control went no further than the power to appoint its heads should be found deficient.

The proposition that Congress must allow some scope for presidential oversight of any law-administrator does not preclude its choosing by statute to place the responsibility for decision in a department rather than the president. Thus, primary responsibility, as well as the capacity, for detailed decisionmaking properly lie in the agencies to which rulemaking is assigned. Both agency decisionmaking and presidential oversight of an agency's performance may take into account only those factors that the statutes establishing them permit as relevant. Since the Constitution does not itself resolve the tension between president as administrator and president as political counterweight to Congress, a court would be hard pressed to deny Congress the choice in a particular setting of treating him as administrator. That assignment is itself a part of the law to the faithful execution of which the president must see. If, for example, EPA or OSHA is statutorily forbidden to consider economic cost as an element in a safety decision, that statutory preclusion is part of the law to be executed and consequently a constraint on its execution—for the president, as well as for the responsible agency. Presidential participation, like agency decision, seems possible only within the area of discretion that a particular law establishes. In any event, the president has only those resources Congress chooses to appropriate to him, and these are insufficient for him to make the detailed substantive decisions required by the statutes.

However, the presidential oversight function must in some sense be recognized; even an administrator has power and is not merely a clerk. Significant possibilities for legitimate presidential involvement can be identified and in fact appear to make up the bulk of presidential intervention efforts: seeking and providing information to promote coordination and awareness of national policy issues affecting law execution, requiring analysis or responses on matters suggested by national policy concerns relevant to the agency's charge, or directing that given perspectives be further considered within the framework of relevance established by the agency's organic statutes. All these actions are consistent with the statutory assignment of decisional responsibility to the agencies, the president's resources and vantage point, and his obligation to "take care" that *all* laws are faithfully executed. None need compromise the ultimate decisional authority of the responsible agency or disregard the factors that Congress may have identified as relevant to its decision by implying a substitution of presidential for agency judgment.

The Requirement That Congress Observe Parity in Political Oversight

The Constitution and the structural judgments it embodies require, at a minimum, that Congress observe a rule of parity in providing for political oversight of any government agency it creates. Congress cannot favor itself in providing for political oversight of an agency that administers, as well as assists in the formulation of, its laws. A rule that presidents may not, but members of Congress may, seek to bring political influence to bear on the policymaking of any agency directly affronts the framers' purposes, and serves no apparent function beyond aggrandizement of congressional power at the expense of the president's. Members of Congress are as capable as presidents of making excessive telephone calls or passing on private views under the guise of policy guidance, and often have done so; congressional hearings, for example, are used at sensitive stages of policymaking as instruments of coercion as well as of inquiry. Yet Congress's constitutional *raison d'être* is not to oversee the execution of laws; it is to enact new laws as required. Political pressure from the Congress or its members, is, if anything, *more* objectionable than similar pressure coming from the White House. At the least, the judgment that policymaking should not be subject to political direction must apply for congressional as well as presidential pressures. To conclude otherwise is to reverse the Convention's central decision about the presidency. . . .

This proposition about parity is not an especially strong one. In particular, so long as Congress does not treat itself more favorably than it does the president, the parity notion does not preclude a congressional judgment that some agency policymaking should not be subject to political direction by the president. From this perspective, the "independent agency" may simply reflect an ordering of governmental function that Congress would be free to extend to most if not all government. Government officials who are not heads of departments may be placed in the Civil Service, remote from presidential discipline, even though Congress could not insist on having an opportunity to approve their removal. Nonetheless, the acceptance of a requirement of parity in political oversight would establish an outer limit susceptible of both understanding by the Congress and application against it. That congressional understanding, in itself, might be thought likely to influence congressional judgments about the extent to which political controls would be warranted in a given situation. In creating "independent" agencies, Congress would have to give up the notion that "they are ours." . . .

The President's Claim to Effective Communication

One of the requisites of even a minimal oversight relationship, confirmed as we have seen in the opinion in writing clause, is a channel of communication with the agency being overseen, communication which—at least absent special justification—may be conducted in private and maintained as confidential. . . . [B]road features should be noted in preparation for a limited discussion of congressional prerogatives of control.

Executive privilege as a claim applicable to all law-administrators. The claim to executive privilege, often enough disputed in particular applications, is nonetheless generally recognized. *United States v. Nixon* acknowledged its constitutional dimension, even while holding that in specific instances it might be overcome. A similar proposition underlies long-standing judicial assertions—again, susceptible of being overcome—that the decision processes of law-administrators are not to be inquired into on judicial review, and that the fifth exemption to the Freedom of Information Act, which in effect embodies the more general notions of executive privilege, should be read as accepting that privilege. If executive privilege is the president's, and it is in the presidency that it has its constitutional source, then privilege is a relationship that must extend throughout all law-administration.

Congressional regulation of executive privilege. As *United States v. Nixon* underlines, the fact of executive privilege does not in itself establish the president's right to hold communications to him (or within the executive branch) confidential. Claims to acquire the information subject to the privilege or to deny confidentiality to consultations may equally be based on the Constitution, and in particular cases may be stronger than the privilege claim. The *Nixon* Court indicated that balancing judgments were required, finding there that a special need urged on behalf of criminal defendants overcame what was presented as a blanket, generalized claim of privilege made on the president's behalf. It was argued above, along similar lines, that a presidential claim of constitutional prerogative to communicate in private with a decisionmaker could be overcome by a congressional judgment that a given decision must be made "on the record" to assure its objectivity.

. . . Are there limits on Congress's authority to make and enforce such judgments? . . . Certainly, given the president's constitutionally based claim for confidential consultation with law-administrators, Congress could not expect to require openness for presidential communication only. If it required all presidential communications to be open without equally requiring that of congressional communications, the suspicion would be inescapable that an assault on the presidency, rather than an assurance of fair or acceptable procedure, was at the root of the measure; and accordingly, it ought not survive. . . .

The President's Constitutional Claim to Direct Agency Judgment

The discussion to this point has largely assumed the proposition sometimes thought to have been settled by *Kendall v. United States* that Congress can if it wishes place administrative duties wholly in law-administrators, not subject to direction by the president. But *Kendall* resolves that issue only for settings in which administrators have no discretion. Granted that the agencies and the president are each bounded by law in their exercise of discretion, what directory authority does the president have *within* these boundaries? As in the inquiry into executive privilege, it may be useful to consider the issue in two aspects: first, whether in the face of congressional silence, presidential authority to direct or at least shape the exercise of law-administering discretion may be inferred; second, if so, whether any limits on congressional ability to override that direction by statute may be found. As in the case of executive privilege, it is not possible to be more than suggestive about a subject on which many have written.

Presidential direction of the exercise of administrative discretion as a claim applicable to all law-administrators. Issues attendant to the president's claim to direct the exercise of administrative discretion are well posed by the increasing level of White House involvement in major agency rulemaking through the imposition of requirements to engage in economic analysis of the likely impact of proposed rulemaking, subject to possible White House review.

The stated rationale for these orders was not that the president might wish to substitute his judgment for the agency's on particular issues delegated to it, but that they would permit him to bring to bear balancing and coordination perspectives inseparable from the notion of a single chief executive. . . .

The power to balance competing goals—and the concomitant power to influence at least to some degree the agencies' exercise of discretion—can only be the president's; leaving the balancing functions in the agencies would create multiheaded government, government with neither the capacity to come to a definitive resolution nor the ability to see that any resolution is honored in all agencies to which it may apply. This outcome does not vary with whether or not the agencies are denominated independent. . . .

There are other practical reasons for granting the president the power to shape administrative discretion. Individual agencies almost necessarily lack the political accountability and the intellectual and fiscal resources necessary to achieve such balancing and coordination. Relying on them to do so would produce more costly and less responsible government, in which interagency disputes would be more likely to arise and less amenable to abiding resolution. . . . These claims made for presidential direction do not necessitate complete displacement of agency judgments or an assertion of White House expertise. Coordination and substantive suggestions—albeit backed up with the influence of the presidency—are not incompatible with the agency remaining ultimately responsible for the decision. The OMB supervision of the impact-analysis process under Executive Order No. 12,291, for example, recognizes (as it must) that the authority to issue the rules subject to the impact-analysis process remained in the agency head, subject to whatever political discipline the president might bring to bear. . . .

Congressional regulation of presidential direction of the exercise of administrative discretion. Given that the president's political accountability was foreseen as both a principal check upon him and a source of his authority in poiltical struggles with Congress, the risk that political considerations might enter into the president's oversight of administrative discretion seems a limited rationale for congressional control of that oversight. At the least, for the reasons already indicated, one would think the rationale unavailable wherever Congress had not been equally careful of its own political involvement.

Just as with executive privilege, the key is to identify those congressional statutes that might unbalance the distribution of authority at the apex of government by crippling existing informal controls; there is no a priori need for statutory authorization of presidential direction. Although the recent proposals for presidential override of some important agency policy decisions appear to be a dramatic innovation, the president's informal influence over the daily operation of government is already pervasive. In fact, these proposals would restrain rather than confer executive authority; the resolution of interagency dispute and coordination of agency activity in conformity with centrally determined executive policy are already daily activities. The activities are informal because informality makes for more efficient deployment of the president's limited resources; formalities mean time, people, and added disputes to be resolved. Such activities are not highly visible, because of the danger that, if they were, Congress would encumber them in ways restricting the effectiveness of the president's coordinative apparatus. The White House does not have to be explicit with an agency about results; it has enough power to act much more indirectly using the points of contact already described: the opportunities to persuade and cajole; the loyalties that may arise from appointment; and the probable shared sense of respect and of national mission.

Conclusion

. . . The imprecision of the relationship between president and Congress is at the heart of the mechanism by which the balance of the constitutional structure is maintained. It not hard to *say* that the central issues are the unitary direction of law-administration, political responsibility for the whole within the constraints of law, and a tension between Congress and president in which neither is able to become dominant. Determining when the presidential capacities necessary to maintain that tension have been threatened will rarely be other than a difficult act of judgment.

OMB Interference with Agency Rulemaking: The Wrong Way to Write a Regulation

ALAN MORRISON

Over the last decade, the Office of Management and Budget ("OMB") has played a greater and greater role in the issuance of agency regulations.

This commentary focuses not on the legality of this practice, but on its wisdom.

A Little History

To understand how a president, who campaigned in part on his abhorrence of centralized power, came to place the entire federal regulatory process under the control of a single office, it is necessary to review briefly the origins of the urge to centralize and control agency action. Throughout the late 1960s and early 1970s, Congress enacted comprehensive statutes creating new agencies, granting new powers, and expanding existing agency authority to provide greater protection for the environment, workers, consumers, and the general public. These statutes, for the most part, required agencies to enforce their broad mandates by issuing prospective rules, instead of giving them the choice to proceed by adjudication. By insisting on rulemaking, Congress sought to ensure greater public participation, examination of problems as a whole, added flexibility for the phasing-in of new requirements, and the efficient concentration of the resources of both the agency and the public in a single proceeding.

. . . Not surprisingly, there arose a feeling, often fostered by those who opposed the regulatory legislation in the first place, that the agencies had run amuck and were not doing the jobs that Congress thought it had assigned them. The charge was that many agencies were administering their laws with no consideration of other interests or the economic effect of their decisions. . . . But for some critics, improved coordination was not enough. They sought, and eventually obtained, increased supervision and centralization of executive agency rulemaking, culminating with OMB's complete dominance of the rulemaking process under the Reagan Administration.

The process began under President Nixon with his so-called "Quality of Life Review," and was followed by President Ford's requirement that agencies consider the inflationary impact of major rulemaking proposals. Under President Carter the process of centralization continued, with the establishment of a Regulatory Analysis Review Group within the White House and the requirement of a detailed regulatory analysis of every "major" rule before it was issued. In addition the administration also supported the Paperwork Reduction Act of 1980, under which the OMB gained greater control over the ability of all agencies, including the independents, to gather information through surveys and to impose recordkeeping and reporting requirements. Each of these changes was a step toward greater centralization and particularly toward insuring a greater decisional role for persons outside the agency, including various economic advisers in the White House. Yet with very few exceptions—the cotton dust rule at the Department of Labor being one—the ultimate policy decisions remained both legally and practically in the hands of the individual agencies.

When the Reagan administration took office, it recognized that White House control over administrative agencies was not yet complete. Therefore, one of its first acts was to issue Executive Order 12,291, which dramatically changed the relation between the agencies and OMB. First, the order requires that agencies promulgate only those regulations that are the product of cost-benefit, least-cost analysis, which all persons involved understand as meaning, "you can still regulate, but only as a last resort." Second, the order authorizes OMB to review virtually all proposed rules for consistency with the substantive aims of the Executive Order before an agency can even ask for public comment on the proposal. Finally, at the conclusion of the rulemaking process, the matter is sent once again to OMB, which may delay issuance of the final rule until the agency has considered and responded to OMB's views.

In one sense, Executive Order 12,291 is merely an extension of the prior system of control over agency rulemaking. In another sense, however, the order creates a very different system, because the professed aim of this administration is to cut back significantly, if not actually to destroy, the regulatory system established by Congress. The administration believes that it has a mandate to do so, despite public opinion polls showing the contrary, and it has proceeded under that assumption. The difficulty with its position is that the basic laws have not been amended, and to the extent that the administration has tried the legislative route to bring about change in the basic federal regulatory framework, it has not succeeded. . . .

In January 1985, just before the start of the president's second term, the president issued Executive Order 12,498, which created a new type of OMB control. Its principal feature is the implementation of what its supporters might characterize as an "early warning system," which requires agencies to notify OMB virtually as soon as an idea for a regulation is conceived, and to obtain OMB approval before undertaking any significant steps. The means chosen to give teeth to the order are found in an OMB Bulletin, issued six days after the order, which requires detailed submissions to justify all significant information gathering and research regarding a problem, even when the agency may only suspect that a problem exists and simply wants to determine whether any federal action is needed.

The Good and the Bad

. . . In practice, however, neither Executive Order is being used to further any of these theoretically beneficial ends. Rather, . . . they operate in a manner inconsistent with the basic principles underlying informal rulemaking. Contrary to the ostensible aim of the original Executive Order, the system of OMB control imposes costly delays that are paid for through the decreased health and safety of the American public. This system also places the ultimate rulemaking decisions in the hands of OMB personnel who are neither competent in the substantive areas of regulation, nor accountable to Congress or the electorate in any meaningful sense. In addition, the entire process operates in an atmosphere of secrecy and insulation from public debate that makes a mockery of the system of open participation embodied in the Administrative Procedure Act (APA). Finally, the new Executive Order allows OMB to cut off investigations before they even begin, making it nearly impossible to attack OMB's decision that a potential rule is "unnecessary." . . .

Finally, because Executive Order 12,498 gives OMB unbridled discretion to cut off an investigation before it even begins, the public may not learn of the dangers posed by the next "asbestos" tragedy before even more lives are lost than happened this time. Once again, the issue is not whether planning and allocation of resources are sensible ideas; everyone agrees they are, although they can probably be accomplished without the heavy hand of OMB. The real concern is that OMB is cutting off investigations at the earliest possible stage, before an agency can even hope to meet the burden that OMB has placed on it in order to commence a study of the problem. It is clearly more difficult for critics of the administration to prove that it is neglecting health and safety when there is no accumulated data. Aside from the judiciary's lack of willingness to intervene at the preproposal stage, there simply will be no data to prove that OMB has acted unreasonably in cutting off an investigation. And of course, when the government does not collect the information, it is more difficult for private parties to take the initiative, either by petitioning for a rule, suing the offending companies, or, in the case of workers, insisting upon protections through collective bargaining. For OMB, without any public input, to prevent an agency from conducting a preliminary investigation is the height of irrationality—unless one defines rationality as abhorrence of all regulation. . . . It is because the desk officers at OMB are operating at the direction of the political people at the White House, who have their own agendas and little concern for what Congress has mandated, that so many of these decisions cannot withstand even the limited scrutiny of judicial review under the APA.

What Should Be Done

To end this process, Congress should step in and prevent OMB intervention in agency rulemaking by precluding OMB from carrying out the functions that it has assumed under both Executive Orders. Thus, either by an amendment to the APA, or through a rider to OMB's appropriations legislation, Congress should prohibit all OMB involvement in the substance of agency rulemakings, except through on-the-record comments that any interested person, either inside or outside government, could submit. Congressional oversight, judicial review, and the president's power to discharge executive branch appointees already present sufficient checks to insure that agencies do not act contrary to law or to the wishes of Congress. . . .

Even though Congress may be unwilling or unable to completely remove OMB intervention from the rulemaking process, the president should amend the Executive Orders to make clear that OMB's function is advisory only and to assure that OMB does not overstep its authority by limiting OMB's role considerably. First, the president should limit OMB review to a few major rules each year. . . .

Second, for those rules that OMB does review, it must have a staff adequate to do the job. . . .

Third, all OMB communications on pending rulemakings should be in writing. . . .

Fourth, the president should amend Executive Order 12,498 to prevent it from being used to shut off an inquiry before the relevant facts concerning potential problems can be developed. . . .

Fifth, if despite all the problems with outside review of the substance of rulemakings, the president is unwilling to end such review, this responsibility should be removed entirely from the budgetary function and placed in a separate office with a separate budget. Only in this way will Congress, not OMB, have the final say over how much is spent on reviewing agency rules. The head of such an office should be appointed by the president with the advice and consent of the Senate, report directly to the president, and be answerable to Congress as well. . . .

Finally, it should be clear that the cabinet officer to whom Congress has delegated the power to administer the law is the lawful decisionmaker.

White House Review of Agency Rulemaking
CHRISTOPHER DEMUTH AND DOUGLAS GINSBURG

The Benefits of White House Review

Apart from specific statutory reforms (such as the Airline Deregulation Act of 1978), the establishment of White House review of agency rules was the most important political response to the growing popular and academic criticism of federal regulation. The characteristic failings of regulation that economists and other scholars have identified are twofold. First, regulation tends to be excessively cautious (forcing investments in risk reduction far in excess of the value that individuals place on avoiding the risks involved). Second, regulation tends to favor narrow, well-organized groups at the expense of the general public. These failings are, at least in part, a consequence of the institutional incentives of regulators and the nature of the rulemaking process.

We all know that a government agency charged with the responsibility of defending the nation or constructing highways or promoting trade will invariably wish to spend "too much" on its goals. . . . This tendency is reinforced by the "public" participation in the rulemaking process, which as a practical matter is limited to those organized groups with the largest and most immediate stakes in the results. Although presidents and legislatures are themselves vulnerable to pressure from politically influential groups, the rulemaking process—operating in relative obscurity from public view but lavishly attended by interest groups—is even more vulnerable. A substantial number of agency rules could not survive public scrutiny and gain two legislative majorities and the signature of the president.

Centralized review of proposed regulations under a cost-benefit standard, by an office that has no program responsibilities and is accountable only to the president, is an appropriate response to the failings of regulation. It encourages policy coordination, greater political accountability, and more balanced regulatory decisions. . . .

Responses to Some Criticisms of White House Review

Critics have attacked the White House review programs as being inconsistent with the Administrative Procedure Act, inconsistent with particular regulatory statutes to which they have been applied, and even unconstitutional. But the strictly legal questions raised by the review programs are not very difficult, and none of the legal attacks has been or is likely to be successful. The president is the federal government's chief executive officer, and informal rulemaking is an important method of executing domestic policy. To be sure, most regulatory statutes vest decisionmaking authority not with the president but with "the secretary," "the administrator," or other agency heads. But the same is true of virtually all statutes that create programs administered by a government agency. Agency heads exercise their statutory authority at the president's pleasure and have done so since the beginning of the Republic; it is his constitutional responsibility, not theirs, to take care that the laws are faithfully executed.

The tension between an agency head's statutory responsibilities and his accountability to the president is not resolved in Executive Order 12,291 or in the earlier regulatory review orders. Nor is it resolved in any statute or in the Constitution itself. It is a political question that can be "answered" only through the tension and balance between the president and Congress—that is, the political branches—in overseeing the work of the agencies. Members of Congress and private groups opposed to a president's policies naturally use legal arguments in their efforts to limit the president's influence, and on occasion appeal to the courts for assistance; and of course it is possible that, in any particular case, OMB or the president may exercise discretion in a way that conflicts with statutory requirements, just as an agency itself may do. But the interesting *general* questions presented by White House review of agency rulemaking are not questions of law, but rather those of politics and of policy.

Critics charge that the OMB regulatory review staff is not and cannot be as intimate with the subject matter of a complex regulation as is the staff of the originating agency. That is beside the point, however. OMB does not itself need to be an "expert" rulemaking agency; its role is to serve as the eyes and ears of the president and to advance generally the set of policies (or just "attitudes") that brought the president to the head of the government. The agency staff may be familiar with the content of scores of studies and documents that bear on their general subject area and that, in the case of a final rule, appear in the rulemaking record. The OMB staff is rarely able to bring new knowledge of a field to the attention of the agency. Yet the OMB staff is routinely able to ask hard questions, both substantive and methodological, to which an agency should be expected to have good answers before it proceeds to regulate. . . .

The OMB staff is more expert than the agencies in one field—the field of regulation itself. A great deal has been learned about the techniques of regulation in the last decade, particularly about the use of market-like incentives to accomplish regulatory goals more efficiently than "command and control" government directives can. Because the OMB staff reviews regulations coming from a variety of agencies and contexts, it is often in a position to draw on its own experience and that of another agency in a different field to inform the way in which an agency proposes to approach a new subject. . . .

The greatest benefit of OMB review, however, may result from the agency mechanisms established to respond to the kinds of questions that OMB raises. In response to Executive Order 12,291, agencies either established or enhanced their in-house capabilities to analyze their regulatory decisions.

The necessity to proceed privately has been a weakness only because it has put OMB at a disadvantage in responding to allegations that it does, or at least could, act as a "conduit" for information or influence to be introduced illicitly into the agency's decision calculus. These concerns are, however, misplaced. First, there are no statutory prohibitions of ex parte contacts by agencies engaged in informal rulemaking. Moreover, criticism focusing on ex parte contacts by OMB misses the point because communications that remain secret cannot determine the outcome of the regulatory process. The ultimate result of the rulemaking process must be a decision for which there is a rational and reasoned basis in the record. Nonrecord evidence cannot be used to support a rule, and any decision not anchored in the record will be over-turned. Consequently, ex parte contacts are irrelevant as a legal matter.

As a matter of practice, however, OMB generally does not meet with outsiders when it undertakes review of proposed or final regulations. . . . Consider the example of the Deputy Administrator's meeting with Canadian government representatives concerning EPA's proposed asbestos ban. The Canadian government requested a meeting with OMB, presumably in order to discuss not the scientific or technical issues, in which EPA is expert, but the foreign policy implications of the proposed rule. In these circumstances, an agency with a broader perspective is better suited to represent the administra-tion than the program agency in which rulemaking authority is lodged.

Some congressmen have joined with so-called "public interest" groups to complain that the OMB process might be a source of illicit influence for regulated industries. The fact is, however, that the only action Congress has taken in this regard reflected its frustration with OMB's *unresponsiveness* to a regulated industry. A rider to OMB's appropriation bill forbids OMB to review agricultural marketing orders under Executive Order 12,291. This measure was enacted at the insistence of agricultural interests that were angry at OMB's application of the order's economic principles to modify or disap-prove their marketing orders.

Some critics have also argued that OMB's review of preregulatory initia-tives under Executive Order 12,498 is especially inappropriate. They allege that such review gives OMB the ability to influence regulatory action before the program agency has had an opportunity to assess the dimensions of a problem through advance notices of proposed rulemaking, conferences, sur-veys, contract research, and the like. The allegation is correct—but we count it as a benefit rather than a cost. Scarce government resources must be allo-cated according to some set of priorities; the question is whether those priori-ties will be set unilaterally by each agency or by the president's administration as a whole through a process that reflects conflicting agency demands and the president's policies. . . .

Finally, some have questioned whether the process itself is cost beneficial. Because the administrative cost of running the program is trivial compared to the social cost of even a single ill-advised major regulation, most criticism has focused instead on the delay that OMB review entails. . . . Eighty percent of the regulations reviewed by OMB are cleared without change, and almost all

of these spend fewer than ten days at OMB. The overall average time for
regulatory review is just sixteen days. . . . [A] regulation that remains under
review for many months is not languishing in the bottom of someone's in box.
If the rule remains under review for such a time, it is typically because OMB
has asked for additional information necessary to resolve the cost-benefit
issues and is waiting for the agency to supply such information.

"Micromanagement by Congress: Reality and Mythology"

LOUIS FISHER

Micromanagement is a relatively new word to express a very old complaint:
intervention by Congress in administrative details. The problem is a real one,
but telling Congress to "stay out" has never been very effective. Congress
oversteps at times; on other occasions the executive branch conducts itself in a
manner that invites, if not compels, Congress to intervene.

We cannot hope to provoke significant changes in the behavior of either
branch. We can, however, encourage a deeper understanding of why Congress
is involved in administration. Avoiding slogans about micromanagement is
one step in appreciating the complexity of executive-legislative relations. I
make two assumptions: it is easier to tolerate what one understands; understanding can improve the performance of government.

The Framers' Design

The Framers were no doubt concerned about congressional involvement in
executive details. They were familiar with the inefficiencies of the Continental
Congress from 1774 to 1787 and hoped to devise a constitutional system that
would ensure energy and accountability in the executive. It is helpful to recall
the major developments over this period.

From 1774 to 1787 the Continental Congress experimented with various
techniques for discharging the duties of government. Until a permanent executive was installed in 1789, Congress had to handle all three functions: legislative, executive, and judicial. During this interval Congress first delegated
managerial responsibilities to a number of committees. This system failed to
work, as did the later system of boards staffed by men recruited from outside
Congress. When departments run by single executives were eventually cre-

ated in 1781, it was not until delays and makeshift arrangements had imperiled the war efforts. . . .

. . . On the eve of the Philadelphia convention, George Washington wrote to General Henry Knox about the need for a system with greater energy than the Continental Congress. . . . Shortly after the Constitution had established three branches for the national government, Washington explained that this separation of powers was neither derived from doctrine nor an expression of timidity toward power. Rather, it reflected the search for a more reliable and effective government. "It is unnecessary to be insisted upon," he wrote, "because it is well known, that the impotence of Congress under the former confederation, and the inexpediency of trusting more ample prerogatives to a single Body, gave birth to the different branches which constitute the present general government."

Comments by Alexander Hamilton and Thomas Jefferson reinforce these observations by Washington. . . .

The outrage routinely directed at congressional meddling seems a little less than principled, however, when we recall how often the executive branch is chest-deep in legislative affairs. Some scholars give Hamilton credit for drafting the bill that created the Treasury Department in 1789. . . .

The Jefferson administration followed the same liaison methods that had been pioneered by Hamilton. Jefferson and his secretary of the treasury, Albert Gallatin, took responsibility for initiating the main outlines of party measures. Gallatin, who had served as chairman of the House Ways and Means Committee, remained in close touch with his former colleagues and attended committee meetings in the same manner as Hamilton.

It is one of the anomalies of constitutional law and separated powers that executive involvement in legislative affairs is considered acceptable (indeed highly desirable) while legislative involvement in executive affairs screams of encroachment and usurpation. That does not seem quite fair. Even the Supreme Court in recent years has adopted this one-way philosophy.

Supreme Court Doctrines

In the legislative veto case of 1983 and the Gramm-Rudman-Hollings decision of 1986, the Supreme Court adopted a highly formalistic model of the relationship between Congress and the president. The net effect is to instruct Congress that it has no business in executive affairs. Predictably, Congress has ignored the Court's preaching, as it should.

In declaring the legislative veto unconstitutional in all its forms, the Court held that future congressional efforts to alter "the legal rights, duties, and relations of persons" outside the legislative branch must follow the full lawmaking process: passage of a bill or joint resolution by both houses and presentment of that measure to the president for his signature or veto. The Court lectured Congress that it could no longer rely on the legislative veto as "a convenient shortcut" to control executive agencies. Instead, "legislation by the national Congress [must] be a step-by-step, deliberate and deliberative process." According to the Court, the framers insisted that "the legislative power of the federal government be exercised in accord with a single, finely wrought and exhaustively considered, procedure."

What legislature is the Court describing? Certainly not Congress, where even the most casual observer can watch proceedings that fall short of being finely wrought and exhaustively considered. There is nothing unconstitutional about Congress's passing bills that have never been sent to committee. Both houses regularly use shortcuts: suspending the rules in the House of Representatives, asking for unanimous consent in the Senate, and attaching legislative riders to appropriations bills. Not pretty, but not unconstitutional either.

The Court also indulged in a rewriting of what the Framers intended by the separation of powers doctrine. The Court claimed that "it is crystal clear from the records of the Convention, contemporaneous writings and debates, that the Framers ranked other values higher than efficiency." "Convenience and efficiency are not the primary objectives—or the hallmarks—of democratic government." These assertions are not documented; indeed they could not be. The Framers placed a high value on efficiency and wanted a government more effective and reliable than the Continental Congress.

By misrepresenting the Framers and the theory of separated powers, the Court issued dicta that in no sense capture the subtleties and dynamics of executive-legislative relations. Both branches have entered into agreements and accommodations that are directly contrary to the Court's ruling. I shall give some choice examples, but first I need to comment on the Gramm-Rudman-Hollings case.

In *Bowsher v. Synar* (1986) the Court promoted an even more rigid model of separation of powers. The Constitution, said the Court, "does not contemplate an active role for Congress in the supervision of officers charged with the execution of the laws it enacts." The fact is that the Constitution does not contemplate a number of things, including the president's ability to make law unilaterally by issuing executive orders and proclamations. The Court insists that if Congress wants to legislate it must sedulously follow each and every rule, including submission of a bill to the president, while the president can make law on his own without any congressional involvement. Again, not exactly even.

The Court also stated that the "structure of the Constitution does not permit Congress to execute the laws; it follows that Congress cannot grant to an officer under its control what it does not possess." Although Congress cannot execute the laws, there are circumstances where it is reasonable for congressional committees to share in the execution of the laws. I shall give some examples. Moreover, congressional oversight of executive activities is a legitimate constitutional responsibility. The Court seems to forget what it has said in the past. In 1957 the Court noted that the power of Congress to conduct investigations "comprehends probes into departments of the federal government to expose corruption, inefficiency or waste."

Grounding itself on the legislative veto case, the Court claims in *Bowsher* that "once Congress makes its choice in enacting legislation, its participation ends. Congress can thereafter control the execution of its enactment only indirectly—by passing new legislation." That, of course, is nonsense. Congress controls the execution of laws through hearings, committee investigations, studies by the General Accounting Office, informal contacts between members of Congress and agency officials, and nonstatutory controls. Congress controls the execution of laws when each house invokes its contempt power and when committees issue subpoenas. These actions do not conform to the legislative veto ruling (passage of a bill by both houses and presentment to the president), and yet the Court has recognized the legitimacy of both congressional contempt citations and committee subpoenas.

In seeking precedents to support the Gramm-Rudman-Hollings ruling, the Court reached back to a passage from *Humphrey's Executor v. United States* (1935). In that case the Court upheld the right of Congress to place limits on the president's removal of commissioners of the Federal Trade Commission. In the Gramm-Rudman-Hollings decision the Court quoted this language from Justice George Sutherland in the 1935 case:

> The fundamental necessity of maintaining each of the three general departments of government entirely free from the control or coercive influence, direct or indirect, of either of the others, has often been stressed and is hardly open to serious discussion. So much is implied in the very fact of the separation of the powers of these departments by the Constitution; and in the rule which recognizes their essential coequality.

What would we say if a student of American government, in a final examination, wrote that the separation of powers doctrine kept the three branches of government "entirely free from the control or coercive influence, direct or indirect, of either of the others"? Would we not wonder where that student had been throughout the semester? Had the student read nothing of Madison's essays in the *Federalist* Nos. 37, 47, 48, and 51, where Madison emphasizes again and again that branches will not stay separate unless they have some control over one another? Was the student absent or asleep when the class discussed checks and balances? Did the student ignore Hamilton's essay in *Federalist* No. 66, where he said that the separation maxim was "entirely compatible with a partial intermixture" and that this overlapping was, in fact, not only "proper, but necessary to the mutual defense of the several members of the government, against each other"? Hamilton drove home the same point in *Federalist* Nos. 71, 73, and 75. . . .

The Court also garbled Madison's position. Although Madison quoted Montesquieu that there can be no liberty where the legislative and executive powers are united in the same person, immediately afterward he explained that the French philosopher did not mean that "these departments ought to have no *partial* agency in, or no *control* over, the acts of each other." The meaning of Montesquieu, said Madison, "can amount to no more than this, that where the *whole* power of one department is exercised by the same hands which possess the *whole* power of another department, the fundamental principles of a free constitution are subverted." Gramm-Rudman-Hollings had many constitutional defects, and I identified several when I appeared before the House Committee on Governmental Affairs. The deficit-reduction scheme, however, which delegated to the comptroller general a role in the sequestration process, was never an effort to place the whole of the executive power within the legislative branch. The writings of Madison give no support for the Court's decision in *Bowsher.*

The Constitution contemplates an overlapping, not a separation, of powers. . . . Some of the states wanted to add a separation clause to the national bill of rights. The proposed language, which Madison submitted to the House in 1789, read as follows:

> The powers delegated by this constitution are appropriated to the departments to which they are respectively distributed: so that the legislative department shall never exercise the powers vested in the executive or judicial [,] nor the executive exercise the powers vested in the legislative or judicial, nor the judicial exercise the powers vested in the legislative or executive departments.

Here was a good opportunity to go on record against intervention, meddling, overreaching, and micromanagement. Congress rejected the proposal, as well as a substitute amendment to make the three departments "separate and distinct."

The Supreme Court has not always been as doctrinaire on separation of powers as in the [*Immigration and Naturalization Service v*] *Chadha* and *Bowsher* cases. . . .

The Legislative Veto

Probably no congressional action in recent years matches the reputation for intrusiveness and micromanagement so much as the legislative veto. This device was attached to hundreds of statutory provisions, allowing Congress or one house or its committees and subcommittees to control executive activities. All these devices had a common quality: none went to the president for his signature or veto. Some had the additional quality of requiring action by only one house. A well-known article in 1975, suggestively called "Congress Steps Out: A Look at Congressional Control of the Executive," concluded that the committee veto "should be considered per se invalid." Moreover, this study said that the use of most simple (one-house) and concurrent (two-house) resolutions "should be considered invalid as inconsistent with the Framers' intentions and their views of the separation of powers."

In *Chadha* the Court agreed with this assessment by striking down the legislative veto because it violated two constitutional principles: bicameralism and presentment of bills to the president. What the article in 1975 and the Court decision in 1983 never came to terms with is that the legislative veto was not a case of Congress's "stepping out." It was not a tool invented by legislators to invade the executive branch or dabble in micromanagemernt. It was a mechanism to further the interests of the *executive*. Unless you understand that, you cannot comprehend either why the legislative veto began or why it persists even to this day, despite the Court's pronouncement.

The legislative veto emerged in the 1930s primarily as a way to broaden executive authority. President Herbert Hoover wanted to reorganize the executive branch without having to pass a bill through Congress. He realized that Congress would never consent to delegating that authority without being able to check his actions with something short of passing a law. Thus a pact was born: Hoover could submit a reoganization plan that would become law within sixty days unless either house of Congress disapproved.

This was the essence of the legislative veto: a simple quid pro quo that allowed the executive branch to make law without any legislative action but gave Congress the right to recapture control without having to pass another public law (and attract a two-thirds majority in each house to override a presidential veto). The legislative veto survives because, using Jackson's term, it helps make government "workable." It persists, notwithstanding *Chadha,* because it is better than the alternatives.

From the Court's decision on June 23, 1983, to the adjournment of the Ninety-ninth Congress in October 1986, Congress included an additional 102 legislative vetoes in bills and President Ronald Reagan signed them into law. A shocking case of Congress failing to comply with a high court edict? Another example of Congress willfully violating constitutional boundaries? Before we satisfy our moral appetite to issue condemnations, let us look at the record.

Most of the new legislative vetoes require agencies to obtain the approval of the Appropriations committees before taking certain actions. The agencies comply with this arrangement because it gives them a substantial amount of discretion to move funds around and initiate other managerial decisions. If they refused the quid pro quo, Congress would withhold the authority and force agencies to do what the Court in *Chadha* said is necessary: come to Congress and seek approval through the regular legislative process. Neither side wants to go through the entire legislative hoop—passage by both houses and presentment of a bill to the president—to make these midyear adjustments.

Take a look at what happens when these arrangements collapse. In 1984 Reagan received the housing appropriations bill and observed that it contained a number of committee vetoes. In his signing statement he asked Congress to stop adding provisions that the Court had held unconstitutional. He also said that the administration did not feel bound by the statutory requirements that agencies seek committee approval before implementing certain actions. In short, he invited agencies merely to notify the committees and then to do whatever they wanted.

It did not take the genius of a Nostradamus to predict what would happen. The House Appropriations Committee reviewed an agreement it had reached with the National Aeronautics and Space Administration. The accommodation had placed dollar caps on various NASA programs, allowing the agency to exceed the caps if it first obtained the approval of the Appropriations committees. Because of Reagan's signing statement, the House Appropriations Committee said that it would remove the constitutional objection by repealing the committee veto. At the same time it would repeal NASA's authority to exceed the dollar caps. If NASA wanted to exceed those levels, they would have to comply with *Chadha:* get a public law.

Here we leave the abstract world that neatly assigns "executive" and "legislative" powers to separate compartments and begin the search for workable solutions. The administrator of NASA, James M. Beggs, wrote to both Appropriations committees to suggest a new accommodation. Instead of putting dollar caps and committee vetoes in a public law, he proposed that the caps be placed in the conference report that accompanies the appropriations bill. He then promised that the agency would not exceed those caps "without the prior approval of the Committees." What could not be done directly by statute would be done indirectly by informal agreement. *Chadha* does not affect these nonstatutory "legislative vetoes." They are not legal in effect. They are, however, in effect legal. If agencies violate agreements with their review committees, they can expect onerous sanctions and penalties. Is this micromanagement? You be the judge.

I will give another example of a committee veto that was challenged on constitutional grounds. After the fight was over and the dust cleared, it remains part of public law. For a number of years the following language has appeared in the appropriations bill for foreign assistance: "None of the funds made available by this Act may be obligated under an appropriation account to which they were not appropriated without the prior written approval of the Committees on Appropriations." For some reason the Reagan administration decided to challenge this procedure in 1987. James C. Miller III, director of the Office of Management and Budget (OMB), raised numerous objections to the foreign aid bill, including the provision for the committee veto. He said it "violates constitutional principles" established by the Court in *Chadha.*

The response from Congress was both predictable and bipartisan. David Obey, chairman of the Foreign Operations Subcommittee of the House Appropriations Committee, along with Mickey Edwards, the ranking minority member, told OMB it would delete *all* the language, removing not only the committee veto but the authority to obligate funds under a different account (transfer authority). Obey said that the OMB letter "means we don't have an accommodation anymore, so the hell with it, spend the money like we appropriated it. It's just dumb on their part." Edwards added that OMB "has not had a history of being very thoughtful or for consulting people" and that the provision was an example of "the spirit of cooperation between the executive and legislative branches, which the administration is not very good at." OMB backed down, realizing that it had shot itself in the foot. The language appears in the continuing resolution signed by Reagan on December 22, 1987 (P.L. 100-202). Another case of micromanagement? Judge for yourself.

Why Does Congress Intervene?

Instead of issuing broadsides about Congress's "meddling" or indulging in micromanagement, we need a better understanding of what Congress does and why it does it. Surely one of the deepest penetrations by Congress into administrative matters is the Iran-contra affair. . . . Why did this level of congressional involvement occur? I suggest two reasons.

First, President Reagan failed to see that the laws were faithfully executed. . . .

Second, even when the scandal became public in November 1986, Reagan failed to respond adequately. . . .

Iran-contra involves congressional intervention at the highest level and for high stakes. Other examples of congressional intervention are less visible and less earthshaking. Nonetheless, they illustrate how Congress can be drawn into "micromanagement."

Consider the following language, which appears in section 626 of the agriculture appropriations bill, included in the continuing resolution signed by President Reagan on December 22, 1987:

> None of the funds provided in this Act may be used to reduce programs by establishing an end-of-year employment ceiling on full-time equivalent staff years below the level set herein for the following agencies: Farmers Home Administration, 12,675; Agricultural Stabilization and Conservation Service, 2,550; Rural Electrification Administration, 550; and Soil Conservation Service, 14,177.

Why would Congress descend to that level of minutiae and circumscribe the flexibility of officials to manage their agencies? Every provision like this has a history. During the Nixon years, when the administration was impounding large amounts of appropriated funds, it was recognized that personnel ceilings were a more subtle form of impoundment. OMB could simply place a ceiling on the number of employees an agency could have, making it difficult for the agency to spend all its funds. Personnel ceilings could also be used by the administration to dictate priorities: give preferred programs full staffing and limit employees for programs out of favor. The Appropriations committees warned the administration, in committee reports, that continued use of personnel ceilings would result in statutory controls.

The congressional concern became more pronounced when it was learned that personnel ceilings had led to waste and inefficiency. The "ceiling game" meant that thousands of employees would be separated from their agencies just before the end of the fiscal year and rehired when the new fiscal year began. Ceilings also forced agencies to use contractors for activities that could have been handled less expensively within the agency. Language in congressional reports became more specific, again raising the specter that these nonstatutory directives would be replaced by statutory controls unless the administration respected legislative priorities. When it appeared that agencies would not make good-faith efforts to carry out the congressional will, Congress resorted to statutory controls. Personnel floors were established for a variety of agencies, and these floors remain in place today.

These collisions between the branches and the failure to reach an acceptable accommodation leave behind a residue of statutory controls on what agencies can and cannot do. Agencies that maintain trust with their oversight committees can be expected to retain substantial discretion and relative freedom from legislative intervention. . . .

The Item Veto

Congress is criticized for inserting an increasing amount of detail into appropriations bills, forcing the president either to accept or to reject a bill as a whole. Many of these details, provisos, qualifications, and conditions give close guidance to agency activities. It is argued that these omnibus measures undermine the president's general veto authority. That argument is even more compelling when the president receives a giant continuing resolution in the closing days of a fiscal year, or perhaps even after the fiscal year has expired. Under these emergency conditions the president is under heavy pressure to sign the bill to prevent the government from closing down. . . .

Of course, the budget process can be changed to permit the president to get at these items. Merely take all the detail in the conference report and include it in the appropriations bill. Instead of a single sum of $31 million for special research grants, identify the sixty programs.

There are two drawbacks. First, President Reagan complained about the size of the continuing resolution. He said it weighed fourteen pounds. Add all the detail from the conference report and it would probably weigh more than thirty pounds. Perhaps the opportunity to exercise a veto over these items might offset this disadvantage. The more serious drawback is that itemization of appropriations bills substantially reduces the discretion of agency officials. The present practice of appropriating in lump-sum amounts gives them latitude to move money around within a large account. In some cases they have to obtain approval from their oversight committees, but they are willing to do that because of the flexibility they get in return. Itemization would lock all the details in. If agency officials wanted to change any of the items, they would need to pass another public law. Neither branch wants that.

Conclusions

This has been a rather brisk tour of a very complicated subject. I have not meant to champion micromanagement. Neither do I denigrate it. If I am shown what is being micromanaged, I will be able to analyze the reasons and come to some kind of judgment. Obviously there are risks and costs to congressional intervention. If agencies want to minimize what we blithely call meddling, they can play the game straight. If they choose to sabotage statutory programs and decide to implement White House policy instead of public law, the light for congressional intervention turns green.

The larger question is what can be done to foster a climate of good faith and trust that permits agencies to carry out laws with minimal congressional interference. When trust is broken and promises are not kept, Congress is stimulated to acquire new staff and impose new statutory restrictions. To regain control of the programs it has authorized and appropriated, Congress creates a rival bureaucracy. It is not the best of all worlds. We can all agree to that. The more difficult issue is how to make it better. Constructive steps can be taken. Decrying micromanagement by Congress is not one of them.

NOTES AND QUESTIONS

1. Strauss insists that the effort to locate agencies in one or another of the three branches is misguided and futile, and that so long as separation of powers principles govern the competition among the president, Congress, and the courts at the apex of government, agencies' powers and their relationship to each of the three branches at the apex should instead be evaluated according to the principles of separation of functions and checks and balances. How useful are Strauss's formulations of these principles in resolving specific conflicts either at the apex or at the agency level? For example, how would his analysis help the Court to decide the constitutionality of a statute creating an independent prosecutor not under the control of the Justice Department, the issue addressed in Morrison v. Olson, 487 U.S. 654 (1988)?

What does it mean to say, as Strauss does, that the principle, derived from article II of the Constitution, that the president must direct, coordinate, and balance the agencies does "not necessitate complete displacement of agency judgments or an assertion of White House expertise" and that the agency "remain[s] ultimately responsible for the decision"? Why doesn't the principle of presidential direction imply that the president may indeed substitute his judgment for that of the agency so long as the decision stays within the zone of discretion granted by Congress and the president is prepared to take political responsibility for it? Given the direction principle, why should the president's power to implement it be confined to disciplining agency officials after the fact? For proposals along these lines, see Lloyd Cutler and David Johnson, "Regulation and the Political Process," 84 *Yale Law Journal* 1395 (1975); and American Bar Association, Commission on Law and the Economy, *Federal Regulation: Roads to Reform* (Washington, D.C.: The Commission, 1979).

2. Who has the better of the exchange between Morrison and DeMuth-Ginsburg? Does Shapiro's article in Chapter III capture the essence of their disagreement, or do other issues divide them? Do they differ, for example, on the nature of rulemaking proceedings under the APA? How would Morrison respond to the DeMuth-Ginsburg argument that "communications that remain secret cannot determine the outcome of the regulatory process. The ultimate result of the rulemaking process must be a decision for which there is a rational and reasoned basis in the record." Could Congress, as Morrison suggests, require that all communications between OMB and executive branch agencies concerning pending rulemaking proceedings be reduced to writing and placed on the public record? Does your answer depend on the nature of the communi-

cation? The agency's use of it? Other factors? See, for example, Harold Bruff, "Presidential Management of Agency Rulemaking," 57 *George Washington Law Review* 553 (1989).

3. In the Bush administration, similar controversies arose concerning not only OMB review but also the activities of the Council on Competitiveness, chaired by Vice-President Quayle, which sought to provide regulatory relief to hard-pressed industries. See, for example, Kirk Victor, "Quayle's Quiet Coup," *National Journal,* July 6, 1991, p. 1676; Graeme Browning, "Getting the Last Word," *National Journal,* September 14, 1991, p. 2194.

4. In criticizing *Immigration and Naturalization Service v. Chadha,* Fisher likens the legislative veto to other congressional "shortcuts" such as legislating under a suspension of the rules or unanimous consent, and to other forms of indirect congressional controls on execution such as committee investigations. Are these analogies apt? Do these techniques contradict any constitutional text comparable in its explicitness to the Presentment Clause? Is it correct to argue, as Fisher seems to do, that congressional "micromanagement" is not a problem because it only occurs when a president fails to do his job properly? To return to the *Chadha* case itself, no one has ever figured out why Congress subjected Chadha, the immigrant, to a legislative veto; certainly the veto did not seem to implicate any policy concern in that case.

5. In effect if not form, the legislative veto remains alive and well—a fact that Fisher uses to argue that the now-invalid legislative veto actually strengthened the executive branch by increasing Congress's willingness to allow it discretion and flexibility. Indeed, although Fisher does not make the point, these post-*Chadha* improvisations between the two branches may be even *more* objectionable to opponents of the legislative veto because they effectively confer the veto on single committees rather than on Congress as a whole.

Barbara Hinkson Craig, a political scientist specializing in the politics of the legislative veto, disagrees with Fisher's generalized defense of the veto and his claim that the president benefited from it. She emphasizes that the veto's effects differed depending on its form (one-house, two-house, committee-level), congressional motives for enacting it, the policy arena, the intended target in the executive branch, and the site of congressional review. Craig contends that the legislative veto reflects Congress's mistrust of its own law as much as mistrust of the executive branch, and she points out that of the 125 actual exercises of the legislative veto, 66 rejected presidential impoundment of funds, 24 rejected reorganization plans, and only 35 rejected agency regulations, projects, or decisions. Finally, she notes that agencies, in alliance with congressional committees, could exploit the veto to oppose the president's policy. Craig, "Wishing the Legislative Veto Back: A False Hope for Executive Flexibility," in *The Fettered Presidency: Legal Constraints on the Executive Branch* (L. Crovitz and J. Rabkin, eds.) (Washington, D.C.: American Enterprise Institute, 1989).

For a proposal to create what the author regards as a functionally equivalent substitute for the legislative veto, see Stephen Breyer, "The Legislative Veto After Chadha," 72 *Georgetown Law Journal* 785 (1984).

6. Not surprisingly, political scientists have produced an enormous literature on the dynamics of congressional–executive branch relationships. Some of their recent findings may be quite surprising to the uninitiated.

For example, David Mayhew has shown, contrary to the conventional wisdom even among political scientists, that whether rule is divided (a president of one party, a Congress controlled by the other) or unified makes little difference to the frequency with which Congress enacts major legislation and conducts major investigations of the administration. Mayhew, *Divided We Govern: Party Control, Lawmaking, and Investigations, 1946–1990* (New Haven: Yale University Press, 1991).

Studies during the 1980s of congressional controls over administration shifted from the previously dominant paradigm, which had emphasized the ability of agencies to resist both congressional and presidential controls, to one emphasizing that elected officials can and do shape agency behavior in systematic ways, especially through the appointment power shared by the president and the Senate. For a review and synthesis of this work, see B. Dan Wood and Richard Waterman, "The Dynamics of Political Control of the Bureaucracy," 85 *American Political Science Review* 801 (1991).

Congressional oversight of agencies—its inquiry into whether agencies are complying with legislative goals—received much attention from political scientists during the last decade. In an influential article, Mathew McCubbins and Thomas Schwartz criticized the traditional view that Congress sadly neglected its oversight responsibilities. In fact, they argued, congressional behavior reflected a rational institutional preference for "fire-alarm" oversight of agencies over another, less effective form, called "police-patrol" oversight. In their model, Congress does not ordinarily conduct a centralized search for problems in the agencies but instead creates decentralized structures and incentives so that citizens and groups which can identify violations at the lowest cost will take the initiative to alert the appropriate committees. McCubbins and Schwartz, "Oversight Overlooked: Police Patrols versus Fire Alarms," 28 *American Journal of Political Science* (1984). A more recent study finds that in addition to fire-alarm oversight, police-patrol oversight is far more common and effective than earlier commentators had believed, and that oversight activities increased markedly since the early 1970s. Joel Aberbach, *Keeping A Watchful Eye: The Politics of Congressional Oversight* (Washington, D.C.: Brookings Institution, 1990). See also, Lawrence Dodd and Richard Schott, *Congress and the Administrative State* (New York: Wiley & Co., 1979), especially chapters 5 and 6.

7. During the 1980s, Congress devised some innovative mechanisms and administrative entities designed to overcome the well-known strategic behaviors that facilitate logrolling and often prevent Congress from giving coherent, disciplined policy direction to agencies. Perhaps the most notorious example is the system of controls established by the budget legislation. See generally, Kate Stith, "Rewriting the Fiscal Constitution: The Case of Gramm-Rudman-Hollings," 76 *California Law Review* 593 (1988); Paul Kahn, "Gramm-Rudman and the Capacity of Congress to Control the Future," 13 *Hastings Constitutional Law Quarterly* 185 (1986). Congress's imposition of extensive agency reporting requirements is another. See D. Roderick Kiewiet and Mathew McCubbins, *The Logic of Delegation: Congressional Parties and the Appropriations Process* (Chicago: University of Chicago Press, 1991), chapter 10. A third innovation is exemplified by the Defense Base Closure and Realignment Commission, which relocates the traditional burden of inertia to those who, for parochial or other reasons, seek to overturn the administrative recommendations. One commentator contends that such commissions are "an attempt to recreate outside the public eye the informal bargaining mechanism among branches that existed with the [political] parties." Michael Fitts, "Can Ignorance Be Bliss? Imperfect Information as a Positive Influence in Political Institutions," 88 *Michigan Law Review* 917 (1990), at 952–53. One legal scholar, impressed with the diverse investigative, reportorial, and even supervisory activities performed by criminal grand juries in the past, has suggested the creation of "administrative grand juries" consisting of lay citizens. Ronald F. Wright, "Why Not Administrative Grand Juries?" 44 *Administrative Law Review* 465 (1992). Is Congress likely to be sympathetic to such bodies?

4

Internal Controls: Management, Culture, and Professional Norms

The Management Side of Due Process

JERRY MASHAW

It is necessary to attempt to specify what is meant by "accuracy," "fairness," and "timeliness" of "adjudications" in the context of the social welfare claims process. "Adjudication" encompasses any determination of eligibility or amount of benefits at any stage of a social welfare claims process. This is a reasonably straightforward usage, but it is considerably broader than the lawyer's customary image of an adjudication as a decision made after a trial-type evidentiary proceeding. "Accuracy" involves the correspondence of the substantive outcome of an adjudication with the true facts of the claimant's situation and with an appropriate application of the relevant legal rules to those facts. Accuracy is thus the substantive ideal; approachable but never fully attainable. "Fairness" is the degree to which the process of making claims determinations tends to produce accurate decisions. That a decision is "timely" simply means that it was made within a reasonable or a statutorily prescribed period of time after presentation of the claim. . . . It is not, therefore, a significant overstatement to suggest that, from the traditional perspective of the legal system, adjudicatory processes which contain adequate procedural safeguards in hearings and appellate checks on initial decisions are considered self-correcting mechanisms for the accurate finding of facts and the authoritative application of law to fact.

This article argues to the contrary that the purposes, necessary modes of operation, and clientele of social welfare programs so severely limit the value of procedural safeguards and appellate checks in assuring accurate and timely adjudication of social welfare claims that there is a need for additional safeguards on the integrity of this very important segment of the administrative process. One such additional safeguard—a management system for assuring adjudication quality in claims processing, sometimes called a quality control or quality assurance system—will be described here in broad outline. The remainder of the discussion is concerned with the due process implications of such a management system and with the possibilities for its judicial imposition on certain social welfare programs as a matter of constitutional due process or statutory construction. The purpose of the discussion is not to provide a detailed analysis of the issues raised, but rather to stimulate new ways of thinking about due process of law in the context of social welfare programs.

Limitations of Trial-Type Hearings and Appeals in Ensuring Accurate, Fair, and Timely Adjudication of Claims

In adversary judicial proceedings, procedural safeguards and appellate review are generally viewed as the guardians of fairness and accuracy. In this context, problems of accuracy and fairness tend to cluster around two dominant issues. The first is the problem of designing systems in which a fair opportunity to contest does not result in an equal opportunity to obfuscate and to delay. The solution to difficulties of this type has been sought largely through adjustments in either the procedural rules or the evidentiary system, the adjective law governing the process of adjudication. . . .

The second problem has been the development of cost-allocation principles to ensure that the adversary process is generally available. . . .

There are, however, limits on the extent to which courts and legislators can refine adjective law without making it a disproportionate concern in litigation that purports to deal with matters of substance. And the wisdom of increasing the potency of adversaries through public subsidies is always open to serious question, unless litigation is viewed as an ultimate rather than as an instrumental end. Hence, it is not surprising to find that in recent years a search has begun for alternatives or additions to adjective law reform and subsidies as devices for improving the quality of adjudicative justice. This search is evidenced by a concern for the development of techniques of judicial administration to make the process of adjudication more efficient, and by an increasing willingness to view the adjudicatory process as one in which the positive management of cases and case flow to achieve accurate and fair results is an appropriate role for the adjudicator. This new focus suggests some movement away from passive judicial reliance on adversary processes and toward positive judicial management of adjudication. When dealing with adjudications of social welfare claims, a posture of positive management of the adjudicatory process to ensure quality is not only appropriate, as in the judicial system, but essential.

The Positive Focus of Social Welfare Claims Adjudication

Perhaps the most general consideration which supports a management strategy for assuring accuracy, fairness, and timeliness in social welfare adjudications is the positive focus inherent in the administration of programs involving benefits and compensation. . . .

The notion that claims adjudicators are engaged, not in providing a forum for the resolution of conflicts, but rather in the systematic and affirmative implementation of certain prescribed legislative policies is reflected in the nonadversary and informal procedures of most social welfare claims processes. . . . Even when the claimant is exercising appeal or de novo hearing rights after an initial denial of his claim, programs involving the payment of public funds employ a nonadversary procedure in which the government is not specially represented and recognize an obligation to aid the claimant in presenting his case.

The Necessity for "Informality"

. . .

Hearings as a means for monitoring the quality of initial eligibility determinations.   A threshold problem in using adversary process to produce high quality adjudication of social welfare claims is the necessity for relying upon claimant appeals or requests for de novo review as the mechanism for triggering the adversary process. Indeed, there are really two types of problems here. First, in such a system there will be no quality check through hearings when favorable action is taken on the claimant's request because, unless the claimant disagrees with the amount of the award, no appeal will be taken.

[Second, e]ven when the decision is negative, appeals leading to hearings are highly and mysteriously selective. . . .

. . . Administrative appeal or de novo hearing provides a very unsystematic check on the quality of initial adjudications of claims. The process of social welfare claims adjudication therefore necessarily places heavy reliance upon the initial claims adjudicator as a developer of facts, as a formulator of syntheses which subsume facts under relevant standards, and as a counselor to both the claimant and the unrepresented interests of the program he administers. Accuracy, fairness, and timeliness depend largely upon the competence and vigor with which the initial adjudicator performs these functions, as well as the function of ultimate decision.

Appeal hearings and adversary process.   To a limited degree the Supreme Court has recognized that a fully adversary hearing may be inappropriate even at the hearing stage of social welfare claims proceedings. In Social Security disability hearings, for example, no one appears to represent the agency. The Administrative Law Judge (ALJ) therefore has the obligation to develop facts adverse to the claimant if he considers such development necessary to a full and fair determination of the merits. Similarly, the ALJ aids an unrepresented claimant in presenting his case. And ultimately, he decides whether the claimant is entitled to benefits. In short, the procedure is inquisitorial rather than adversary, but the Supreme Court nevertheless rejected a combination-of-functions challenge to it in *Richardson v. Perales*. . . . No federal social welfare system currently provides representation for claimants as a matter of right, nor does it seem likely that a public program for providing attorneys (or other types of representatives) for all social welfare claimants at hearings could be instituted at a budget figure low enough to make the program attractive to agencies, legislatures, or the courts. . . .

Other disadvantages may also attend a fully adversary-type of proceeding. . . . Formality and punctiliousness take time—time that can have harsh consequences for the social welfare claimant who is awaiting a decision.

This is not to say that present nonadversary hearing procedures are faultless or that adversary procedure would produce no gains. . . . But there is little evidence to substantiate the proposition that increased formality or adversariness would improve social welfare adjudications. . . .

Two conclusions would seem to follow from this discussion: First, providing procedural safeguards or opportunities for appeal may be of limited value in ensuring fairness, accuracy, and timeliness in social welfare claims adjudications; and second, because increased or continued reliance on formal adversary procedure has limited value in this context, the assurance of accuracy, fairness, and timeliness should be pursued by other available means.

Quality Assurance in Adjudication as a Management Technique

A system for monitoring the performance of personnel and the quality of end products is such an obvious necessity in any large business enterprise that the failure to employ some method of quality control would be considered desperately poor, if not irresponsible, management. Moreover, a quality control system can be adapted to virtually any type of enterprise or end product for it involves merely the development of standards, the evaluation of performance against those standards, and action to upgrade substandard performance.

As straightforward and sensible as such a management device might appear, quality control programs having all these attributes have been used in only two federally funded or administered social welfare programs—the Social Security System (OASDI) and veterans' benefits programs. The experience in these programs provides a basic model for a quality control or quality assurance program and permits identification of the problems and limitations inherent in quality control techniques.

Standards and Techniques for Evaluation

What is an error? Development of standards and methods for the evaluation of accuracy, timeliness, and fairness in adjudication is not a simple matter. . . . The correctness of factfinding and the appropriateness of policy application often involve questions of judgment. . . .

The problem of second-guessing cannot be completely eliminated, but there are strategies for dealing with it which ameliorate its impact on the reliability of the evaluation system. One such strategy is to refine the evaluation format so that it differentiates between relatively clear errors and what might be termed "judgment deficiencies." A second is to attempt to force agreement between the initial adjudicator and the evaluator on whether an error has occurred. As explained below, the VA [Veterans Administration] uses both strategies. . . .

The system operates as follows: There is a daily first-line review of the total work product of most adjudication units. A random sample of all claims on which any action was taken by the unit on a given day is reviewed for both procedural and substantive error. The reviewer, who is attached to the regional office, corrects any error found, whether it involves the particular action on a claim taken that day (e.g., adding a new dependent to a veteran's file) or any action taken perviously which comes to his attention (e.g., the initial determination that a veteran had been honorably discharged and was therefore eligible for benefits). He also enters the numbers and types of errors found into a monthly report to the regional office. The report is then sent both to the national Office of Appraisal and back to the regional office.

The Office of Appraisal monitors these monthly reports, looking for trends. While the national office does not conduct a separate statistical review of a region's monthly output, it does review in depth the total operation of a regional station. As part of this review, the national office conducts another random sampling of the station's work product, looking for the same types of errors that the station checked for each day. The findings in the station's monthly quality control reports are then checked against the results of the national office's own review. In this way the office in effect reviews the station's statistical quality control (SQC) operations. If the variation between results of the office sample review and the findings expressed in the station's monthly report is statistically significant, that fact becomes part of the national office's report on the management performance of the regional station.

The portion of the first-line and national office reviews which is relevant for present purposes is the qualitative review. Each case which is selected for review is evaluated for: (1) substantive error (errors leading to an incorrect result), (2) judgment deficiency (errors in the development of a claim file and cases in which the reviewer thinks a different result more tenable than the one reached), and (3) procedural discrepancies (errors which do not affect basic entitlement). Each of these categories is broken down on a standard form into a series of subheadings and specific inquiries. . . . Although the problem of second-guessing judgments is not eliminated, a quality assurance system like the VA's has the capacity to identify variance from the institutional judgment of the agency. And within a closed system, uniformity and accuracy tend to merge.

A variation on this approach may also help to deal with the possible divergence between record and objective reality. The categories of errors available for assignment by VA quality reviewers under the rubric of judgment deficiency include such things as failure to request a needed medical examination or piecemeal development of the claims record. In short, the reviewer is saying that for one or several reasons the case record has been so inadequately developed that a substantive error may well have occurred. This sort of evaluation can be done without a costly and time-consuming de novo development of the case by the reviewer.

The agency should not be wholly content with this sort of record review. An attempt should be made by the agency to validate its claims policies and procedures by checking actual claims adjudication performance against an external standard. This approach is embodied in the Social Security Administration's Evaluation and Measurement System. . . .

How many "inaccuracies" are too many? Standards for accuracy, such as "permissible errors per hundred cases," are also difficult to develop. . . . Regardless of approach, the important point is that supervisory personnel must have accurate information on error rates and types of errors before they can identify problems and take action to make improvements.

The evaluation of fairness. . . . When adjudications in a particular program involve marginal cases or matters of judgment, and when quality assurance evaluations are made solely on the basis of record evidence, management evaluation and control of accuracy are necessarily tenuous. Fairness in these contexts must be appraised by an independent evaluation of the process elements of adjudication. A supplementary check on fairness should be directed at those adjudicatory procedures and routines which are meant to place the relevant facts, policies, and arguments before the adjudicator and to facilitate sound decisionmaking—things such as case development effort, articulation of the bases for decisions, adequate notification of actions to the claimant, and explanation of opportunities for appeal. . . .

Timeliness as a quality factor. . . . The possibility of creating incentives for speedy but otherwise poor quality adjudications is obvious. However, this is a problem which can be dealt with by formulating a sensible and sensitive personnel policy, by adopting an evaluation system which reviews all the elements of adjudication quality, accuracy, fairness, and speed, and by refining the statistical analysis of processing time so that it reveals "creaming" of easy cases to meet timeliness goals.

Continuous Evaluation

Statistical reporting systems. Because effective management requires the ability to perceive trends in adjudication performance and to relate quality data to program changes and exogenous factors which influence program performance, positive case load management implies a continuous monitoring function. . . . When properly employed, statistical sampling improves a quality assurance program. Sampling reduces the cost of continuously monitoring quality and tends to focus attention on important concerns, e.g., the delineation of distinct elements in the adjudication process and the contribution of each element to high quality end products. The assembly of sample data discloses patterns of errors and permits the agency to distinguish random and essentially uncontrollable errors from recurrent errors of a similar type or made by a particular adjudication unit. . . .

Special studies. A well-designed statistical quality assurance [SQA] system can provide a continuous flow of information concerning the quality of adjudications. But the system cannot provide all the information necessary for effective management of the adjudicatory process. For one thing, the data collected by such sampling techniques must often be assembled by computer.

Hence, the information must be limited to that which can easily be encoded. This is likely to produce tabulations of the incidence of error but little information on its causes. Hence, the SQA system often merely alerts the agency to apparent problems which must be investigated further in order to determine whether a problem is real and if so, what should be done about it.

Moreover, the reliability of sample data decreases with the size of the sample. For example, a sample that is drawn to generate reliable information concerning a regional office may be unreliable with respect to a particular adjudication unit within that region. . . .

The Social Security Administration's use of special studies to validate its case development procedures and reveal errors which are effectively hidden when reviewing only the case record has already been described. A similar issue that should be of concern to agencies is whether inaccuracies, unfairness, and tardiness are randomly distributed among claimants. . . . The only technique for analyzing this aspect of adjudication quality is the special study.

The Use of Quality Assurance Information

There are two major requisites for a successful quality assurance program. The first is that the collection of information on the quality of adjudications not be subject to the control of the adjudicators whose product is being evaluated. The second is that the information be developed in such a way that it is useful to and used by those in charge of improving adjudicative performance. These considerations suggest that considerable care must be taken to ensure the independence of the quality assurance staff without pushing them into a detached position in the agency from which they, and their evaluations, have little influence on policy.

Quality Control and the Constitution

The Potential Effects of Good Management on the Requirements of Due Process

. . . Proof of the existence or nonexistence of a sound quality control mechanism should be influential when a court is called upon to balance claimant need for the protections of trial-type hearings against the costs of imposing adversary process at a particular stage of a social welfare claims process. The availability of the alternative protection of good management, which should, of course, show up in the error rate identified by the quality control mechanism, both lessens the need for the protections afforded by a hearing and tends to substantiate the administrative claim that the use of adversary process is an unnecessary, and a potentially costly and time-consuming addition to a process which is already carefully structured to implement a positive program for the protection of the claimant's substantive economic interests.

Quality Control as a Judicial Remedy to Ensure Due Process of Social Welfare Law: Emphasis on AFDC

If, as seems sensible, the underlying integrity of the social welfare claims process should have a bearing on the need for trial-type hearings as a means for assuring "fundamental fairness" to claimants, it seems equally sensible to suggest that the realistic prospects that such hearings or appeals will protect claimants should also affect the appreciation of what due process requires in the context of social welfare programs. *Goldberg v. Kelly* made much of the critical situation of the claimants and emphasized the need to tailor the hearing process to their capacities. The logical and limited extension of that principle is that . . . if hearings cannot provide reasonable assurance of accurate adjudication of claims in a social welfare program—and in AFDC [Aid to Families with Dependent Children] there is substantial reason to believe that they cannot—then there should be judicial imposition of a comprehensive quality assurance program to bolster that program's capacity for high quality adjudicative performance.

The limited value of trial-type hearings to AFDC claimants. To summarize, hearings protect claimants against unfair and inaccurate decisions only on the assumptions: (1) that the claimant is aggressive, knowledgeable about the program, and skillful in developing and presenting facts (or has access to those who are); and (2) that the hearing system works. Neither of these assumptions has proven to be realistic with respect to AFDC hearings, and there are reasons to have grave doubts about both. If these doubts have substance, due process requirements which seek to provide fair opportunities for claimant-initiated challenges to agency determinations are focusing on only a small part of the problem. . . .

Judicial imposition of management-oriented remedies. Due process should require greater protection for the claimant, and such protection should include the application of systematic management techniques which will discover errors, identify their causes and implement corrective action. This is suggested, not only by the inadequacies of the hearing process, but also by the protective purposes of AFDC and other social welfare programs. There is, indeed, a sense in which the imposition of a basic quality control program as a part of due process might be viewed as a lesser constraint on legislative and executive judgments concerning the administration of these programs than is the imposition of judicially delineated hearing requirements. The former remedy is at least consonant with the avowedly paternalistic objectives of the program, suggests solutions that are managerial rather than legalistic, and avoids adversary postures that are generally considered inappropriate by program professionals. . . .

Yee-Litt [*v. Richardson*] suggests that at some point a court will be willing to examine questions of whether poor practices or management are a serious impediment to the realization of constitutional rights and to take action which limits administrative discretion about the effectiveness of program policies.

Although the judicial response in *Yee-Litt* merely reinforced the traditional requirement of trial-type hearings, courts have been more management-oriented when reviewing the performance of welfare program functions other than fair hearings. Judicial orders have, for example, required the development of affirmative action plans or procedures to deal with the loss or theft of welfare checks, to advise clients of their rights under the NOLEO provisions, to eliminate sexual discrimination in the Work Incentive program (WIN), and to prevent delayed delivery of welfare checks.

. . . In the process of adjudicating claims of violations of the Eighth and Fourteenth Amendments, the courts have been exposed to the detailed working of prisons, youth correction centers, and even homes for the mentally retarded. Moreover, having once found custodial care in these institutions to be inadequate or improper under the Constitution's broad prohibitions against cruel and unusual punishment or the denial of due process and equal protection, the district courts have issued orders of remarkable scope and specificity.

Assuming that innovative remedial responses are possible, the outline of a lawsuit requesting the imposition of standards of administration which have a reasonable prospect of producing fair and accurate results in AFDC eligibility determinations might look something like this: (1) demonstrate the tendency of the program to make a substantial number of errors; (2) establish a duty to avoid those errors; (3) demonstrate that current management practices are inadequate in light of the error rate and the administrative duty; and (4) propose remedial requirements which will help alleviate the problem. . . .

Conclusion

It may properly be objected that this article looks at management solutions to problems of adjudication and the prospects for their judicial imposition on wayward administrative agencies through rose-tinted glasses. Even so, there are a number of advantages in thinking about due process in social welfare systems in management terms.

The first advantage is that such thinking begins to focus attention on the realities of the protection that can be expected from hearings. . . .

Additionally, viewing due process as potentially requiring a management system for assuring the quality of social welfare claims adjudications begins to translate the legal issue of fundamental fairness into terms which are meaningful to the people who administer social welfare programs. . . .

Finally, a management approach to due process suggests that the arguments against the introduction of adversary procedure in social welfare claims systems do not necessarily lead to the conclusion that the judiciary should treat these claims processes as peculiarities which do not respond to constitutional commands for due process. Rather, those arguments may lead to evaluation of what process is due the social welfare claimant in the social welfare system's own terms.

"Culture" and "Compliance"
JAMES Q. WILSON

Organizational Culture

Every organization has a culture, that is, a persistent, patterned way of thinking about the central tasks of and human relationships within an organization. Culture is to an organization what personality is to an individual. Like human culture generally, it is passed on from one generation to the next. It changes slowly, if at all. . . . Philip Selznick likened the creation of "organization character" to character formation in an individual: A viable organization is not merely a technical system of cooperation (any more than an individual is merely a mechanism processing food and sensations); it is an institution that has been "infused with value" so that it displays a "distinctive competence" (or a distinctive incompetence!). An organization acquires a distinctive competence or sense of mission when it has not only answered the question "What shall we do?" but also the question "What shall we be?"

There are some difficulties with viewing organizations as culture. One mistake is to assume that an organization will have *a* culture; many, perhaps most, will have several cultures that are often in conflict. The culture of the United States Navy is very different depending on whether you are assigned to submarines, aircraft carriers, or battleships. Another is to give so much emphasis to culture—that is, to the subjective states of organization members—that one loses sight of the objective conditions of organizational participation. . . . Organizational culture admittedly is a vague concept, but no less real than concepts such as national culture or human personality. We find it hard to explain how exactly Italians differ from Germans or introverts from extroverts, but we do not doubt that there are important differences.

The predispositions of members, the technology of the organization, and the situational imperatives with which the agency must cope tend to give to the organization a distinctive way of seeing and responding to the world. This is especially the case when the organization's stated goals are vague. When, as is often the case, these factors produce different definitions of core tasks for different people (or, more typically, different subunits), the organization will have several cultures. . . .

Culture and Mission: A Summary

Every organization has a culture, many have several. When a single culture is broadly shared and warmly endorsed it is a mission. The great advantage of mission is that it permits the head of the agency to be more confident that operators will act in particular cases in ways that the head would have acted had he or she been in their shoes. There are fewer distortions in the flow of information because both the sender and the recipient of the message share common understandings. FBI agents behaved as if J. Edgar Hoover were looking over their shoulders in part because the agents believed that was the right way to behave. Field representatives of the Army Corps of Engineers, office managers in the Social Security Administration, and district rangers in the Forest Service also act as they do in part as if they felt the ghosts of Sylvanus Thayer, Arthur Altmeyer, and Gifford Pinchot looking over their shoulders. Of course, there are no ghosts; indeed, many members of these agencies may never have heard of these people. But the ethos shaped by these men has been embodied in the expectations of living coworkers and superiors, thereby giving organizational life to long-dead and dimly remembered founders.

The importance of that ethos becomes clearest when it begins to decay. This may happen because of the arrival of groups in the organization that have a different occupational or professional culture (for example, the Forest Service), or it may result from the decisions of new leaders who by plan or by accident destroy the old sense of mission or attempt to replace it with one that may not be well-adapted to organizational needs or cannot elicit the enthusiasm of the operators. . . .

The advantages of a clear sense of mission are purchased at a cost. Tasks that are not defined as central to the mission are often performed poorly or starved for resources. Subordinate cultures may develop around these peripheral tasks, but promotional opportunities for members of these cultures may be so restricted that the ablest members will avoid assignment to these subunits because service in them is "NCE" (Not Career-Enhancing). . . . A strong sense of mission may blind the organization to changed environmental circumstances so that new opportunities and challenges are met with routinized rather than adaptive behavior. But even short of occasions for major organizational change, the perceptions supplied by an organizational culture sometimes can lead an official to behave not as the situation requires but as the culture expects. . . .

Outputs—work—may be hard to observe because what the operator does is esoteric (for example, a doctor performing a diagnosis or a physicist developing a theory) or because the operator acts out of view of the manager (e.g., a police officer handling a family quarrel or a ranger supervising a forest). If operator actions are esoteric or unobserved, the problem of moral hazard arises: the operator may shirk or subvert. Outcomes—results—may be hard to observe because the organization lacks a method for gathering information about the consequences of its actions (e.g., a suicide-prevention agency may actually prevent suicides but it has no way of counting the number of potential suicides that did not occur); because the operator lacks a proven means to produce an outcome (e.g., prison psychologists do not know how to rehabilitate criminals); because the outcome results from an unknown combination of operator behavior and other factors (e.g., a child's score on a test reflects some mix of pupil intelligence, parental influence, and teacher skill); or because the outcome appears after a long delay (for example, the penalty imposed on a criminal may lead to a reduction—or even an increase—in the offender's behavior five years later). I realize, of course, that what constitutes an outcome is a matter of judgment. Is the outcome of the work of the U.S. Employment Service referring an unemployed person to a job, having any employer actually hire the person, or helping the person develop a meaningful, long-term career? In what follows I shall refer chiefly to effects that approximate the most operational (or least vague) statement of the agency's goals.

Observing outputs and outcomes may be either difficult or easy. Taking the extreme case produces four kinds of agencies: Agencies in which both outputs and outcomes can be observed; agencies in which outputs but not outcomes can be observed; agencies in which outcomes but not outputs can be observed; and agencies in which neither outputs nor outcomes can be observed. For reasons that I hope will become clear as we proceed, I have called the first kind of agency a *production* organization, the second a *procedural* organization, the third a *craft* organization, and the fourth a *coping* organization.

Production Organizations

Where both outputs (or work) and outcomes are observable, managers have an opportunity to design (within the limits established by external constraints) a compliance system to produce an efficient outcome. . . .

The existence of observable outputs (the activities of auditors, letter sorters, and claims processors) and observable outcomes (taxes collected, mail delivered, checks received) simplifies the managerial problem. But that is not to say that managing these agencies is easy. Though the OASI [Old Age and Survivors Insurance] work of the SSA [Social Security Administration] in principle is simple and measurable, the law and rulings that it administers are incredibly complex. . . .

A problem that confronts the managers of all production agencies is that by plan or inadvertence they may give most of their attention to the more easily measured outcomes at the expense of those less easily observed or counted. There is a kind of Gresham's Law at work in many government bureaus: Work that produces measurable outcomes tends to drive out work that produces unmeasurable outcomes. . . .

The employees of production-oriented agencies are not indifferent to the management systems with which they must cope. They will try on occasion to fudge the numbers by which they are evaluated, either out of a desire to shirk (that is, minimize effort) or subvert (that is, produce outcomes other than the measured one). . . .

In no agency was the "stat game" played with greater zeal than in the FBI during J. Edgar Hoover's tenure as director. . . . By the 1970s, a bureau survey disclosed that 60 percent of the cases that the FBI presented for prosecution to local U.S. attorneys were being declined. In many cases the reason for the declination was that the case was too trivial to warrant prosecutorial effort. . . .

The error made by . . . the FBI was to define outcome too narrowly, so that only some but not all of the desired results were being observed and counted. . . . This error is made all too easily in any organization, but especially in government agencies where outcomes are not measured by sales in a market of voluntary transactions. . . .

Procedural Organizations

When managers can observe what their subordinates are doing but not the outcome (if any) that results from those efforts, they are managing a procedural organization. . . .

Perhaps the largest procedural organization in the government is the United States Armed Forces during peacetime. Every detail of training, equipment, and deployment is under the direct inspection of company commanders, ship captains, and squadron leaders. But none of these factors can be tested in the only way that counts, against a real enemy, except in wartime. This is especially true of those weapons systems designed to deter nuclear war. . . .

The conditions that define a procedural bureaucracy seem to make it ripe for management in ways that encourage the development of professionalism. What better way, one might ask, to manage organizational activities, the outcomes of which cannot be observed from any administrative perch, than by recruiting professionals to do the work in accordance with the highest professional standards? These standards would constrain the practitioners to put the client's interests ahead of their own and to engage in behavior that is most likely to produce the desired outcome. This sometimes happens (as in the case of better-run mental hospitals), but more often it does not. The reason, I believe, is that a government agency cannot afford to allow its operators to exercise discretion when the outcome of that exercise is in doubt or likely to be controversial. . . . Putting the fig leaf of professionalism over the nakedness of unknown outcomes will not fool anybody.

In short, because it is constraint driven, management becomes means-oriented in procedural organizations. *How* the operators go about their jobs is more important than whether doing those jobs produces the desired outcomes. . . .

In a procedural organization, standard operating procedures (SOPs) are pervasive. Popular accounts of service in the peacetime army or navy are replete with stories about rules and procedures. ("If it moves, salute it; if it doesn't move, pick it up; if it is too big to pick up, paint it.") . . .

Craft Organizations

In wartime, many army and navy units change from procedural to craft organizations. Whereas formerly their members acted under the direct gaze of managers (marching on parade, practicing on the rifle range, maneuvering in convoys), now they fight in the haze, noise, and confusion of distant battle-fields. . . . But wartime commanders do learn (usually rather quickly) whether those battalions won their engagements. A craft organization consists of operators whose activities are hard to observe but whose outcomes are relatively easy to evaluate. . . .

Many investigative agencies practice goal-oriented management. Detectives in a police department are evaluated on the basis of crimes solved, not procedures followed. Attorneys in the Antitrust Division of the Justice Department exercise substantial independence in initiating and developing cases. The division can afford to give them this freedom because higher-level managers review the final report and decide whether it constitutes grounds for prosecution. . . .

There are scarcely any operators in any federal agency whose daily work is harder to observe than the field representatives of the Army Corps of Engineers. They work for a year or two in remote locations, overseeing the building of military bases or other facilities. . . . Outputs—the daily work of the engineers in the field—could not be centrally directed. But outcomes could be evaluated: It was easy to learn whether the air base was built on time and within budget, and it was not too difficult to decide whether it was built according to specs. . . .

It is just this combination of self-taught or professionally indoctrinated skills and group- or profession-induced ethos that justifies calling such agencies craft organizations. . . .

Coping Organizations

Some agencies can observe neither the outputs nor the outcomes of their key operators. A school administrator cannot watch teachers teach (except through classroom visits that momentarily may change the teacher's behavior) and cannot tell how much students have learned (except by standardized tests that do not clearly differentiate between what the teacher has imparted and what the student has acquired otherwise). Police officers cannot be watched by their lieutenants and the level of order the officers maintain on their beat cannot readily be observed or, if observed, attributed to the officers' efforts. Some of the activities of diplomats (for example, private conversations with their counterparts in a foreign government) are not observed and many of the outcomes (for example, changes in foreign perceptions of U.S. interests or in foreign attitudes toward U.S. initiatives) cannot easily be judged.

The managers of these agencies must cope with a difficult situation. They can try to recruit the best people (without having much knowledge about what the "best person" looks like), they can try to create an atmosphere that is conducive to good work (without being certain what "good work" is), and they can step in when complaints are heard or crises erupt (without knowing whether a complaint is justified or a crisis symptomatic or atypical). . . .

Where both outputs and outcomes are unobservable there is likely to be a high degree of conflict between managers and operators in public agencies, especially those that must cope with a clientele not of their own choosing. The operators will be driven by the situational imperatives they face—the teachers' need to keep order in the classroom or the of cers' desire to create order on the street or restore order in the quarreling fa ily. The managers will be driven by the constraints they face, especially the need to cope with complaints from politically influential constituencies. Complaints can be rejected when the manager can show that the complained of behavior did not occur or they can be partially deflected when the manager can argue that the outcomes achieved justified the action in question. But coping agencies are precisely those that do not know with confidence what behavior occurred and cannot show with persuasiveness what outcomes resulted. And so managers, depending on their personal style, cope with the complaints as best they can. . . .

In coping organizations as in procedural, management will have a strong incentive to focus their efforts on the most easily measured (and thus most easily controlled) activities of their operators. They cannot evaluate or often even see outcomes, and so only the brave manager will be inclined to give much freedom of action to subordinates.

The subordinates, of course, are not without their own resources. Some will conform their behavior to whatever is being measured ("they want stats, we'll give 'em stats"); others will subvert the management strategy by ignoring the measured activities (thus jeopardizing their own chances for advancement) or by generating enough stats to keep management happy while they get on with their own definition of what constitutes good work.

Legality, Bureaucracy, and Class in the Welfare System

WILLIAM SIMON

When lawyers confronted the welfare system in the 1960s, they charged it with oppressive moralism, personal manipulation, and invasion of privacy. They focused attention on the "man-in-the-house" rules that disqualified families on the basis of the mother's sexual conduct and the "midnight raids" in which welfare workers forced their way into recipients' homes searching for evidence of cohabitation.

When I represented welfare recipients from 1979 to 1981, the workers showed little interest in policing their morals or intruding on their private lives. The "man-in-the-house" rule and the practice of unannounced or night-

Reprinted by permission of The Yale Law Journal Company and Fred B. Rothman & Company from *The Yale Law Journal* Vol. 92, pp. 1198–1269.

time visits had been repudiated. Yet the pathologies emphasized by the lawyers of the 1960s seemed to have been mitigated at the cost of exacerbating others that were in some respects their mirror images: indifference, impersonality, and irresponsibility. The new pathologies were typified by cases in which newly arrived Cuban refugees were denied assistance because they could not produce appropriately certified copies of birth certificates for their children or in which people who sought assistance from the wrong worker were sent away without explanation, thinking mistakenly that they were entitled to nothing. . . . If the literary personification of the pathologies of the old regime was Dostoyevsky's Grand Inquisitor, with his relentless intimacy and psychological omnipotence, the personification of the pathologies of the new regime is Kafka's Doorkeeper, who stands, passive and inscrutable, before the door to the Law and announces only when it is too late, "this door was intended for you. I am now going to shut it." . . .

. . . Since the 1960s, the social work profession has ceded to the professions of law and management both its ideological custody of the welfare system and its preeminent role in its administration. The concerns of law and management have converged in the three basic themes of the recent literature and practice of welfare administration: first, the formalization of entitlement, by which I mean the formulation of the eligibility norms as rules; second, the bureaucratization of administration, by which I mean the intensification of formally hierarchical organization; and third, the proletarianization of the work force, by which I mean the diminution of the status, skill, education, and reward associated with the front-line welfare worker's job. . . .

Although its primary purposes are historical and critical, there is a prescriptive theme running through this essay. I suggest that a vision of legality and organization opposed to the dominant vision is more plausible as an ideal and as a guide to reform in welfare administration. This opposed vision involves a legality of relative informality, a decentralized enforcement structure, and a corps of enforcers with some of the attributes of skill, education, and status associated with professionalism. This vision is in many respects similar to the understanding of welfare legality and enforcement of the New Deal social workers who founded the modern public assistance system. . . .

The Transformation of Public Assistance

The trends of formalization, bureaucratization, and proletarianization in public assistance are exemplified by the recent history of the AFDC [Aid to Families with Dependent Children] program. . . .

The Reformulation of Substantive Norms

The substantive eligibility norms have been revised to increase their substantive formality, that is, to give them the form of rules rather than of standards. There are three aspects to this distinction.

First, rules are more explicit than standards. . . .

Second, rules constrain judgment more than standards by limiting the number of factors to be considered in decisionmaking. . . .

The third aspect of the formalization of substantive norms concerns the style of judgment and the attitude prescribed for the decisionmakers. . . .

. . . The formalization of recent years . . . embraces mechanical judgment and literalistic interpretation. At the same time, it implicitly prescribes an attitude of impersonality. It seeks to alienate the worker from the purposes of the norms she enforces. There are two versions of this alienation; in one the worker is unaware of the purposes underlying the norms she enforces; in the other the worker feels bound to apply the norms in spite of their purposes. . . .

A cumulative effect of the three aspects of formality is that the system appears to require fewer and simpler substantive judgments and to require less information from the client about her personal circumstances. Another effect is that assistance is less sensitive to the circumstances of the applicant.

The Reconception of Proof

Under the old regime, the dominant view of the factfinding phase of the eligibility process focused on the worker's assessment of the credibility of the claimant's oral statements. . . .

Since then, however, this view has been rejected in favor of one that emphasizes the claimant's responsibility for verifying eligibility through documents and that portrays the worker's role as passively policing the claimant's compliance with documentation requirements.

The Advent of Quality Control

Another important development is the change in the character of organizational review and supervisory procedures. Under the old regime, two such procedures were emphasized. One was relatively constant and direct personal supervision of the worker by a supervisor in the same office. The relationship was conceived as a collegial one between professionals, albeit professionals of unequal experience and authority. Supervision involved both review of written case records by the supervisor and oral consultation in which the supervisor explained her judgments and the worker was encouraged to participate actively and respond to the supervisor's comments. The goal was an informal consensus regarding the appropriate disposition of the case. The process was supposed to involve an intense scrutiny of both the applicant's situation and the worker's response to it. . . . Review was done by a person with extensive knowledge of the worker, the case, and the immediate context of decision, and it was done in a manner facilitating participation by the worker.

Supervision was supplemented by a variety of procedures, the most important of which was the federal audit. Periodically, HHS [Health and Human

Services] would review the records of a sample of the caseload. It would disallow reimbursement for the federal share of improper payments to the reviewed cases. When the review disclosed a pattern of mistake or lawlessness, HHS pressed the state to change its policy and instruct its workers appropriately. Efforts to affect the program through the audit appear to have been largely concerned with major policy matters.

Beginning in the 1960s, the federal audit was supplemented and eventually replaced by procedures called Quality Control. Aside from their zealous invocation of the jargon of business management, the new procedures differ from the old in three respects. First, they use statistical methods to designate the sample of cases for review and to derive from the reviews an estimate of the total number of the state's cases and the total amount of its payments in "error." Second, since 1973, these estimates have been backed by the threat of "fiscal sanctions" in the form of withheld federal reimbursement for a portion of the estimated erroneous payments in the entire caseload, not just the cases reviewed. Third, Quality Control involves a more intensive and detailed review of the sample cases than the old procedures. In response to the federal system, the states have developed their own supplementary quality control processes, which make similar but less intensive reviews of larger samples of cases. These supplementary reviews are designed to lower error rates in advance of the federal reviews. . . . Whether or not AFDC Quality Control has made any contribution to the measurement of the welfare system's performance is still an open question, but it has clearly contributed to the system's transformation.

First, Quality Control has reinforced the trend toward formalization of eligibility norms. Relatively complex eligibility schemes relying on standards generate more quality control errors than relatively simple ones relying on rules. . . .

Second, Quality Control has intensified organizational hierarchy. It has created corps of elite officials charged with supervision of lower-level decisionmaking. . . . Quality Control reviewers make judgments with little opportunity for participation by the worker or supervisor and with little or no explanation to them.

Third, Quality Control has reinforced the trend toward increased documentation requirements. The enforcement of some documentation requirements reduces error by providing more accurate information regarding substantive eligibility. However, a major portion of the documentation efforts of some states in recent years has been devoted to procedural or "paper" errors unrelated to substantive eligibility. . . .

Fourth, by emphasizing "positive" errors (overpayments and payments to ineligibles) and deemphasizing "negative" errors (underpayments and wrongful denials or terminations), Quality Control may shift costs of the transformed eligibility process to recipients. . . .

Perhaps the most important adverse impact on claimants arises from the fact that Quality Control does not measure failures of compliance with rules that provide important protections for claimants. . . .

The Expulsion of the Social Workers, the Advent of the Clerks

The principal personnel goal of the old regime was administration at both upper and lower levels by professional social workers. The old regime emphasized three aspects of the front-line worker's job which called for the abilities of a professional social worker. First, as we have seen, financial assistance was assumed to involve relatively complex, particularized judgment. Second, financial assistance was linked to counseling, a task requiring intimate and concrete knowledge of the client and her circumstances. AFDC was originally seen as a program to enable mothers to stay home with their children, and counseling was prescribed to improve the recipient's ability as a parent and homemaker. Later, the program's purposes were reconceived, and counseling was prescribed to enhance recipients' capacities for self-support in the labor market. Third, an important part of the worker's job in this conception was the mobilization of community resources available outside the public assistance program to serve the particular needs of the recipient. The social worker was "among other things, a guide through a kind of civilized jungle, made up of specialized agencies and service functionaries the citizen can hardly name, let alone locate." At a minimum, this involved advice and referral; more ambitiously, it involved active participation in securing assistance for the client from outside agencies. The social work profession developed an elaborate body of psychosocial and pedagogical theory to educate people for these tasks.

This personnel goal was never achieved. Only a small minority of front-line workers under the old regime had any social work training, and only a handful had attained the basic professional credential, the master's degree in social work. Nevertheless, social work professionalism had a significant influence on the system. . . .

In the 1960s and 1970s, the ideal of social work professionalism in public assistance administration was repudiated, and the actual influence of the social work profession was almost entirely eliminated. . . .

The most important event in this development was the administrative separation of social services from financial assistance, which HHS encouraged in the late 1960s through federally sponsored demonstration projects, and mandated by regulation in 1972. Proponents explained this reform as a way of reducing the danger that services would be coercively imposed on the client as a condition of financial assistance and as a way of increasing administrative efficiency through specialization. A consequence of this separation was that nearly all front-line workers with qualifications or aspirations as professional social workers left financial assistance to work in social services, which involved the professionally most prestigious and challenging jobs. Financial assistance jobs were then restyled as clerical. . . . Analogous changes in the upper tiers of state welfare administrations increased the influence of technocrats at the expense of social work professionals.

A critical result of these developments has been the expulsion from the system of people socialized to think of the front-line worker's role as involving complex, particularized judgments and direct responsibility to claimants as individuals. They have been replaced by people socialized to think of the role as characterized by routine, unreflective judgment and responsibility only to hierarchical organizational authority.

Division of Labor

The old regime's conception of the welfare worker's job has been decomposed into many more specialized functions—"minijobs" as one management consultant puts it. The separation of social services and financial assistance is merely one example of this trend. . . .

The general tendency of these developments is to limit workers' understanding of the way the welfare system works and of claimants and their circumstances.

Mechanization

A variety of mechanical and computer systems have been introduced to mechanize the eligibility process. . . . For our purposes, the most significant effect of mechanization has been to reinforce or facilitate some of the trends described above.

First, mechanization requires formalization. Eligibility determination can only be automated when norms are explicit and judgment, constrained. Second, mechanization facilitates centralization. A variety of functions previously performed at the local office level have been centralized through computers. . . . Third, mechanization has encouraged intensification of the division of labor, since the machinery brings new work roles with it. Existing work must be redivided and reapportioned among the old and new roles.

Productivity Enforcement

Quality Control has been supplemented by productivity regimes designed to measure the quantity of work and facilitate the sanctioning of low producers. Under the old regime, appraisal of productivity was relatively informal and decentralized. It was integrally related to the appraisal of quality and was largely the responsibility of the local office supervisor. The basic measure of productivity was the caseload, the number of recipients or families for whom the worker was responsible. The managers of the new regime have rejected this approach for a number of reasons. For one thing, the caseload measure depended on the system of work assignment by which workers retained responsibility for particular claimants throughout their stay in the program. This system limited the flexibility of managers in assigning work. Second, the breadth of the caseload measure limited the possibilities of hierarchical control since it did not enable the manager to monitor specific activities within the case handling job. Third, the semicollegial nature of the supervisor worker relation was thought to inhibit the supervisor from vigorous application of coercive sanctions.

The newer productivity measurement systems are more elaborate. . . .

The basic purpose of such regimes is to enhance centralization and limit worker and local office autonomy. . . . The productivity requirements contributed to the further attenuation of the worker's sense of responsibility to the client by encouraging workers to threaten clients in order to induce them to comply more promptly with documentation requirements and by implicitly penalizing time spent responding to client requests.

Effects

. . . Putting aside for the moment a variety of conceptual problems and acknowledging the limitations on information, two tentative generalizations seem plausible.

First, there is no reason to believe that the administrative transformation of the system has made any major contributions either to the liberals' goal of increasing the participation rate—the percentage of the substantively eligible population that actually receives benefits—or the conservatives' goal of reducing inappropriate expenditures.

There does seem to have been a substantial increase in the participation rates in public assistance programs, and particularly AFDC, in recent decades. Much of the increase, however, seems attributable, not to administrative reform, but to the increased availability of information and advocacy resources outside the welfare system through informal transmission, community groups, and subsidized legal services and to declining cultural inhibitions about accepting welfare. . . .

The achievements of administrative reform with respect to cost reduction also remain to be demonstrated. Payment error rates have decreased substantially, but as we have seen, reducing payment error does not necessarily lower grant expenditures (as in the case of "paper" errors), and error reduction has been achieved only at substantially increased administrative expense. Efforts to compare these added expenses with savings from error reduction have been insubstantial and inconclusive. And although a few specific reforms, such as some versions of periodic income reporting, appear to have reduced grant expenditures by more than their administrative costs, these savings have sometimes been achieved only by imposing large costs on eligible claimants.

Second, the transformation of the system seems to have had an important effect on the way claimants and workers experience the system and their relation to each other. The reforms seem to have reduced the claimant's experience of oppressive and punitive moralism, of invasion of privacy, and of dependence on idiosyncratic personal favor. But they also have reduced their experience of trust and personal care and have increased their experience of bewilderment and opacity.

As far as the worker's role is concerned, the reforms have consistently reinforced experiences of alienation and subordination. Moreover, they appear to have encouraged the worker to view her own interests as in constant conflict with the claimant's. . . .

The Sociological Jurisprudence of Welfare Reform

. . . The dominant vision of contemporary welfare jurisprudence associates social goals such as efficiency, freedom, and fairness with a legality of formality and in turn with a bureaucratized enforcement organization staffed at the lower tier by proletarians. . . .

Rules, Hierarchy, Proletariat

In the early 1960s, the dominant vision was principally associated with liberals concerned with the interests of the poor; more recently it has been co-opted by conservatives concerned with cost-cutting and disciplining the public work force. The rhetoric of both groups has converged on the term "discretion." The dominant vision contrasts legality with discretion and prescribes the elimination or minimization of discretion. The meaning of discretion in this literature, however, turns out to be elusive. Sometimes discretion is defined simply as the absence of formalization, bureaucratization, or proletarianization. Thus, discretion can mean informality in eligibility norms (standards rather than rules), decentralized administration, or administration by professionals. Sometimes, however, discretion is defined in two more general senses.

The first sense refers to decisionmaking without regard to norms, or more simply, arbitrariness. Although the dominant vision frequently speaks of the problem as discretion in this sense, it seems unlikely that many of the phenomena to which it reacts are accurately characterized this way. It is extremely difficult for people to act in any setting without regard to norms, let alone in a public bureaucracy. . . .

The second general sense of discretion is the practical ability to make a decision that violates the applicable norm. This is the most frequent use of the term, and much of the rhetorical power of the dominant vision derives from its tendency to associate discretion in this sense with informality, decentralization, and professionalism. Yet this general notion is often hard to apply. First, in order to identify discretion in this sense, one needs to know what the applicable norms are. But it is often, and perhaps typically, the case that there is a basic disagreement over what the applicable norms are. A further difficulty is that, on a formal level, discretion of this sort can never be reduced. There is a law of conservation of discretion: one limits the discretion of one set of actors only by increasing that of others. Thus, while the dominant vision spoke generally of reducing discretion, its proposals tended to reduce only the discretion of lower-level officials and to increase correspondingly the discretion of others. Conservative proposals tended to increase the discretion of upper-level officials; liberal proposals tended to increase the discretion of the federal judiciary.

The rhetoric of discretion raises two related questions: the question of value and the question of trust. The question of value is how norms are to be derived to assess the conduct of public officials. The question of trust is under what circumstances officials can be relied upon to comply with applicable norms.

The dominant vision responds to these questions in terms of notions of society and law that resemble in important respects those of Max Weber. The starting point is a vision of dissensus and distrust as basic conditions of modern social life. In these circumstances, the central goals of social order are efficiency, freedom, and fairness. The goal of efficiency is to coordinate the activities of individuals so as to attain the greatest aggregate satisfaction; the goal of freedom is to enlarge each individual's autonomy to pursue her subjective goals; the goal of fairness is to protect people from the danger that collective activity will favor some citizens at the expense of others. . . .

The Weberian answers to the questions of value and trust are founded on this basic notion of the legal system insulated from the surrounding society by substantive and procedural formality. . . .

Bureaucratic, proletarian organization is also related to procedural formality. Procedural formality expresses a commitment to order society in accordance with the will of the sovereign. Hierarchical bureaucracy with the sovereign at the top and a proletarian work force at the bottom purports to maximize the sovereign's control over enforcement. . . . The range of problems of the dominant vision is apparent in the current efforts of the Reagan administration to come to terms with the widely varying but generally high reversal rates in disability appeals to Administrative Law Judges (ALJs) of the Social Security Administration. For many years, reversal rates have averaged around 50 percent or more, but they have varied widely among individual ALJs; some judges decide for claimants in a high percentage of cases; other judges applying the same substantive norms decide for claimants in a low percentage of cases. How does one decide whether either or both groups are performing improperly? The Reagan administration responded to this challenge by having the Social Security Administration's Appeals Council review the files of a sample of cases decided by ALJs. The Reagan administration reasoned that since by law the Appeals Council has the highest authority within HHS with respect to disability decisions, its decisions on review are presumptively correct. Because the Appeals Council disagreed with the ALJ's reversals more frequently than with their affirmances and with the high reversal rate ALJs more frequently than the low reversal rate ALJs, the administration interprets the study to show that the ALJs are reversing too much. It has used the study to justify a series of reforms designed to decrease reversals, including instructing the ALJs to follow the voluminous and relatively formal rules of the Program Operating Manual System designed for front-line workers and focusing supervisory efforts on the high reversal rate ALJs.

This program exemplifies the problems of the Weberian view. First, there are problems of communication, particularly imprecision. As the administration admits, the Program Operating Manual System's instructions are substantially underinclusive. They deny eligibility to a large number of applicants with disabilities, such as psychiatric ailments or ailments that cause disabling pain, which cannot be established with simple physical observations or quantitative medical tests. They do this despite the fact that nearly everyone permitted to make a judgment on the matter believes that many of these claims are within the substantive legislative norm.

Second, there are problems of supervision. Because the Appeals Council's review was conducted on a paper record, it lacked some information available to the ALJs who made the decisions under review: physical appearance and demeanor of the claimant. American law has always asserted the importance of this type of information in doctrines such as those limiting the scope of appellate review of trial court findings of fact and the right of confrontation. Appearance and demeanor are often dispositive in two of the largest categories of cases that come to the ALJs: cases involving psychiatric impairments and those turning on assessments of the degree of pain.

Third, there are problems of alienation. By reducing the capacity of the ALJ to respond to the contingencies of the particular case, the reforms reduce the capacity of the system to treat the claimant as a concrete individual. From the claimant's point of view, the reforms would make the appeal less an inquiry about whether he can legitimately be expected to earn his own living and more an inquiry about whether he fits a set of formal categories that have no perceptible relation to his experience and needs.

And fourth, there is the organizational paradox of formality. The same factors that make the administration doubt the authority of the ALJs—their administrative insulation from the immediate control of the sovereign and their ethos of decisional independence—from another point of view inspire confidence in them. And the same factor that gives the administration confidence in the authority of the Appeals Council—its proximity to the sovereign—raises questions about it. The Appeals Council's lesser independence and its closer association to policymaking executives associated with the Reagan administration's budget-cutting commitments undermines the claim to authority from its superior position in the hierarchy. These factors suggest that the review judgments may represent, not more accurate applications of the disability norms, but the betrayal of those norms to the Reagan administration's ad hoc policy commitments.

Standards, Decentralization, Profession

. . . The view of legality and society that competes in contemporary America with what I have called the dominant vision in welfare jurisprudence . . . resembles the program of the social work profession under the old regime of welfare administration, but in recent decades its most prominent influence has been in the literature on the judiciary. The competing view has been developed in theories of reasoned elaboration and irrebuttable presumptions, of due process and process values, of judicial administration, and of federal habeas corpus, class actions, and structural remedies. . . .

In the competing view, law is seen as both autonomous of society and a function of it. Substantively, legal norms are less formal than rules but more formal than organic social norms: their characteristic form is the standard, a norm that requires an indivdiualized assessment of how a social goal can best be furthered in a particular case. Procedurally, the norms are legitimated through enactment by specialized institutions, but participation in enactment is broadened by delegation of power to social groups and by the incorporation of social norms into the promulgated law. The distinction between enactment and enforcement is attenuated; it is expected that substantive norms will be elaborated and developed in the process of enforcement.

The most characteristic feature of the organizational structure that the competing vision associates with this type of legality is the profession. The profession plays the role in the opposed vision that the aristocracy played in classical political theory and the civil service played in Hegel's political theory; it mediates between legality and society. . . .

This organizational structure differs from the dominant vision in that it is more decentralized and prescribes a higher degree of education, skill, status, and reward for front-line decisionmakers. Decentralization is associated with the relatively diffused character of sovereignty in the opposed vision. Since sovereignty is not embodied in a unitary actor or institution, it cannot be organized as the apex of a hierarchical bureaucracy. Moreover, the commitment to relatively indivdiualized judgment under standards requires autonomy on the part of the decisionmaker to respond to the contingencies of the particular case. Hierarchical review is limited by the inability of reviewers to know the full range of information on which the initial decisionmaker acts. In addition, decentralization is required to make possible meaningful participation by citizens affected by the decisions. The importance of such participation arises from the attenuation of the differentiation of law and of the distinction between enactment and enforcement. Since citizens have information and views that are valuable to the decisionmaker, participation secures the benefit of this information and these views. Moreover, such participation gives the decisionmaker an opportunity to justify the decision to the citizen as an expression of legal norms rather than of individual will; it also enables the legal system to express a respect for the citizen that mitigates the socially centrifugal effects of adverse decisions. Such participation requires limited review because the more review restricts the decisionmaker whom the citizen actually confronts, the more attenuated becomes the citizen's participation in the decision. . . .

Class

If the preceding argument is correct, one cannot account for the course of welfare jurisprudence and welfare reform in recent years in terms of the general goals of efficiency, freedom, and fairness and the general conditions of modern society. One has to look to more concrete social and historical contingencies. The most interesting of these contingencies are variations on the theme of class. . . . To put it crudely but not unfairly, organization premised on shared values is considered possible as long as the values are those of elite professionals; organization premised on trust is considered possible as long as those to be trusted are elite professionals.

Thus, one way to interpret recent welfare jurisprudence is as an expression of the efforts of professionals to define their place and relationships in the course of the recent expansion of the welfare state.

The issue of class with which welfare jurisprudence has been most concerned is situating the large number of service roles created by the welfare expansion of the 1960s and 1970s. This expansion involved roles that, at various points since the beginning of the welfare state, have seemed ambiguous both in their functional definition and their position in the social hierarchy. Many of them implicated occupations such as social work, nursing, and teaching, which have long claimed professional status but with mixed success. In these fields, the gap between aspiration and reality has at times inspired attempts at redefining the functional content and social status of occupational role. In the 1960s and 1970s, the expansion of welfare activity, the demand for institutional reform, and the creation of new contexts of practice temporarily unsettled the definitions of a variety of these service roles. Jurisprudence contributed to the resolution of this indeterminacy in favor of a proletarian definition of the welfare worker's role.

There are at least two promising approaches to interpreting this development; one emphasizes the constraints of social structure, the other the limited goals of the liberal welfare politics of the 1960s and 1970s. The two approaches are not really distinct. Social structure is simply congealed politics. . . .

Social Structure

The approach based on social structure explains the triumph of the dominant vision by asserting that social structure precludes the downward extension of professionalism that would be required by the application of the opposed vision to welfare administration. This interpretation rests on the view of class implicit in contemporary public law discourse.

At the center of this view stands the professional class. The professional class is distinguished from the class above it (the upper class) by relatively lesser wealth, power, and status, and from the classes below it (the working class and the poor) by relatively greater wealth, power, and status. It is distinguished from all the other classes by a distinctive ethical orientation toward work. This orientation is expressed in the premise of universality in Hegel's notion of the "universal class." It views work as expressing and implementing values that rise above the competing ends of antagonistic individuals and groups. It repudiates the ideal of work as instrumental to the satisfaction of private ends or as a means of imposing one's will on the world and asserts an ideal of work as an intrinsically satisfying form of participation in the life of the community. It is this transcendent, universalistic orientation to work that requires and makes possible autonomy and responsibility in the organization of professional work. . . .

The argument from social structure asserts that the possibilities for the expansion of professionalism are subject to severe limits.

The argument from social structure against the downward expansion of the opposed vision thus has a depressing implication: At least for the poor, the experience of dignified, responsible, just treatment at the hands of the state must be a rare and random event. Except in those few and unpredictable moments when claims are precipitated into the privileged forums of the opposed vision, the experience of such people in their everyday dealings with the state must continue to be alienation and bewilderment.

Politics

The political approach interprets the dominant vision as ideology. By ideology I mean discourse that tends to portray as functions of fixed social structure what are in fact contingent political choices. . . .

The liberal professional reformers of the 1960s sought both to expand the welfare system and to make it more responsive to the poor. They exercised their influence through the federal judiciary and the upper levels of the federal executive. From this perspective, the expanding corps of lower-level service roles of somewhat ambiguous status presented them with a political choice analogous to one faced by some of the reforming monarchs of modern Europe. The choice is between professionalizing and proletarianizing strategies of transformation of the state. In the professionalizing strategy, the reformer creates an organizational structure designed to foster autonomous, responsible activity by people socialized in the reformer's own perspective; in the proletarianizing strategy, the reformer creates an organizational structure designed to subject subordinates to her direct control. The appeal of the professionalizing strategy is that it promises to overcome some of the deficiencies of Weberian bureaucracy as an instrument of control and, by extending the reformer's own mode of life and work, to create valuable allies for her. The disadvantages are, first, that if the strategy fails to inculcate the reformer's perspective, organizational autonomy may be used in ways that will frustrate her ends; and, second, that the expansion of professional status dilutes the exclusivity of the positions and perhaps the privileges of the reformer and her present allies. Thus, one explanation for the embrace of the proletarian strategy by the liberal reformers of the 1960s is that downward professionalization might have seemed to pose unacceptable risks to their own positions. . . .

A related explanation looks at the problem from the perspective of the lower-tier public work force. The roles this group occupies are not entirely the creatures of elite professionals; the workers themselves have participated in designing them. The choice between professionalizing and proletarianizing strategies of elite reform correspond to a choice between strategies of worker politics. . . .

To the workers, the professional strategy held the promise of a nobler and more satisfying life at work and of political solidarity with a wider segment of the society. But it was also a riskier strategy. It was a strategy that, if implemented broadly, would have made much greater redistributive demands on the society. And it was a strategy that would have depended on alliances with professionals and poor people from whom many of the new service workers were divided by ethnicity, education, and status, and whose reliability they might well have suspected.

From the point of view of the liberal reformers' goal to make the system more responsive to the poor, the proletarian strategy had severe drawbacks. The dominant vision obscured the difficulties of translating influence over the upper levels of the system into control of a lower-tier work force alienated from the purposes of those at the top. Moreover, the liberal strategy depended on liberals retaining influence over the upper levels. When in fact liberal influence waned and control passed to conservatives relatively indifferent to beneficiary interests and more intent on curbing expenditures, the proletarian strategy was easily co-opted for quite different purposes. (To be sure, the proletarian strategy has some of the same disadvantages as a means of controlling expenditures that it has as a safeguard of beneficiary interests, but the political liabilities of the alternative professionalizing strategy would have seemed even greater to the conservatives than to the liberals.) As the conservatives implemented the strategy in a way that reallocated enforcement resources away from beneficiary interests, its adverse consequences to recipients were exacerbated. . . .

If one adopts this political explanation of recent welfare jurisprudence, then the future prospects of the opposed vision would seem to depend on political factors, in particular, on the possibilities of reestablishing the New Deal alliance of professionals, workers, and the poor. From this perspective, an interesting feature of the opposed vision is its capacity to link the interests of the public work force in dignified, satisfying working conditions with the interests of the poor in dignified, responsible treatment by the welfare state.

The Ambiguity of Professional Culture

There is a further objection to the opposed vision as an ideal of public administration. This objection focuses, not on the premise of exclusivity, but on the premise of universality. It denies that the public culture of the professional class has been or could be anything but an expression of the contingent views or interests of the professionals themselves or their allies or patrons. . . .

The case for the opposed vision suggests that the implausibility of the professional project lies less in its aspirations to universality than in the premise of exclusivity. . . . [D]ownward professionalization would have to mean, not simply exposure or indoctrination of new recruits in the currently prevailing doctrines of the professions, but the downward extension of the social circumstances associated with the limited but significant amount of trust that is currently accorded professionals: education, reward, status, collective work experience, peer culture, and work roles incorporating autonomy and responsibility. . . .

Conclusion

The opposed vision of welfare administration could be implemented in many ways at varying levels of ambitiousness. For example, one modest reform would involve the abolition of the AFDC Quality Control system and its replacement with an expanded system of quasi-adjudicatory review. Review for all purposes would be superintended by an expanded corps of hearings officials. Two changes in current practice could be made to remedy the problems that arise from the current hearing system's dependence on recipient initiation. First, in addition to cases initiated by recipients, review of a representative sample of cases could be administratively triggered. Centrally triggered cases would be reviewed by an auditor and by either an ombudsman-type official charged with protecting the claimant's interests or a representative of the claimant's choosing. If the reviewers disagreed with each other or the worker who made the initial decision, the disagreement could be resolved by decision at a hearing in which all participate. Second, review could also be triggered by eligibility workers who disagreed with their supervisors about any eligibility decision. There could be selective publication of hearing decisions and the workers would be taught to regard adjudicatory decisions as precedent to be followed in similar cases. . . .

. . . The more modest ambition of the reform is to increase the control of review procedures structured in accordance with the opposed vision; the less modest ambition is to change, however marginally, the way the eligibility workers view their own work and interests by encouraging them to see themselves as autonomous, responsible participants in the implementation of a public program designed to alleviate individual need.

More thoroughgoing reforms would require more adventurous decentralization and redistribution of wealth, power, and status.

NOTES AND QUESTIONS

1. Compare Mashaw's benign view of quality controls with Simon's disparaging view. Do their differences reflect disagreement over the reliability of such controls, their effects on administrators and clients, the normative status of efficiency in AFDC administration, or other factors? Note that Mashaw's definition of "fairness" as accuracy lacks any dignitary component directed at the effects of process on the client's dignity and self-respect. (Mashaw would later develop such a theory in "Administrative Due Process: The Quest for a Dignitary Theory," 61 *Boston University Law Review* 885 (1981)). Would Simon accept this definition? Quite apart from a concern

for the client's feelings about process, wouldn't Simon also want any control system to protect the caseworker's dignity and self-respect? Can any procedural system achieve all of these goals? Do you share Mashaw's skepticism about the possibilities of procedural rights to protect AFDC clients' substantive interests?

2. Mashaw's management due process norms are meant to control the quality of agency adjudications. Might they nevertheless have some application to the informal rulemaking process structures analyzed by McGarity in the article summarized in note 3 to Chapter IV above, or is the notion of due process in the rulemaking context too "weak" to support such constraints?

Social security claims adjudication, the subject of Mashaw's articles, is in Wilson's terms a production organization. Should the same due process norms apply to the four kinds of organizations? How does current due process doctrine deal with the differences in organizations that Wilson highlights?

3. Wilson's organizational typology is constructed around the interaction of two variables: the observability of outputs (work) and the observability of outcomes (results). Clearly, the usefulness and application of this typology depend in part upon how one defines an agency's outputs and outcomes and how one distinguishes among them. Wilson exemplifies this point when he notes that accurate claims processing in welfare offices is more easily observed than being helpful to clients. Each of these two activities, moreover, might plausibly be viewed as both an output and an outcome.

4. What does Simon mean when he says of a profession that "it mediates between legality and society"? Note that Simon does not define professionalism (as is customary) in terms of the possession of an esoteric body of technical knowledge or the norm of selfless devotion to the client's interests. Instead, he defines it in terms of a higher degree of education, skills, status, and reward than those possessed by nonprofessional decisionmakers, making decentralized decisions possible.

Simon discusses the riskiness of what he calls the professional political strategy, which depends upon an alliance between people who are divided by class, ethnicity, and other social markers. Does he consider the possibility that well-educated, high-status professionals would be *less* responsive to the needs of clients than nonprofessionals who are more similar to the client in those respects? Would the unionization of caseworkers advance or retard Simon's "competing" (or "opposed") vision? Is professionalism in bureaucracies inherently inconsistent with hierarchical controls? Are the norms of the profession determinate enough to provide the necessary curbs on arbitrariness—the dark side of decentralized, individualized judgment—by professionals?

5. In considering the internal controls on agency discretion, we should recall the insight that "[t]he most pervasive control of all—that is, self-control, reinforced by professional attitudes within the public service—is difficult to depict; but assuredly it must not be overlooked, for without it the external controls would be of small moment." Walter Gellhorn, Clark Byse, Peter Strauss, Todd Rakoff, and Roy Schotland, *Administrative Law: Cases and Comments* (Mineola, N.Y.: Foundation Press, 8th ed., 1987), p. 1. What is the content of this self-control? Can it be taught to officials? Is there a tension between its implicit conception of the role of bureaucracy and other normative conceptions of bureaucracy in a democratic polity, such as responsiveness to elected officials?

5

Public Participation

The Reformation of American Administrative Law
RICHARD STEWART

Administrative Law as Interest Representation

The expansion of the traditional model to afford participation rights in the process of agency decision and judicial review to a wide variety of affected interests must ultimately rest on the premise that such procedural changes will be an effective and workable means of assuring improved agency decisions. Advocates of extended access believe that an enlarged system of formal proceedings can, by securing adequate consideration of the interests of all affected persons, yield outcomes that better serve society as a whole. The credibility of this belief must now be considered. . . .

The Provision of Representation

The threshold problems in assuring representation of all affected interests in the process of administrative decisions are determining which interests are to be represented and the means by which such representation is to be provided. The limitations imposed by the developing law of standing and intervention on the potential interests entitled to be represented are, in practice, relatively insubstantial. . . . Thus the practical extent of standing and participation rights turns on the means for providing representation to the multitude of interests affected by agency decisions, rather than on doctrinal limits to access rights.

Broad participation rights do not, by any means, ensure that all relevant interests will be represented before the agencies. Representation of these interests is especially unlikely in what may be a frequent situation in administrative law—where the impact of a decision is widely diffused so that no single individual is harmed sufficiently to have an incentive to undertake litigation, and where high transaction costs and the collective nature of the benefit sought preclude a joint litigating effort, even though the aggregate stake of the affected individuals would justify it. "Public interest" advocacy is aimed at providing representation for such widely scattered interests. Today, such advocacy is undertaken primarily by the private bar through "public interest" law firms funded primarily by foundations, or by private firms who subsidize pro bono work out of their regular business. In addition, there are some membership organizations, particularly in the environmental field, that retain lawyers at less than going market rates, although they too are often dependent on foundation support. In a few important instances the government has provided "public interest" representation before administrative agencies. . . .

"Public interest" advocates, however, do not represent—and do not claim to represent—the interests of the community as a whole. Rather they espouse the position of important, widely shared (and hence "public") interests that assertedly have not heretofore received adequate representation in the process of agency decision. . . . Reliance on expanded public interest advocacy as a solution to the problem of agency discretion raises a number of troubling issues.

First, problems are raised because the resources presently available for private representation of fragmented "public" interests fall woefully short of those necessary to ensure adequate representation of all those interests significantly affected by agency decisions. . . .

The fact that public interest lawyers often select the interest to be represented points to another difficulty. The lawyer is often not subject to any mechanism of accountability to ensure his loyalty to the scattered individuals whose interests he purports to represent. . . .

These difficulties are only somewhat alleviated where the plaintiffs represented are an organization and its members.

Given the considerable discretion and independence enjoyed by public interest advocates, we must regard with scepticism claims that a scheme of interest representation will relieve "citizen" frustration and apathy toward government and serve to legitimate agency policies. Most individual members of the class of interests assertedly represented will probably be completely unaware of the "participation" in their behalf. Alternatively, such individuals may see no tangible connection between their interests and the litigation, or they may feel that their putative advocate is ignoring their real needs or actually working against them.

Similar difficulties arise with proposed government advocate agencies.

The Costs of Interest Representation

The resort to formal procedures and judicial review to effectuate a more equal representation of interests affected by administrative decisions is likely to entail serious costs both in resources and in possible impairment of the quality of such decisions. Formal rights of participation have immediate significance only to the extent that agency policy must be implemented through formal proceedings. Because bias in agency policies is often attributed to informal decisions, courts have imposed requirements that force agencies to adopt formal procedures for hitherto informal decisions. To the extent that trial-type procedures are required, with the right of participants to introduce evidence and to cross-examine, a considerable measure of delay and increased expenditure of resources will be involved. These characteristics will be aggravated with the expansion of the number of parties entitled to participate. Liberalized grants of standing to seek judicial review will promote further delay and require additional commitments of resources.

Unorganized interests, by threatening to invoke new rights to demand formal decisionmaking procedures and judicial review, may enjoy a measure of derivative bargaining power in informal agency procedures because of the ability thus to impose the high costs of litigation and delay on agencies and other affected groups. Nonetheless, those representing unorganized public interests may, for a variety of reasons, prefer to require that agency decisions be subject to formal proceedings. . . .

The resource and delay costs of formal proceedings are incurred by the agency as well as private parties and may seriously undermine the effective discharge of agency responsibilities. . . . So long as formal participation rights remain at the heart of the pluralist solution to the problem of agency discretion, the possibilities for streamlining multiparty adjudicatory proceedings on complex issues are distinctly limited.

Nor is more extensive use of notice and comment rulemaking an especially promising means of providing for effective representation of all affected interests without an excessive commitment of resources. Although notice and comment rulemaking has been termed the "most democratic of procedures" because all may participate, and has been urged as an alternative to multiparty adjudication, public interest advocates have tended to scorn resort to rulemaking proceedings on the ground that participation in such proceedings may have little impact on agency policy determinations. In notice and comment rulemaking the agency is not bound by the comments filed with it, and many such comments may be ignored or given short shrift. Indeed, the content of rulemaking decisions is often largely determined in advance through a process of informal consultation in which organized interests may enjoy a preponderant influence.

The Indeterminacy of Interest Representation

The difficulties and costs of furthering the representation of all affected interests through formal proceedings might be thought tolerable if such representation substantially improved the quality and fairness (however those terms may be defined) of the resulting decisions. But the impact of such representation on agency decision is at best problematic.

Representation of unorganized interests may have some impact on administrators' decisions by providing additional inputs of data and argument, and by calling attention to aspects of particular problems that might otherwise be overlooked. . . . But agencies will continue to be exposed to intensive pressures from regulated or client groups, on whom the agencies must rely for information, political support, and other forms of cooperation if the agency is to survive and prosper. Efforts to secure greater participation rights cannot eliminate these pressures or change the institutional factors that make for agency dependence on such groups.

Accordingly, the expansion of participation rights at the agency level is unlikely to resolve the fundamental problem of asserted bias in agency choice under broad legislative delegations. By multiplying the range of interests that must be considered, by underscoring the complexity of the issues involved, and by developing a more complete record of alternatives and competing considerations, expanded participation rights may reduce the extent to which procedures will effectively control agency discretion in decisionmaking. Indeed, by emphasizing the polycentric character of controversies, expanded representation may decrease their tractability to general rules and exacerbate the ad hoc, discretionary character of their resolution. . . .

Substantive Standards for Judicial Review

The adequate consideration requirement. . . . Given the diversity and multiplicity of interests that may be represented as a result of increased participation rights, preparation of opinions that demonstrate "adequate consideration" of parties' contentions may impose a heavy burden on the agency (or its opinion writers). Moreover, exhaustive discussion of the parties' contentions may not be adequate to sustain the agency decision, because the court may require the agency on its own motion "to seek out experts representing varied and opposing . . . views." In effect, the "adequate consideration" principle may become a convenient formula for reversing agency decisions which the court finds unpalatable, since it can almost always find some aspect of the controversy that has been overlooked or some contention that arguably has not been given its due.

Moreover, completeness in agency records and opinions alone may not be sufficient; what counts as "adequate consideration" of an issue or interest necessarily turns on the weight or value to be assigned to that issue or interest. . . . The reliance on ad hoc judicial interest balancing in the application of the "adequate consideration" requirement will tend to prevent the development of generalized substantive rules as grounds of decision in particular cases. Since it is unlikely that principles or guidelines can be developed for weighing particular interests, agencies attempting to solve complex problems will be largely unable to anticipate what a subsequent reviewing court may demand of them. . . .

The Overton Park technique. . . . The technique of construing unclear statutes to control agency policy biases may not be feasible or justifiable in more than a small proportion of cases. Given the open-ended quality of many legislative delegations, considerable (and questionable) judicial reconstruction of the statute may be required in order to extract significant, generally applicable constraints on agency discretion. In addition, efforts to constrain agency choice over a broad class of cases may unduly restrict administrative flexibility in future controversies.

Moreover, the very process of enlarging the number of interests represented tends to multiply the issues for decision in a way that diminishes the odds of finding a clear statutory directive to resolve the controversy. . . .

Judicial interest balancing. . . . Some advocates of public interest representation have forthrightly asserted that it "is the job of the court" to decide "where the public welfare in balance lies." In addition to straining courts' competence to resolve policy questions involving complex scientific and economic issues, such an expansion of judicial power would give the courts a degree of across-the-board responsibility for social and economic policymaking that is wholly inconsistent with our received constitutional premises, under which the legislature remains free, subject to the minimal requirements of the doctrine against delegation of legislative power, to delegate a discretionary power of choice to agencies so long as the agencies stay within statutory bounds and observe appropriate procedural safeguards. . . .

The protection of "underrepresented" interests. . . . Perhaps reviewing courts should give special weight to those interests that are likely to be "underrepresented" in the informal agency process and hence have a lesser impact on the agency's policy decisions. Although some of the leading decisions in the transformation of the traditional model suggest judicial receptivity to such a rationale, there appears to be no adequate principle to define and justify any such general judicial power of revision over agency policies. The reviewing court must have some basis for determining which interests are "overrepresented" and which are "underrepresented" in agency decisions, which implies the existence of some accepted means of defining relevant interests and of ascertaining an initial distribution of weights, in order to determine whether agency decision has been distorted. . . .

Judicial "remand" to the legislature. . . . The proposal might be generalized to require invalidation of agency action that seriously harms a relevant interest and is not clearly authorized by statute. . . .

Implementation of such a policy of clear statement would pose difficulties in defining the class of interests protected, the degree of invasion required to raise a question of legislative authorization, and the degree of specificity in legislative authorization required to validate the agency's action. More importantly, such a doctrine would threaten to paralyze the administrative process. Given the inevitable breadth of agency discretion, a great many administrative decisions would be struck down. . . . Judicial efforts to redress disparities in interest-group resources and to stimulate agency attention to hitherto neglected interests by superimposing a formal system of participation rights are therefore likely to have a quite modest effect on the administrative process as a whole. Perhaps, however, a model of interest representation can be achieved more directly and more effectively by explicitly political mechanisms.

Political Modes of Interest Representation

. . .

Political Systems of Interest Representation: Interim Conclusions

The foregoing discussion is unlikely to kindle enthusiasm for a political system of interest representation as a means of fostering greater administrative responsiveness to the entire range of interests affected by agency action. The direct election of agency members or their selection by designated interest groups would constitute a radical departure from established principles and practices, which dictate that governmental powers should ordinarily be exercised by public officials who are not formally accountable to any identifiable private organization or group. The potential dangers in an interest representation system are obvious: heightened conflict over policy choices leading to domination or deadlock; the fragmenting of governmental authority and responsibility; the impairment of administrative efficiency and impartiality; the erosion of government's ability to lead and innovate. If, in an effort to meet some of these dangers, power is shared between interest-group representatives and other agency organs, the virtues of a system of representation may be undermined. Moreover, there is no assurance that the processes of election or selection will produce a weighting of affected interests in the membership of a representative council—much less in the policies of the agency—that is appropriate and just by whatever criterion one may apply. . . .

In view of the potential hazards in a political interest-representation system, and the inertia in accepted principles of political organization, it seems improbable that such a novel system of administrative organization would ever be adopted. Even in the case of advisory committees, which would seem to provide fertile ground for experimentation, there appears scant prospect for any serious attempt to develop explicitly representative mechanisms. . . .

Evaluation and Prospect

. . . Such criticisms of the interest representation model do not necessarily imply that the principle of representation of affected interests has no place in the continuing effort to control and legitimate administrative decisions. Representational mechanisms may well play a useful and significant role in the future evolution of administrative law, but they do not represent a comprehensive solution to the problem of agency discretion for the forseeable future. . . .

This is not necessarily grounds for despair. It may be persuasively argued that the ideal of a unitary theory of administrative law is untenable and is likely to distract us from the world's complexity and hinder the development of realistic solutions to the variety of problems that confront us. If this argument were accepted, the interest representation principle could be viewed, not as a general model for dealing wth agency discretion, but as a technique for dealing with specific problems of administrative justice. Selective use of such a technique could be warranted in cases where the desirability of fully and formally assessing the effect of alternative policies on various affected interests clearly outweighs the burdensome delays and other costs involved, especially if there is reason to believe that the agency will take an unduly narrow view. In other situations control techniques such as deregulation, application of the clear statement principle, requirements of adherence to previously established substantive rules of decision, and application of the allocational efficiency criterion may appropriately be applied.

From this perspective, occasional judicial resort to the machinery of expanded participation rights in order to focus attention on "underrepresented" interests could be acceptable as a limited part of a more general effort to redress deficiencies in contemporary agency performance. Given judicial selectivity and the likelihood that funding will never be provided in sufficient abundance to afford representation to every relevant affected interest, the resource costs of such a limited system of interest representation may not appear excessive.

In addition, the aggrandizement of judicial power resulting from court enforcement of the interest representation model may not be large or seriously troubling. . . . In the end, the array of doctrinal techniques utilized by courts to expand participation rights may prove acceptable, not because we really believe in the interest representation principle, but because they represent useful judicial levers for the redress of clear failures in the operation of specific agencies. But if we are thus willing to countenance selective judicial intervention in discretionary agency policy choices, it would be preferable for the judges explicitly to set aside policy choices as unsound rather than resorting to indirect and costly procedural stratagems. . . .

If this analysis is accepted, a central problem becomes how to determine the occasions on which selective judicial effort to promote consideration of a greater variety of affected interests should be applied, and how such efforts should be meshed with other techniques for controlling agency discretion. Like the notion of interest representation, no one of these alternative techniques offers a total solution, yet each can make a significant contribution. However, merely to acknowledge the limited validity of these various control techniques provides little guidance as to the circumstances in which each should be applied. Conceding the lack of any single "solution" for administrative discretion, one might seek to order these diverse control techniques through a differential analysis.

Administrative agencies might be classified by their function, structure, powers, environment, and the nature and quantities of discretion exercised. The tasks and policy issues involved in welfare administration might be found to be completely different from those presented in the regulation of the airline industry. The need for judicial supervision might vary between agencies (such as the Federal Energy Office) so hastily and recently created that they can scarcely be recognized as independent bureaucratic units, and long established agencies (such as the ICC [Interstate Commerce Commission]) whose operations have for decades moved in well-worn grooves of their own making. Some agencies that nominally enjoy enormous statutory discretion (such as the FCC [Federal Communications Commission]) will be found in practice to be closely confined by political pressures, while other agencies with narrower statutory mandates (such as the Internal Revenue Service) may enjoy a greater measure of operational independence. These and other vital differences—which are likely to be obscured by any single conception of administrative law—invite comparative classification.

Such a classification of agency functions and institutional contexts might be paralleled by a similar classification of the various techniques for directing and controlling administrative power, including judicial review, procedural requirements, political controls, and partial abolition of agency functions. The two systems of classification might then be meshed to determine the most harmonious fit between the purposes and characteristics of particular agencies and various control techniques. Any design quite so grandiose is of course unlikely to be achieved in full, but it marks out a potentially rewarding line of inquiry that may represent our best hope of realistic future progress in administrative law.

Bureaucratic Justice: Managing Social Security Disability Claims

JERRY MASHAW

A Modest Proposal: Bureaucracy with a Human Face

However sensible the bureaucratic model of adjudication may be in general, the problems of perspective that we have identified—the tendency to ignore subjective evidence, the susceptibility to transitory political concerns, the suppression of claimant (and other outsider) involvement in claims processing, the submerging of off-budget costs—are nevertheless real. Moreover, they suggest the need for an effective counterforce. Otherwise, systematic rationality may play out a progressive logic of control and objectivity that satisfies the Nietzschean definition of the ultimate stupidity—forgetting what it is that we were trying to do.

If a counterforce is needed, and ALJ [Administrative Law Judge] hearings and judicial review cannot provide it effectively, where might it be found?

Here, it seems to me, there are two plausible reform strategies. One is to make the claimants themselves the counterforce; that is, to structure the state agency process so that claimants are as real to adjudicators as are the medical consultants, unit supervisors, bureau chiefs, QA [Quality Assurance] staffs, medical listings, DOT [Dictionary of Occupational Titles], DISM [Disability Insurance State Manual], and disability determination forms that make up their daily work environment. The examiners could be forced to talk to claimants, to treat them as important sources of information, to explain their eligibility decisions.

What would happen as a result of face-to-face contact? To some degree SSA [the Social Security Administration] already knows the answer. It has experimented with the use of interviews at the reconsideration stage of the process. Those tests were indeed premised on a series of hypotheses that respond to the concerns we have previously expressed. If presenting one's case or information in person and getting immediate feedback are important elements in the perception of fairness, claimants should be better satisfied with such a procedure. This in turn would presumably reduce appeal rates. If, in at least some cases, seeing the claimant adds important information tending to support a claim, the number of claims granted by state agencies should go up. And this result need not imply a loss of appropriate objectivity. It would occur in part from a change in the interpretive perspective carried to the available documentary materials and in part from pursuit of additional objective evidence in cases that seemed stronger on personal contact than the medical evidence of record indicated.

If the latter effects were true, the reversal rate on appeal should go down. Where personal contact is the key to an allowance, the claim would have already been awarded. And when denying the claims of people who have more than a documentary existence, examiners would probably make a better evidentiary case, one that will be more likely to stand up on appeal.

Program costs could be expected to go up because some claims that would not have been appealed to a hearing would be granted at the reconsideration stage. Administrative costs would probably go up. Each case would take longer to process at the state agency level, and the reduction in appeal rates would probably not offset these costs. On the benefits side, a number of "nonappealing" but eligible claimants, who would not have received benefits at all under the present system, would receive them. Moreover, successful claimants who would otherwise have succeeded only after appeal would be spared considerable delay.

In its experiments with reconsideration interviews, SSA found all these propositions to be true. The information from the experiments is not nearly as complete as one would like. *In particular, the tests did not have a sensitive measure of either claimant satisfaction or of decisional accuracy.* One cannot tell why the interview made a difference in either dimension (satisfaction or accuracy) when it did. Hence it is not really possible to tell from the studies whether the hearing stage could be eliminated with no substantial loss or whether the interview substantially interfered with management control of the system. SSA's opinion on the latter question must have been that it did not, for in 1976 the administration recommended adding a reconsideration interview to the current process nationwide. Until recently, however, this position has not been supported at the departmental level (Health and Human Services), and the costs of moving to the proposed system have never been included in the HHS budget.

A second counterforce idea is representation of claimants, an attractive notion for several reasons. First, neither the present system nor the personal interview process really comes to grips with the limitations on claimants' understanding of the program. A system that provided claimants with specialized representatives whenever they were initially denied benefits could instill somewhat more confidence both that informed choices were being made concerning whether to request reconsideration and that relevant evidence was not being overlooked. Representatives could also be expected to filter out frivolous claims. Representatives who viewed themselves as participating in the system over an extended period would be discreet. Rather clear "losers" would be counseled to accept defeat, and energies would be devoted to worthier claims. The number of claims persisting past an initial denial might well decrease.

If properly constructed, representation could also play a mediating role between the claimant's perceptions of distress and the program's policies. Much of what lawyers do in civil and criminal claims can quite properly be viewed as educating the client about the legal system rather than educating the legal system about the client. Both the availability of relevant expertise and the counseling of a personal representative should increase claimants' sense of fair treatment.

There are obviously problems with structuring representation into the reconsideration process. First, who would these representatives be? In my view there is no need for them to be lawyers. Not only are there few, if any, legal issues involved, most lawyers do not understand the disability system well enough to be of much use to the claimant (or to the decisionmaker). The VA [Veterans Administration] system of claims representatives, operating through veterans' organizations, is an attractive model; but there are no obvious representational substitutes for veterans' organizations such as the VFW [Veterans of Foreign Wars] or American Legion. Nevertheless, it seems critical that, as in the VA situation, representatives have essentially the same qualifications and training as disability examiners. Indeed, they might *be* disability examiners who sometimes work as deciders in the state agency and sometimes as claimants' representatives in a separate bureau established for that purpose.

But if the representatives are also government employees, how can they be given the proper incentives to provide vigorous assistance to their "clients"? In part, I would suggest, through the same devices currently used to monitor examiner behavior. If examiner development and judgment can be supervised, so can a representative performing a similar task. To be sure, there are additional functions and problems involved, but they hardly seem insurmountable. Moreover, we should not forget that the current system is one in which claimants, save for a tiny percentage who are represented at hearings (say 60,000 out of 1,300,000 claimants entering the system each year), are now almost totally at the mercy of government employees (examiners, ALJs) who have no personal obligation to them for the development of their cases. Assigning a government employee the explicit role of representing claimants will certainly increase the incentives to protect the claimants' interests. It might also be possible to generalize the representation idea by establishing a separate bureau of paraprofessional claims representatives (perhaps as an offshoot of the Legal Services Corporation) to handle other types of benefits claims as well.

In my view, a combination of face-to-face reconsideration interviews and representation provided at the time of any initial denial notice is worth trying in a carefully controlled test. By "carefully controlled" I mean a test that is sensitive to the need for data on correctness and satisfaction and that will, therefore, provide a basis for well-informed policy choice. My intuition is that on most of the dimensions of adjudicative quality, the combined method would yield results so superior to the current system that the costs of the change could be partially recouped by eliminating ALJ hearings and judicial review. If this were the case, there would have been developed a bureaucratic model of administrative justice that substantially improves upon both the traditional adversary process model and the existing model of bureaucratic rationality that has emerged as a response to SSA's need to obtain control over the disability decision process.

For the interview-plus-representative idea responds both to the shortcomings and to the strengths of the bureaucratic model. Although bureaucratic rationality deemphasizes intuitive judgment and personalized concern, it also seeks continually to improve upon its technical or scientific mode of thought. Merely introducing the claimant into the process without providing him or her with access to technical resources is to introduce a foreign body that is likely to be rejected by the host organism. The claimant's participation in the development of a claim must be structured in a way that orients the facts of his or her functioning to the stylized reasoning process of a manageable, and managed, adjudicatory system. That orientation requires expertise about the system, as well as sympathetic association with the claimant's situation.

The interview-plus-representative proposal can also be uncoupled. Claimants who prefer privacy to personal contact can be content with the aid of a representative and a decision based wholly on the claims file. The fiercely independent could represent themselves or reject SSA's offer of a representative in favor of employing their own.

The Role of Courts in Regulatory Negotiation—A Response to Judge Wald

PHILIP HARTER

Most environmental disputes are highly complex, involve significant technical questions of risk assessment and its management, are based on or establish issues of public policy, and affect many parties. As a result, they have all the earmarks of a typical rulemaking proceeding. It is, therefore, a relatively short step from the successful use of ADR [Alternative Dispute Resolution] in environmental matters to their application in developing regulations, albeit the techniques would need to be adapted to the peculiarities of the regulatory process. To that end, the Administrative Conference of the United States commissioned a study of the use of direct negotiations among the parties in interest in developing regulations, then study formally recommended procedures for doing so. Several agencies have used the process, and several more are well along in planning to use it for new regulations.

The Federal Aviation Administration used ADR to revise its rules governing the amount of time a pilot can fly without taking a rest and the minimum periods between duty assignments. The original rule had proved enormously controversial, and the FAA had tried unsuccessfully several times before to update it. Its complexity had generated more than 1,000 pages of interpretation. Significantly, especially given its history, once the rule was issued, it was not challenged judicially. The Environmental Protection Agency used regulatory negotiation to develop two rules, one involving the nature of penalties imposed for truck engines that fail to comply with the dictates of the Clean Air Act and the other to establish the procedures for granting emergency exemptions from pesticide registration requirements. As with the FAA, the comments submitted in response to the Notices of Proposed Rulemaking were remarkably sparse, and no judicial challenge has been filed against either rule. The Occupational Safety and Health Administration used a variant of the process to address the occupational exposure of benzene. That rulemaking had extended for nearly ten years and had reached the Supreme Court, and thus had become a particularly difficult regulatory proceeding by the time the negotiations began. The discussions involved four industries and four unions. The parties made significant headway in developing a consensus, but for complex reasons largely beyond the negotiations themselves no final agreement was reached. Nonetheless, the participants viewed the process as worthwhile since it elucidated the needs and concerns of the other parties. That they made it that far in that rulemaking demonstrates the potential of the process. . . .

Parties are interested in negotiating a regulation, as opposed to relying on a more adversarial mode, and hybrid rulemaking is surely that, because it allows them to share in making the actual decision and to help develop a better, more informed rule. They would not incur the time, expense, and anguish of making hard choices, however, if their work were to be easily undone. Inappropriate forms of judicial review could do just that. On the other hand, judicial review can also play a vitally important role not only for ensuring the integrity of the process but also for providing positive incentives to make the rulemaking process work well.

Undoubtedly the best approach to judicial review would be to obviate its need because everyone has been sufficiently satisfied with what happened before the agency. This has been our experience so far with *all* of the rules developed by this process. But, alas, that surely will not continue forever. As Judge Patricia M. Wald points out in her article, "Negotiation of Environmental Disputes: A New Role for the Courts?," "negotiation typically does not eliminate court involvement altogether; instead it changes the nature and scope of the judicial role." . . .

Overview of the Regulatory Negotiation Process

. . .

Once the agency, either on its own initiative or in response to a request by an interested party, decides that it *may* be appropriate to use regulatory negotiations to develop a proposed rule, it would appoint a neutral convenor to conduct a feasibility analysis. The convenor would identify those interests that would be significantly affected by the regulation and the issues that need to be addressed by the rule, and in any negotiations, should they be undertaken. The convenor typically does this by starting with the agency and asking it both who would be affected and what issues are important. He or she would then go to the people named by the agency and ask them the same questions. That effort, coupled with independent research, will usually uncover at least the major players and issues. . . .

The convenor must also assess whether the parties believe it is in their overall best interest to attempt to negotiate a proposed rule or to rely on the normal process. Factors likely to influence that decision are:

- Whether the issues are mature and ripe for decision; it is difficult to negotiate a solution to a problem that is only dimly understood or that can be put off until another day.
- Whether there is a reasonable deadline for the negotiations; reaching agreement is a tough task since parties must make hard choices; hence, a deadline is decidedly helpful; this can usually be provided by the agency's committing itself to developing the rule through the traditional process should the negotiations prove unsuccessful after a specified period.
- Whether the outcome of the proceeding is genuinely in doubt; various

parties may each have enough power—clout in one form or another—so that no one party, including the agency, can totally dominate what happens.

- Whether any interest would have to compromise an issue that goes to the very core of its existence; no one compromises issues like that.
- Whether there are a diverse number of issues that are likely to be of different value to different interests.
- Whether there are a limited number, roughly no more than 15–25, of distinct interests that would be substantially affected, and whether they are such that individuals can be selected to represent them. . . .

If, after that analysis, it appears that regulatory negotiation would be appropriate, the covenor would recommend it to the agency. If the agency decides to go forward, it would appoint a senior agency official—generally the one charged with developing a draft rule within the agency—to be its representative to the negotiations. A neutral mediator, who is responsible for aiding the *process* of the negotiations, is usually helpful for any complex or controversial rule. Generally the person who served as the convenor will fill that role since he or she has developed an understanding of the issues and a rapport with the parties. The mediator has no power whatever to decide issues or impose a decision; his or her role is one of helping the negotiators themselves reach an agreement by helping to define the issues, generating creative options, carrying messages back and forth, being an "agent of reality" to each party, making it "safe" for the parties to talk directly, as well as facilitating the plenary discussions.

The purpose of the negotiating committee is to develop a "consensus" on a proposed rule. It is surely preferable if the consensus extends to the actual language of the important provisions of the rule and its supporting preamble, since many policy choices surface in the drafting. "Consensus," as Judge Wald notes, can be a difficult term to define in the abstract and a difficult status for a reviewing court to discern, whatever the definition. The parties can choose to define it however they see fit, but its definition can have a significant bearing on the willingness of parties to participate and perhaps on any subsequent judicial review. Unless the parties agree otherwise, the definition that has proved the most workable is that "consensus" means that each *interest* represented in the negotiations concurs in—or at least does not oppose—the result.

That definition has several important consequences. First, no party can be outvoted; hence, each preserves whatever power it has. This has proved critical to most parties' willingness to participate in regulatory negotiations. Second, it forces the parties to come together to solve a mutual problem—developing the rule. They can no longer act as a group of disparate interests, each of which can dissent from a provision, go back home, and tell their constituents how tough they were. Rather, all participants must decide whether, on the whole, they are better off accepting the agreement or trusting their fate to another process. Third, it means that individual members of an interest group may dissent without destroying the consensus, so long as the interest as a whole concurs.

The agency agrees to use the consensus proposal as the notice of proposed rulemaking unless it is outside the agency's authority or something is significantly wrong with it. That should happen rarely, however, since presumably the senior agency official has signed off on it, and he or she should have done his or her homework to ensure that no other relevant agency officials had problems with the proposal as it was developing and before its adoption.

Following the convenor's analysis, the agency would begin the actual process by publishing a notice in the *Federal Register* and in substantive publications read by mere mortals who may be interested in the subject matter. The announcement would indicate the agency is intending to establish an advisory committee pursuant to the Federal Advisory Committee Act for purposes of negotiating the rule. It would describe the issues likely to be raised and the parties that had been identified by the convenor as being interested in participating. In that way, others can determine whether they would be affected and whether they should participate, or whether their interests are already adequately represented.

If new parties do come forward, the convenor and the agency must decide whether to include them. Because the purpose of the negotiations is to ensure that the relevant issues get raised and resolved, it is far better to err on the side of inclusion. Thus, they should be included unless doing so would result in too many parties and it is not likely that a party seeking admission would raise significant new issues. Even in that case, it may be appropriate to form caucuses of allied interest for purposes of participation. Even if a party does not join the committee, it can express its views during the committee's meetings or in response to the resulting Notice of Proposed Rulemaking. Thus, no one is shut out.

The group needs to decide what information is required to make a responsible decision. One of the major values of this process is that the parties bring with them, or can otherwise obtain, insight and perspectives for developing a workable solution to the regulatory question. In addition, the negotiating group will typically be furnished with technical information developed by the agency. The Environmental Protection Agency has also established a "Resource Pool" the committee can draw on for purposes of the negotiations, such as for conducting short-term research or analysis.

The groups operate under the Federal Advisory Committee Act which requires open meetings, although private caucuses can be held. Most experts on negotiating emphasize the need for closed sessions to encourage candid exchanges. The "sunshine" of the meetings has perhaps made things a bit more difficult at times but not impossibly so, and closing meetings in which public policy decisions are being made has a cost of its own. Therefore, very little effort has been made to close any meeting.

If the group is able to reach a consensus, it submits it to the agency as a recommendation for a Notice of Proposed Rulemaking (NPRM). The protocols developed for three negotiations have explicitly required each participant to sign the consensus and to ensure that that person has the authority to do so. In other instances, the approval has been more informal.

The agreement has meant that the representative concurs in it, but it does not purport to bind all members of that representative's constituency. Thus, for example, an individual member of the interest group—be it a trade association, a labor union, or an environmental organization—who is dissatisfied could file comments with the agency in response to the NPRM or challenge the rule in court. Indeed, the interest as a whole would even disown the earlier agreement and file adverse comments or challenge it judicially. As a practical matter, however, that should rarely happen: if one party disaffects, others will too. Since it is likely that the outcome of the rule was genuinely in doubt when the process began, it probably still is. In that case, trashing the agreement will result in the loss of something that is known in return for a situation where the party could lose even more. Thus, a more desirable condition from everyone's perspective would be for the individual dissidents to challenge while the interest as a whole signals the decisionmaker—be it the agency, OMB [Office of Management and Budget], or a court—that it backs the agreement. In that way, the challenger can attempt to show that it would be affected differently than others or that its concerns were not raised. That will serve as an important check on those who do participate not to ignore the full range of interests. But, if the others lend their support, the decisionmaker will also know that the challenger does not speak more broadly. That in turn will help illuminate whether the challenger has real needs or whether it simply seeks more than it got at the table.

The Needs of Review

Sloughing off a course full of detail, the Administrative Procedure Act (APA) provides that a reviewing court is to "decide all relevant questions of law" and "hold unlawful . . . agency action . . . found to be . . . arbitrary, capricious, an abuse of discretion, or otherwise not in accordance with law." . . .

Scope of Authority

Everyone involved in negotiating a regulation must recognize on a fundamental level that the proposal must be within the agency's authority granted by statute if the agency is to use it as the basis for a rule. That is simply a recognition that neither the agency nor certainly the negotiating group is sovereign. Rather, the agency must act within the bounds of discretion provided by Congress if its actions are to be a binding, official position. Thus, just as in the judicial review of a rule developed by another process, the court's first task must be to determine whether the rule is within the scope of the agency's authority. That process would be very much as it is customarily. Not only is this step required by the APA, it also serves several practical functions with respect to negotiating regulations. . . .

First, the statute, along with any implementing gloss placed on it by the agency and the courts, is the norm against which the negotiations are conducted. As in any negotiation, a party should keep its eye on what it could obtain without an agreement, and it should make a bargain only if it will be better off for doing so. In Fisher and Ury's popular term, the party needs to continually assess its "best alternative to a negotiated agreement" or BATNA. The agency's scope of authority defines the contours of the parties' BATNA in regulatory negotiation. It is important for that reason alone for the courts to enforce that expectation. Were it otherwise, one party could pressure another into giving up something it won legislatively although that party had gained a reasonable expectation of achieving that result.

Second, judicial enforcement of the limitations on an agency's authority also provides a deterrent against that ever present villain of democratic politics—logrolling. Were it not for judicial oversight, it would surely not be inconceivable for the negotiating parties to make impermissible "deals" that are outside Congress's contemplation. . . .

Third, the judicial role also prevents sell-outs in which a party agrees to something short of what it should obtain under the governing law. Perhaps the greatest concern here would be in the agency's hasty acceptance of a negotiated deal in return for some other, intangible benefit.

Thus, the judicial role in defining the scope of the agency's authority not only serves its usual role, but also provides additional incentives for the parties. While this role is an important and necessary one, the courts should be decidedly circumspect in exercising this duty too aggressively. The court should, of course, be on the alert for logrolling and the like, but, that aside, it should be reluctant to second-guess a diverse group of interests—representing a wide spectrum of concerns—that has concurred in the outcome, and whose individual members feel that on the whole the rule meets their needs.

The process itself helps ensure that the proposal is within the agency's authority: if the agreement does not reflect a major provision of the statute or is otherwise outside the agency's authority, presumably some party would be better off if it did and that party would likely insist on compliance. While that may sound a bit utopian, it is a measure of the extent to which the agreement is out of conformance with the statute. It may be that a bit of play is needed in the joints of a statute. The parties are typically better able to determine "what works" within the theory of the statute and hence what is the best way of achieving its overall goal. Or, a provision may be included in a statute to benefit a particular interest. If that interest does not insist on its full exercise as part of the agreement, the rule should still be accepted without that interest's being estopped from insisting upon the full application of the provision in future rulemakings. That they agreed indicates the interest achieved the protection sought in the statute which would otherwise come through a rigorous application of "scope of authority" review. . . .

Thus, the court needs to define the authority of the agency but provide a little leeway to accommodate practical interpretations and implementation.

Arbitrary and Capricious

The Supreme Court directed the reviewing courts to make a "searching and careful" inquiry into the facts and to decide whether the decision is based on a consideration of the relevant factors. More recently, Judge Leventhal's characterization of the judicial process has become the frequently used shorthand. As Judge Wald began her article by pointing out, Leventhal advocated—and as a reviewing judge required—that "the agency . . . take[] a 'hard look' at the salient problems and engage[] in reasoned decision-making; the decision itself must come 'within a zone of reasonableness.' " Whichever term is used, the court is expected to carefully review the factual basis on which the agency based its decision and to carefully analyze whether the agency considered the relevant factors (or relied on an impermissible factor) and ultimately reached a decision within the zone of reasonableness under the statute.

It should be recognized, however, that in developing this standard of review, the courts adapted a more deferential, less intrusive oversight to meet the needs of the times. The former standard was simply not up to corralling agency action in the face of complex technical and social science questions or the pervasive influence of regulation. The courts responded with a new form of review by changing the interpretation placed on twenty-five-year-old statutory commands. This adaptation provides important incentives to agencies to do a thorough job in developing the underlying facts and to exercise their discretion with care. It therefore serves as a powerful check against the arbitrary use of administrative power.

In the same vein, it may be appropriate for courts to alter their means of ensuring that agency action is not arbitrary or capricious if the rule is the product of regulatory negotiation. Thus, the destination of [Citizens to Protect] Overton Park [v. Volpe] would remain the same, but the route taken in getting there can and should be changed. That is not because the standard should be any lower, but because the nature of review developed to supervise one process may inhibit the use of another that has other means of achieving the same end. Judicial review should therefore be tailored to ensure that those means are fulfilled.

Without supervision, the agency may not have adequate incentives to delve into the relevant factors—statutory, factual, or political—that are involved in a rule. The "hard look" form of judicial oversight addresses that lack by providing the stimulus of the pain of reversal unless the agency does its homework. But, the directly affected parties are in a far better position than the agency or a reviewing court to determine what the "relevant factors" are and the weight to be accorded each; they can determine how much information is needed to resolve the outstanding factual issues; they are in the best position to determine the appropriate political trade-offs—all within the context of the agency's authority. And again, if there were a "relevant factor" that would affect the outcome of a rule, it would be in the interest of some party to raise it for discussion and insist that it be adequately considered lest a consensus fail. That the rule reflects a consensus of the affected parties therefore goes a long way toward meeting the goals of *Overton Park* and ensures that the rule is neither arbitrary nor capricious.

The issue then becomes whether the rule does indeed reflect that consensus. The very fact that we are discussing judicial review posits that at least someone is unhappy with it and hence that it does not reflect unanimity, which is, after all, one definition of consensus. What, therefore, is the court to do?

The issue in contest may, of course, have been addressed directly in the preamble to the rule so that the review would be much the same as in a typical rulemaking. This emphasizes that a negotiated rule will still have a preamble that explains the basis and purpose of the rule and the reasons the various choices were made as they were.

If the issue is not explicitly developed, the court could still satisfy itself in large part by a procedural review of the process. The question is whether the parties to the consensus reflect sufficiently diverse perspectives that the full range of issues would be raised in the negotiations.

A beginning point would be to exmaine the process by which the negotiating committee was established. That would surely entail an examination of the issues raised in the Notice of Intent and its list of interests that would be represented on the committee. Thus, the court would determine whether the notice was sufficient to alert an interest as to whether it might be affected and whether someone with allied needs would be present so it could be relatively comfortable that its issues of concern would be raised.

Indeed, one of the major advantages of negotiated rulemaking is that by means of the convening process and the notice of intent, a concerted effort is made to identify the affected parties and to encourage them to participate directly in the development of a rule. Regulatory negotiation is therefore likely to result in broader participation than the traditional process. Thus, more issues are likely to be confronted.

To ensure that a party's concerns are considered by the negotiating committee in the first instance, or by the agency in response to comments filed on the NPRM, the court should not reward a party for sitting out the process and challenging only the final rule. Therefore, if the notice and the preliminary outreach are sufficient to alert those likely to be affected, it would be appropriate for the court to provide an incentive to come forward and participate directly in the discussions by requiring a reasonable explanation of why the challenger believes its interests—those relevant factors—may not have been raised and its failure to come forward. Thus, the court should require a reasonable explanation as to why the petitioner did not "exhaust its administrative remedies" by participating before the agency.

Parts of an earlier article have been construed as my arguing that a party should be "bound" by an agreement so long as its interest was represented at the table. After having seen these things in action and after reflecting on Judge Wald's article, I am prepared to confess error on that score. "Interests" are too slippery for that, and determining just how far a consensus goes can be as elusive as the term itself. It now seems to me that a failure to provide an adequate explanation should not be regarded as a fatal preclusion of review, but the petitioner should be put to a relatively higher burden of knocking down the agreement that others forged. That is certainly close, if not identical, to current standards.

If such an explanation is provided, the court must still determine whether the decision is within the "zone of reasonableness" or whether it reflects an adequate consideration of the "relevant factors." In determining whether the important factors were considered and resolved appropriately, the court would want to look at the range of interests that were represented and that concurred in the result by signing the document. The petitioner should be required to make some showing that *its* specific interest was somehow not dealt with adequately, and hence that a relevant factor was not considered, or that the result was a clear error of collective judgment.

Knowing that a single member of a broader constituency is challenging a rule should help the court focus on the precise issue in controversy. Either the petitioner feels it would be affected differently than the others in its constituency, or it disagrees with the balance that was struck. If the petitioner was affected differently than the others at the table, then the court must decide whether, all things considered, the decision is still within that zone of reasonableness. That decision would turn on the interplay of those interests that were present and the explanation of the decision in the preamble. If, on the other hand, the challenger's interest was represented, so that it is disagreeing with the resolution of the issues, the court should recognize that negotiated rulemaking is like legislation—a method for reaching a political accommodation—and reject the challenge.

The Thorny Problems

In concluding her analysis of the judicial review of negotiated rules, Judge Wald emphasizes three thorny problems: "For any particular agency action, interest-group pluralists have to determine which interest groups should be represented before the agency, what kind of participation is appropriate, and what sort of procedural rules will govern the ironing out of a consensus among those conflicting group interests." These are surely the major issues, plus perhaps a fourth: how much information should the group be required to develop in reaching a decision. The hybrid process thrives on paper, for the quite legitimate and important reason that it is a means of controlling discretion that is needed precisely because parties do not participate directly. But, much of the analysis of the hybrid procedure is largely a surrogate for direct participation; thus, it should not be necessary if the parties are there. Since, however, judicial review has come to focus on the paper record, so here too we may need to adapt. We will, as Judge Wald suggests, have to evolve a response to each concern. In the meantime, there are some preliminary answers or guideposts.

First, to meet the dictates of *Overton Park,* sufficiently diverse interests need to be represented to ensure that the relevant factors will be raised and considered. On a practical level, the convening process needs to be one that will help identify the parties that will be significantly affected and will also provide notice to those who were not picked up by direct contacts. Thus, the answer to the question of which groups should be represented is that all who represent parties that would be significantly affected by the rule are welcome. The process, within the confines of available slots, is one of self selection; if too many wish to participate, it is usually accomplished through caucuses or through encouraging participation at meetings without formal membership on the committee. . . .

Second, each interest should have equal access to the table. It may be that a coalition of allied interests will need to be formed and that the caucuses will need to select designated representatives. The individual representatives may be supported by "team" members who have particular viewpoints or an expertise that can be drawn upon as needed.

Third, a workable definition of "consensus" is that each interest represented at the table concurs in the result and that each representative signs the document. Each party then has the power to insist on sufficient information to make an informed choice as balanced against all other demands on resources and on the full consideration of those factors the party thinks are relevant.

NOTES AND QUESTIONS

1. How promising is the "comparative classification" technique that Stewart suggests for determining when "selective judicial intervention" is warranted? Consider the number of variables he includes and the lack of consensus about how they should be applied in particular cases. Would courts be able to develop principles for deciding when and how to intervene? Would such principles be predictable?

2. At the height of the consumer movement's political influence during the early 1970s, legislation was advanced to establish a federal consumer protection agency that would be authorized to intervene on behalf of "consumer interests" in a broad range of agency proceedings as well as perform certain ombudsman functions such as consumer complaint processing. The bill died in conference in the face of a veto threat from President Ford and the nomination of a Democratic candidate committed to a strong consumer agency bill. The legislation was never enacted. A number of conceptual and practical objections to such an advocacy agency were raised. It was argued, for example, that there is no consumer interest not already well-represented in a competitive marketplace; that such an agency would be dominated by special interests and ideologies masquerading as consumer advocates but committed to policies that actually harm consumer interests; that the existing regulatory agencies bear primary responsibility for identifying and protecting consumer interests; and that such an agency would increase the cost and duration of administrative proceedings. See, for example, "To Establish An Independent Consumer Protection Agency," joint hearings before the Subcommittee on Reorganization, Research, and International Organizations and the Subcommittee on Consumers, of the Senate Committee on Commerce (1973).

At various times, federal agencies such as the Antitrust Division of the Justice Department and the Federal Trade Commission have participated in other agencies' regulatory proceedings to assert an interest in competition and consumer welfare.

3. Is public participation the same as public interest representation? The public

participates in agency policymaking through a variety of mechanisms, of which formal intervention as a party in agency proceedings is only one. Some are indirect and operate through the political process: congressional legislation and oversight; appointment of agency officials; advocacy by organized interest groups; and the like. In agency rulemakings, anyone is permitted to file comments. Media coverage of the most controversial agency policies can influence decisionmakers, as many *60 Minutes* programs have done. Market pressures, which aggregate individual choices, shape and constrain agency decisions. What are the strengths and weaknesses of these techniques? Notwithstanding these techniques, is the system biased in ways suggesting that the public interest remains unrepresented? If so, what do you mean by the public interest? How can it be institutionalized without creating new biases?

4. Negotiated rulemaking (sometimes called "reg-neg") began in the early 1980s as a pilot project sponsored by the Administrative Conference of the United States. The origins of the idea, however, go back at least to 1941 when Dean Acheson, then assistant secretary of state and chairman of the Attorney General's Committee on Administrative Procedure, which developed the legislation that would become the APA, proposed an "informal conference" procedure strikingly similar to negotiated rulemaking. See "Administrative Procedure" hearings before a subcommittee of the Senate Judiciary Committee, May 15, 1941, p. 810. See also, Peter Schuck, "Litigation, Bargaining, and Regulation," 3 *Regulation* 26 (July/August 1979). In 1990 the Administrative Conference approach received the imprimatur of Congress, which amended the APA to provide agencies with formal authority to conduct negotiated rulemakings under specified procedures.

The idea is not without its detractors and skeptics. See, for example, William Funk, "When Smoke Gets In Your Eyes: Reg-Neg and the Public Interest—EPA's Woodstove Standards," 18 *Environmental Law* 55 (1987); Patricia Wald, "Negotiation of Environmental Disputes: A New Role for the Courts?" 10 *Columbia Journal of Environmental Law* 1 (1985). For a review of some agency experiences with negotiated rulemaking, see Henry Perritt, "Administrative Alternative Dispute Resolution: The Development of Negotiated Rulemaking and Other Processes," 14 *Pepperdine Law Review* 863 (1987). In 1990 the Administrative Conference published a Negotiated Rulemaking Sourcebook, which contains extensive background materials and a bibliography.

5. Is "bureaucracy with a human face" possible, as Mashaw hopes, or is it oxymoronic? Might representation of the kind that Mashaw proposes cause the adjudicator to be less attentive to the claimant's interests, relying on the representative to fulfill that function? Would the increased formality of the interview be advantageous or disadvantageous to claimants?

6. Prompted in part by the revived interest in civic republicanism (see, e.g., the Seidenfeld excerpt in Chapter I) and critiques of pluralistic liberalism (see, e.g., the Lowi excerpt in Chapter VI, section 1), political scientists and philosophers have devoted renewed attention to the norms and forms of participtaion in the administrative state. The most provocative treatments of this subject include Benjamin Barber, *Strong Democracy: Participatory Politics for a New Age* (Berkeley: University of California Press, 1984); Carol Pateman, *Participation and Democratic Theory* (Cambridge: Cambridge University Press, 1970); Roberto Unger, *Social Theory: Its Situation and its Task* (New York: Cambridge University Press, 1987); Unger, *False Necessity: Anti-Necessitarian Theory in the Service of Radical Democracy* (New York: Cambridge University Press, 1987). For a review of this literature, see P. P. Craig, *Public Law and Democracy in the United Kingdom and the United States of America* (New York: Oxford University Press, 1990), chapters 10 and 11.

6

Market-Oriented Controls

Privatization: Politics, Law, and Theory
RONALD CASS

Forms and Assumptions

Forms

Privatization proposals vary along several dimensions. They suggest different *routes* to lessen government involvement in a given activity; they bring about different *degrees* of public and private control over an activity; and they leave different *types* of control in public and private hands. The variation in each of these dimensions provides significant information about what is desired and what the consequences of a proposal will be. The simplest starting point, and that which probably coincides most closely with current legal thinking about government powers and obligations, is division of privatization possibilities by the route or manner of disinvolvement.

Divestiture. The first group of privatization possibilities involves formal government ownership of an asset and proposes that government terminate such ownership. . . .

In addition to the sale of a going concern, government may reduce its ambit of authority by relinquishing other assets that may or may not be used to generate income in the manner of a private enterprise. The sale of land is exemplary. . . .

Divestiture also can take the form of abandonment of assets. The government, before 1935, abandoned millions of acres of land to "homesteaders" who paid no money but agreed to improve the land in certain respects.

Contracting. A second group of privatization proposals involves the contractual rearrangement of control over some, but not all, aspects of an activity. These arrangements fall into two categories. One is the lease of government assets to another party. . . . In the other category, known as "contracting out," the government purchases goods or services from another party.

All of the contracting proposals provide for a sharing of decisionmaking responsibility between government and the other contract party. The division of control could take very different forms. . . .

Reprinted by permission from 71 *Marquette Law Review* 449 (1988).

Increasing choices: Deregulation and vouchers. The prior proposals would reduce government control over property or would substitute private inputs for government inputs in particular activities. Another group of privatization proposals is concerned with government restrictions on the range of choices available to private parties and suggests the reduction or removal of current constraints on private choice.

Deregulation addresses situations in which the government directly regulates particular private behavior. . . .

A related form of privatization is available when government relies less on command-and-control mechanisms than on direct provision of benefits to a class of beneficiaries. Various commentators have suggested that beneficiaries could be better off, and government would be more responsive to their needs and interests, if government allowed beneficiaries some choice among goods or services. Thus, instead of providing a publicly funded school for its residents, a locality might provide an "education voucher" redeemable at any school of the parents' (and children's) choice. . . .

Direct dollar choices. A final group of privatization proposals is also concerned with expanding the ambit of individual choices, focusing particularly on the choice to spend a given sum of money in exchange for given benefits. These proposals assert that private choice is restricted whenever government officials are authorized to collect a large block of funds and then, at their discretion, divide the funds over activities of their choosing. This group of privatization proposals calls for a more direct link between the payment of money and the receipt of a good or service. One example of this effort to tie costs more closely to benefits is the call for imposition of user fees, specific charges paid directly by those who elect or use particular government-provided goods or services.

A different route to collapsing individual decisions respecting payment for and receipt of benefits is simply to reduce general government revenues. . . .

Assumptions

The privatization proposals build on several assumptions, positive and normative. No single set of assumptions explains all of the proposals. Indeed, some privatization proposals arguably rest on assumptions which are at odds with those underlying other proposals. In general, the proposals build on some combination of the following five assumptions. The first two assumptions relate to normative goals, the remaining three to positive prediction of the means calculated to accomplish them.

Normative bases—Baseline and burden: The contractarian tradition. . . . The baseline process for any activity is autonomous, private action; government power is presumptively disfavored until its proponents bear the burden of justification.

This normative proposition builds on the Lockean tradition which treats government as though it were the creation of contractual agreement among individuals who are free to choose whether and on what terms to have government. . . .

Utility and Wealth

The second normative assumption underlying much of the movement for privatization may be held along with the first assumption or independent of it. This norm proposes the maximization of aggregate individual utilities as the proper goal for social decisionmaking. . . .

Often it is not clear which version of social utility maximization informs a particular proposal, but many privatization proposals seem to accept two notions that are not uniformly accepted among those who espouse the social utility norm. First, privatizers generally seem comfortable with the proposition, common to non-Paretian social welfare functions, that we can conclude that social utility has increased even where one or more parties have been made worse off; that is, unanimous consent is not necessary to establish a gain in overall social utility. Second, many privatizers appear to accept as a norm the maximization of preferences weighted in dollars—what one would spend to secure or to prevent a given outcome, subject to current tastes and budgetary constraints—possibly in place of, but more likely as a surrogate for measurement of maximization of social utility. . . .

Information positive bases and individual action. The normative bases, while influential, seldom are brought to the surface in privatization arguments. Positive assumptions for privatization, while also often inarticulate, more frequently are explicit bases for argument.

One positive assumption that frequently supports privatization efforts is the informational advantage of decentralized, incremental decisionmaking over centralized, comprehensive decisionmaking.

Incentive Advantages: Private Enterprise and Competition

A second positive assumption, often explicit in privatization proposals, is that most governmental bureaucratic. structures perform their assigned tasks poorly because the bureaucrats are not motivated to do better. Many of the contracting proposals provide no significant change in the government's mission, but posit that the mission can be accomplished better at lower cost—that is, more efficiently—under a different structure. Two different, though often conflated, assumptions support these proposals.

First, it is assumed that private enterprise, and especially private, for-profit enterprise, can provide better performance incentives than can public enterprise.

Government as Rent-Seeking

A final assumption supporting much privatization advocacy is that democratic-representative government inescapably operates as a vehicle for creating economic rents. The argument is that our government structures consistently produce appropriable private benefits at public expense.

This argument is compatible with a variety of normative goals for public decisionmaking. It does not deny the existence of principled justifications for collective action. Rather, the assertion is that even when there is a principled explanation for invoking government processes, those processes are apt to be used to other ends. This argument has received considerable play in recent years and supports the call to reduce the ambit of government action.

Legal Issues: Limits and Lines

Privatization potentially raises a series of legal issues, few extensively litigated in this context. These issues can be organized in three untidy categories: structural concerns, general obligations, and specific entitlements.

Structural Concerns: Constraint and Authority

Delegation. . . . Although there is no basis for a general constraint on delegation of government authority, it is likely that some government powers will be found to be nondelegable, that constitutionally only a specific governmental actor can perform certain given functions. Thus, even if some adjudicatory authority can be exercised by any public or private delegee, whatever adjudicatory authority is deemed to come within the judicial power of article III must be exercised by judges who have lifetime tenure, irreducible salaries, and who do not perform functions which are inconsistent with their judicial roles. These restrictions on delegation, however, will be rare; each must rest on some specific constitutional inhibition or particularized substantive concerns. Even privatization proposals involving activities that intuitively appear to be essentially governmental are unlikely to pose constitutional delegation problems.

Public liabilities. While only the extraordinary activity will be committed solely to government hands, the law may impose other constraints on privatization. The same sorts of concerns that inform decisions on government structure also inform decisions respecting the imposition of special liabilities on, or the grant of special immunities to, government. . . .

It is a truism in our system that government often is asked to eschew conduct that we tolerate from private actors. Government, for instance, cannot engage in certain types of discrimination among practitioners of particular religions that are permitted for private actors. Government often must give its employees procedural protections that private employers are free to offer or withhold. Also, government frequently is disabled from distinguishing among messages in ways that private speakers and owners of private property are free to do.

Each limitation especially imposed on government power has its own particular background and basis, but all share a common sense that government should be treated differently. . . .

Given the Courts' holdings and statements, most privatization proposals—including most contracting out, as well as deregulation, sale of assets, and so on—should allow the privatized activities to be carried on free from the special inhibitions on government action. Where direct, physically coercive power is exercised over individuals involuntarily committed by government to private hands or where clearly governmental actors retain control over the very aspect of the activity to which a special constraint attaches, as in some of the debtor-creditor cases, the public constraint still will bind the public actors. As in *Rendell-Baker,* however, even fairly slight movements away from that posture are likely to support a relaxation of public liabilities.

Public immunities. The obverse of special governmental liability is special governmental immunity. . . . If privatization generally will remove public constraints from certain activities, will it also delete the public immunities? Similar questions have arisen in at least three contexts: the state action exemption from antitrust liability, the market participant exception to constitutional commerce clause constraints, and the government contractor defense to tort liability. Although only the second of these is a *constitutional* constraint, all three are the products of judicial decisionmaking in the "common law" mode, free from reasonably specific legislative instruction. All three doctrines draw fairly tight lines around the public immunity.

General Obligations

The special constraints and authorizations discussed above directly reflect beliefs that particular decisionmaking structures can uniquely threaten or serve public interests.

The concerns that are classed here under the rubric of general obligations relate to the capacity of concentrated private interests to use government to secure special advantage over more diffuse interests. The concerns grouped under the category of particular expectations derive from the prospect for larger groups to impose incommensurate burdens on smaller groups.

Public benefit. The essence of government, in large measure, is the separation of benefits and burdens. By and large, in private activity benefits and burdens are linked; if one seeks a benefit, usually he must pay its costs. . . . In contrast, government is able to break the normal transaction, to disaggregate benefit from burden. The key to this separation is government's capacity to impose burdens by fiat.

The separation of benefit from burden allows government to overcome free-rider problems that often plague private action, but it also raises the possibility that some groups will use the government's coercive power to impose burdens on the public while arrogating special benefits to themselves. This capture problem is alleviated somewhat by the possibility that those who are burdened will exit from the jurisdiction and more by the requirements of successful coalition-building. These constraints still leave substantial scope for effective interest groups to use the political process to their own advantage.

To buttress the practical limitations on the use of government to confer "naked preferences" on well-situated groups, a number of legal constraints attempt to assure that as long as the burdens of a government activity are imposed on the general public, the benefits will also be duly shared. The federal requirement that property taken through the "eminent domain" power be put to "public use" is perhaps the most visible example. . . . Privatization proposals arguably could be challenged as conferring private benefits at public expense in contravention of restrictive legal provisions such as these.

Although these prohibitions on government capture may generate litigation, they offer little in the way of a serious impediment. . . .

There is, however, a set of legislative and administrative constraints, rooted in similar concerns about government capture, that is far more likely to chafe when *administrative* officials endeavor to privatize by contracting out.

The rules respecting administrative contracting will not preclude legislative decisions to use private contracts in place of public provision of particular services. And the opportunities for private gains to contractors may increase the political prospects for such actions. At the same time, the concerns that underlie constraints on administrative contracting should prompt political decisionmakers to weigh possible efficiency losses from the government contracting process against expected gains from private production.

Public guarantees. In addition to the legal requirements intended to assure that government serves general public interests whenever it imposes burdens on the general public, some governments have made explicit commitments to provide minimal levels or types of certain services. An example, well-known in some circles, is the New Jersey Constitution's mandate that the state provide a "thorough and efficient system of free public schools." Unlike the entitlements considered below, these guarantees do not run to any particular individuals or identifiable groups.

Nonetheless, courts at times have found such guarantees both enforceable by individual citizens and contrary to particular state actions. . . .

Guarantees of this sort arguably could pose obstacles to privatization. The reduction in government spending on particular projects or the increased reliance on local or individual choice could be found to violate substantive commitments. . . .

The conflict between substantive state guarantees and privatization, however, is not so great as this example might suggest. First, the existence of general substantive guarantees does not, in itself, mean that privatization of activities to which such guarantees attach is unlawful. . . . Second, there are relatively few enforceable, general substantive commands of this sort in American constitutions. . . .

There are, to be sure, a large number of general *statutory* obligations that might be thought inconsistent with particular privatization efforts. Although courts also may be reluctant to enforce some of these statutory prescriptions for government conduct, in other instances the courts no doubt will prove a willing forum for challenges to administrative privatization.

Private Entitlements: Grated Expectations

The most likely legal impediments to privatization efforts derive from claims of special, private entitlements, rather than from general public obligations of structural constraints. By definition, legal entitlements confer on their possessors the capacity to extract something from the government, or from someone else with the government's assistance. The action, of course, lies in determining when a claimant has an entitlement and to what exactly the claimant is entitled. . . .

The claims that should have the most concrete foundation are those based in contracts with the government. Contractual objections to privatization could arise, for instance, from collective bargaining or other employment agreements with employees who would be displaced by the decision to eliminate or, more likely, to contract out government services that formerly were performed in-house. . . .

Even where there are fairly clear, positive commitments, government largely remains free to revise those commitments. In contrast with judicial zeal to protect tangible, privately held property against even quite modest physical invasions, courts have approved legislative revisions of intangible rights that substantially and, at times, dramatically diminish the value of private property, upholding such revisions against challenges premised on the takings clause of the Fifth Amendment, the Contract Clause, the Equal Protection Clause, and the Due Process Clause.

NOTES AND QUESTIONS

1. Cass asserts that no special legal obligations can—or should, in his view—attach to private firms conducting previously public activities that have now been privatized. As a legal-descriptive matter, Cass's examples are cases involving activities that were previously private (although having some connection to the state), not those that were public and then privatized by the government. As a normative matter, why should this be so? Even if government concludes that both equity and efficiency concerns are best served by privatization, might fairness concerns nonetheless dictate that the private firms be required to employ some of the administrative procedures, especially those that due process requires of public agencies, that applied before privatization?

2. One form of privatization, part of Cass's deregulation category, is the use of economic incentive systems *within* a program of regulation as an alternative to command-and-control systems. The Environmental Protection Agency has been especially active in seeking to develop economic incentive techniques through its program of emissions trading under the Clean Air Act, a program approved and extended by Congress in the 1990 amendments to the act. The arguments for and against this approach are developed in Bruce Ackerman and Richard Stewart, "Reforming Environmental Law," 37 *Stanford Law Review* 1333 (1985) (for); and Howard Latin, "Ideal Versus Real Regulatory Efficiency: Implementation of Uniform Standards and 'Fine-Tuning' Regulatory Reform," 37 *Stanford Law Review* 1267 (1985) (against). For

many other regulatory applications, see Stephen Breyer, *Regulation and Its Reform* (Cambridge, Mass.: Harvard University Press, 1982), excerpted in Chapter I above. For a detailed discussion of an education voucher proposal, see John Chubb and Terry Moe, *Politics, Markets & America's Schools* (Washington, D.C.: Brookings Institution, 1990).

The incentive technique might also be used more extensively in government procurement, where one strategy is to base future purchasing decisions on information on past and current vendor performance, an approach that current procurement procedures impede. See Steven Kelman, *Procurement and Public Management: The Fear of Discretion and the Quality of Government Performance* (Washington, D.C.: American Enterprise Institute, 1990).

3. The current guru in market-oriented public management is David Osborne. Acting as a consultant to governments and political candidates (including Bill Clinton), he has developed a multifaceted program of "entrepreneurial" government based on studies of public institutions that Osborne claims exemplify the entrepreneurial virtues of innovation, efficiency, flexibility, resource leveraging, problem-solving, accountability, independence, competitiveness, and responsiveness to consumers. See David Osborne and Ted Gaebler, *Reinventing Government: How the Entrepreneurial Spirit Is Transforming the Public Sectors* (Reading, Mass.: Addison-Wesley, 1992). Assuming that entrepreneurial agencies can indeed generate savings and improved services, are those gains likely to make all groups better off, or might they come at the expense of those that are most costly and difficult to serve?

4. For a general treatment of market-oriented techniques in public administration by an economist who is also immersed in the political science, public policy, and legal literatures, see Susan Rose-Ackerman, *Rethinking the Progressive Agenda: The Reform of the American Regulatory State* (New York: Free Press, 1992). A particularly interesting application of Rose-Ackerman's approach is her discussion of a strategy of proxy shopping for social services, which uses market choices of unsubsidized clients to ensure high quality for needier, publicly funded clients. Id., chapter 7.

VII

Comparative Administrative Process

Just as travelers often view their native country in a new light when they journey abroad, we can learn much about the nature of our own administration law by examining how other societies have designed their administrative states. Comparative analysis, by casting the American system into bolder relief, can stimulate fresh perspectives on the familiar. Nations define the problems and opportunities of governance in different ways and have come up with different solutions. Often, of course, those differences reflect distinctive national cultures, political systems, and institutions, which cannot readily be altered or transplanted. The Skowronek excerpt in Chapter II, for example, should dispel any illusion that administrative arrangements suitable for a strong-party parliamentary system will be desirable, much less politically feasible, in a state of weak parties and committed to separation of powers principles.

While it is well not to harbor unrealistic expectations about the practical applications of comparative analysis in American administrative law, we should nonetheless mine it for whatever insights it may yield. Despite the many differences among modern democratic societies, they confront a host of common governance problems. All are expected to deal with more or less similar social conditions such as aging populations, labor-management conflicts, environmental pollution, imperfect markets, politically mobilized special interests, and public demands for protection against certain risks of contemporary life. All impose on the state a major responsibility to ameliorate

321

and regulate these conditions, and all expect it to discharge this responsibility largely through bureaucratic forms and techniques. Perhaps most important, all democratic administrative states engage competing values: public participation and accountability; bureaucratic empowerment, expertise, and efficiency; and liberal norms of fairness and due process.

James Q. Wilson's discussion of national differences in the behavior of government agencies draws upon a growing body of empirical comparative analysis at a cross-national level. Wilson's sources, however, are essentially concerned with comparing differences in the politics, policy, cultural and public administration styles of national governments; a comparison of administrative law practices is more peripheral. An excerpt from the leading casebook in comparative law, which focuses on the distinctive public law arrangements in common law and civil law systems, helps to fill that gap.

"National Differences"

JAMES Q. WILSON

When health and safety inspectors enter factories in Sweden and the United States, they come to enforce pretty much the same set of rules. They look for ladders that are unsafe, floors that are slippery, guardrails that are missing, and fumes that are toxic. Many of the standards developed by the two countries to govern these matters are not merely similar, they are identical.

But what the inspectors do in these factories is very different. The American inspectors, all employees of the Occupational Safety and Health Administration (OSHA), tend to "go by the book"—if they see a violation of the rules they write up a formal citation. If the violation is serious, a fine is mandatory and the inspector does not hesitate to levy it. Even if the violation is not serious, the inspector may impose a fine. If the employer does not correct the violation within a specified number of days, further penalties may be assessed. The OSHA inspectors believe that this is the way it must be: When Steven Kelman interviewed them they said that most employers would ignore any violations unless they were penalized. "Teeth are the only way to impress management," one inspector said; without the power to impose penalties employers would "laugh at you when you came into the plant," another remarked. Most American inspectors did not take into account the economic condition of the firm; whether or not it could afford the cost of correcting violations wasn't their concern. OSHA managers reinforced these attitudes with a lengthy field manual that prescribed in great detail every step in a workplace inspection; with the compilation of data on every aspect of the inspection; and by using these figures to compare the productivity of their inspectors (being "productive" meant making a lot of inspections and issuing a lot of violations). The inspectors were keenly aware of their bosses looking over their shoulders.

In Sweden, the inspectors for the ASV (the *Arbetarskyddsverket,* or Work Protection Board) are expected to use their discretion and not go by the book. There is scarcely any book to go by; the procedures they are supposed to follow are outlined in general language in a six-page pamphlet. Often they arrive at a factory only after giving the employer advance notice; while there, they spend much of their time advising the employer on how to improve conditions. If there are violations of the rules, the inspector typically makes oral recommendations; only occasionally will he or she issue a written notice. Fines are not automatic; they are levied only after persistent failure to correct the violation. If a firm complains that it cannot afford to make the changes,

Swedish inspectors are much more likely than Americans to give the firm more time. Moreover, the Swedes have an optimistic view of employer behavior: When Kelman spoke to them they felt most owners were law-abiding and so sanctions were not of crucial importance. The ASV managers gathered far fewer statistics about their inspectors than did their OSHA counterparts and made little use of the few they did collect.

Kelman summarized the differences between these two agencies, each with essentially the same goal, as follows: "American inspections are designed more as formal searches for violations of regulations; Swedish inspections are designed more as informal, personal missions to give advice and information, establish friendship ties between inspector and inspected, and promote local labor-management cooperation." As far as Kelman could tell, the informal and cooperative Swedish system produced a level of compliance with safety and health rules that was as high or higher than that achieved by the formal and punitive American system.

Such striking differences in bureaucratic behavior are not limited to Sweden or to industrial safety. Graham Wilson has shown that British regulation of the industrial workplace has much more in common with Sweden than it does with the United States: Whereas OSHA is ready and willing to file complaints and levy fines, its counterpart in Great Britain, the Factory Inspectorate, is averse to prosecution and like the ASV tends to give assistance rather than enforce regulations. David Vogel found that regulations designed to reduce air and water pollution were administered in a more flexible, informal, and cooperative manner in Great Britain than they were in the United States. A study of how four nations regulate pesticides, food additives, and industrial chemicals concluded that in Great Britain, France, and West Germany the administrative system afforded the bureaucrats more discretion than was enjoyed by their American counterparts and encouraged them to use informal procedures in formulating and enforcing rules. Environmental protection legislation is implemented in Japan by "administrative guidance" rather than by legal enforcement, with considerable flexibility shown in accommodating the needs of particular industries and localities.

There was no clear relationship between how each nation managed its regulatory process and the laws it enforced or the results it achieved; the consensual European administrative practices essentially served the same goals and produced the same outcomes as the adversarial American practices. How can we explain why similar bureaucracies with identical goals behave so differently in different nations? There are two possibilities: politics and culture.

Politics

The nations that enforce their regulatory policies in a consensual (that is, flexible, nonpunitive, or accommodationist) manner have parliamentary regimes; the nation that enforces its policies in an adversarial (that is, rigid, punitive, or legalistic) manner* has a presidential regime. That this should be the case is puzzling in view of the widespread belief that in the United States business has privileges and influence denied to firms in those nations such as Sweden, which have a much larger public sector, many more nationalized industries, and a more egalitarian income distribution.

The puzzle disappears when we realize what political incentives are created by the way in which political authority is organized. A parliamentary regime concentrates almost all political authority in the hands of a prime minister and cabinet chosen from the majority party (or from a majority coalition of several parties) in the legislature. It is only a slight exaggeration to say that what Prime Minister Margaret Thatcher wants she gets. So long as she has a majority in the House of Commons her proposals become law; if that majority ever deserts her, Commons will be dissolved and a new election held. Between elections the legislature can defy her leadership only at the risk of committing political suicide, that is, being told to face the voters. Parliament has little authority to intervene in the conduct of bureaucratic affairs; its members can ask questions of the political heads of these agencies, but ordinarily it cannot obtain information that the agencies wish to conceal or direct their discretion in ways that they oppose, nor can it investigate their conduct over the objections of the prime minister. The courts do not constitute an important check on the exercise of bureaucratic discretion. Though British courts have become somewhat more activist in recent years, it remains the case that no citizen has much hope of getting a judge to tell a government official what policies to adopt or what procedures to follow. Unless the bureaucracy has directly affronted one of the traditional liberties of a British subject, it has little to fear from the courts. It is almost unthinkable that a British judge should give orders to British air-quality regulators that remotely resemble the orders federal judges routinely have given to the Environmental Protection Agency in this country.

Matters are more complicated in other European democracies. Where, as in Italy or West Germany, the prime minister's party has no absolute majority in parliament, he or she can govern only with the support of a coalition of several parties. These coalitions, especially in Italy, often are unstable. Should one party leave the coalition, a new prime minister may take office even without new elections. In France, a directly elected president exists alongside the conventional parliamentary system, leading to the possibility that the president and the prime minister may be of different parties, as happened in the early 1980s. Moreover, special French courts exist to hear citizen complaints against the bureaucracy that have no counterpart in Great Britain.

---

*By saying that the United States has an adversarial enforcement system I do not mean to imply that it implements its regulatory policies in as stern or uncompromising a manner as the most dedicated advocates of those policies would prefer, nor that it does so wholly without fear or favoritism. I mean only that *in comparison with parliamentary democracies* it is more likely to rely on legal sanctions and less likely to grant enforcement operators wide discretion.

Sweden has a special constitutional officer, the "ombudsman," or people's representative, who like the French administrative courts will respond to citizens who believe they have been mistreated by the bureaucracy.

But these complexities do not alter fundamentally the central fact of a parliamentary "democracy," that political authority over both making and implementing policy is concentrated in one set of hands, those of the executive. Neither the French administrative courts nor the Swedish ombudsmen permit citizens to shape policy; at best, they remedy individual grievances about bureaucrats who clearly have overstepped their legal authority. But when that authority is granted in sweeping language, as it often is, there is no boundary to be overstepped.

If authority is concentrated, the incentives to organize politically (other than to contest elections) are weakened. If a group is unable to persuade the majority party to embrace its programs, it has no recourse but to wait for the next election in hopes that a different majority will assume office. It cannot expect to make an end run around the prime minister's decision by getting a legislative committee to do informally (through the conduct of investigations, the exercise of committee "guidance," or the requirement for committee "clearance") what formally the legislature as a whole would not do. It cannot demand that the executive in charge of the program be a person approved by or drawn from the ranks of the group; the direction of programs will be in the hands of career bureaucrats who owe little or nothing to outside interets. It cannot anticipate that a friendly judge will review an agency's decisions to insure that wherever the statutory language is vague they will conform to the "intent" of the legislature as manifested in obscure reports written by like-minded staff members of some congressional committee; judges in parliamentary regimes ordinarily will not substitute their views for those of appointed officials. It cannot assume that the media will embarrass bureaucrats who have acted contrary to the group's interests with real and imagined misdeeds; bureaucrats in parliamentary regimes find it relatively easy to keep things secret from the press.

If a potential group cannot entertain these hopes or make these demands it has little incentive to organize and express them. Of course, there are well-organized interest groups in Europe as there are in the United States, but abroad these groups typically take the form of peak associations—large, nationwide assemblages of workers, employers, or professionals. They are large and national because they must concentrate their resources on influencing the key policy decisions of a few national leaders. A small organization consisting of a handful of activists, a foundation grant, and a catchy name is less likely to appear (or, if it appears, to be successful) in Europe than in the United States because there are far fewer points in the political system abroad at which such a group could gain access and wield influence.

These institutional arrangements contribute to the adversarial nature of bureaucratic politics in this country. All policies are continuously contested. The legislative coalition that enacted a new program will be quickly supplanted by another coalition that influences the implementation of that policy.

No sooner had Senator Hubert Humphrey promised Congress that the Civil Rights Act of 1964 did not contemplate the cutting off of federal aid as its principal enforcement device than civil rights groups were pressing a federal judge to order the Office for Civil Rights to do exactly that. No sooner had Congress passed an Occupational Safety and Health Act authorizing OSHA to use coercive measures than industry was lobbying members of Congress to insure that OSHA followed a conciliatory strategy; and no sooner had this lobbying begun than labor went to court to make certain that OSHA took a tough line.

Policymaking in Europe is like a prizefight: Two contenders, having earned the right to enter the ring, square off against each other for a pre-scribed number of rounds; when one fighter knocks the other one out, he is declared the winner and the fight is over. Policymaking in the United States is more like a barroom brawl: Anybody can join in, the combatants fight all comers and sometimes change sides, no referee is in charge, and the fight lasts not for a fixed number of rounds but indefinitely or until everybody drops from exhaustion. To repeat former Secretary of State George Shultz's remark, "it's never over."

Under these circumstances a prudent bureaucrat will realize that any effort to keep agency proceedings secret, informal, and flexible will entail heavy political costs. Matters can be kept secret only if there are a few key partici-pants (say, one peak business association and one nationwide labor union) who are aware that they must work with the bureaucracy because there is no one else to whom they can appeal its decisions. If there are many participants and easily available appeals, trying to keep anything secret can be portrayed as a "cover-up." Implementation can be informal and flexible only if the interests involved have an incentive to accept the agency's actions, and they will have that incentive only if they do not think they can get a better deal elsewhere—in court, the press, Congress, or the streets. No agency head relishes the idea of being called in the media a "lackey" of some special interest, being hauled into court to explain why he or she failed to follow every detail of the prescribed procedure, or being grilled by a congressional committee as to why he or she interpreted a vague statutory provision in a way that favored "special interests" rather than the "public interest" or that al-lowed Company A to take more time in complying with an order than was given to Company B.

The political base of an American agency is often insecure; the coalition supporting it is frequently weak, short-lived, or shot through with internal contradictions. Under these circumstances a bureaucrat who may wish to adopt informal and flexible procedures first must establish authority so that flexibility will not be construed as a sign of weakness or evidence of a sell-out. Ronald Brickman and his colleagues, authors of an important cross-national study of chemical regulation, came to the same conclusion:

> The fragmentation of political authority leaves U.S. administrators in a peculiarly vulnerable position. They are confronted with contradictory

statutory mandates. . . . These must often be implemented under the critical eye of other government institutions, and in full view of warring private interests. . . . Unable to strike bargains in private, American regulatory agencies are forced to seek refuge in "objectivity," adopting formal methodologies for rationalizing their every action.

It makes sense then for the bureaucracy to take a tough line: Insist that everything be on the public record, insist that everything be done "by the book," insist that "everybody be treated the same," and insist that the full force of the law fall on every violator.

That acting this way is less risky does not mean that every agency will act this way. The political environment of some agencies involves client politics—a cozy, low-visibility relationship between a private interest and its like-minded government bureau. Or a strong-willed (and unusually lucky) president may succeed in getting an agency head to act other than as his or her immediate political interests require. But the institutional incentives to adopt a formalistic and adversarial mode in managing agency-interest relations are very strong (and, as the next section will suggest, are getting stronger all the time) and so we should not be surprised to discover that on the whole American regulatory agencies behave very differently from their European counterparts.

Graham Wilson compares a regulatory agency to a classroom: "A teacher whose authority is assured may be able to adopt a more friendly and relaxed approach than a teacher whose authority seems highly uncertain." In Europe, regulators tend to have a more secure base of authority than they do here. Thus, left-leaning, "antibusiness" regimes abroad actually may have friendlier and more accommodative relationships with business than do conservative, "probusiness" regimes in this country.

Culture

But politics cannot be the whole story. The American political system based on the separation of powers is essentially the same today as it was thirty or forty years ago; however, regulatory agencies at that time were given far greater discretionary authority than is the case today. The British political system based on prime ministerial rule is not very different today from what it was a century ago; but then, in the early years of factory inspections, employers not only complained bitterly about the hostile behavior of the inspectors, they occasionally assaulted them as they made their rounds.

The political costs attached to secret administrative proceedings make secrecy difficult to maintain; there is, as Edward Shils pointed out many years ago, a populist streak in the American character that provides ample ammunition for any politician eager to attack a "cover-up." For decades, the British quietly have accepted a level of governmental secrecy that would have led to rebellion in the United States. It is inconceivable that the United States would ever adopt an Official Secrets Act comparable in scope and severity to that which long has been on the books in Great Britain.

The ability of the courts to intervene in American administration is based on their power of judicial review (that is, on their right to declare executive and legislative acts unconstitutional) and on such laws as the Administrative Procedure Act, but the present-day scope of judicial activism could not be predicted simply from knowing its constitutional or statutory foundation. . . . [T]he federal courts for many years maintained rules of standing and deferred to administrative discretion in such a way as to make judges relatively modest influences on federal administrative agencies. Today those rules and that deference are a thing of the past and so judges, because of decisions they have chosen to make, are major players in bureaucratic politics.

Finally, there are many government bureaucracies in the United States that display in their everyday routines a degree of informality and flexibility that seems to belie the findings about regulatory agencies. No one who has spent even a few hours in a representative French and American schoolroom could fail to observe the looser, less directive, more individualized atmosphere of the American public school. The level of discipline in the American armed services is less than that in many European armies. British sailors are more deferential to their officers on merchant ships than are American sailors. The group-based decisionmaking style of Japanese managers has been contrasted with the individualistic style of American executives so often as to have become a cliché, but it is true.

Michel Crozier has provided us with an insightful look at two French government agencies performing routine clerical and manufacturing tasks. Though he made no explicit comparisons with similar American agencies, the picture he painted of the French agencies left little doubt that their managerial atmosphere was quite different from what one encounters here. In France, the organizational culture was rigidly formal, with little informal communication, few work-based voluntary associations, heavy reliance on written rules and procedures, and elaborate (though often resentful) deference to the outward trappings of status and rank. The easy give-and-take among workers at different hierarchical levels so typical of the daily life of most American bureaucracies was conspicuously absent in the French system.

In short, the way in which a bureaucracy operates cannot be explained simply by knowing its tasks and the economic and political incentives that it confronts. Culture makes a difference. . . . Culture is defined as a set of patterned and enduring ways of acting, passed on from one generation to the next. A national culture consists of those patterned and enduring ways of acting characteristic of a society or a significant part of that society. Culture is to a group what personality is to an individual, a disposition that leads people to respond differently to the same stimuli. Though every traveler is immediately aware of how differently the British or the Japanese or the Swedes respond to meeting a stranger, addressing a clerk, or joining a group, there is no systematic, well-established account of these differences. As with most important things in life, we are aware of more than we can explain. But it is

possible to list a few major cultural factors that from the available research
seem to influence how people behave in formal organizations.**

Deference Versus Self-Assertiveness

In some societies the right of government officials to make decisions is taken
for granted. People may disagree about the substance of the decision but they
do not question the authority behind it. As a consequence they defer to
officials who act for the government. Police officers have authority by virtue
of their badges, teachers by virtue of their positions, bureaucrats by virtue of
the laws that authorize their offices. In other societies the right of government
officials to make decisions frequently is challenged. People usually are disposed to accept a governmental decision but they are always ready to question
the right of a police officer to stop them, a teacher to control them, or a
bureaucrat to decide a matter of importance to them. The formal signs of
authority—badges, uniforms, laws, regulations—do not produce in the citizenry a habitual deference to the persons who display those signs.

Steven Kelman contrasts the deferential Swedish political culture with the
self-assertive or adversarial American political culture to explain why the
Swedish ASV is able to rely on informal, consensual methods not only for
enforcing health and safety rules but also for writing them. Modern Sweden,
like many northern European nations, emerged from a long history during
which power was centralized in the hands of a king or an aristocracy; for
centuries, a wide, unbridgeable gulf separated the rulers from the ruled.
Deference to authority was obtained by a combination of coercion and awe:
the king's power was enhanced by his grandeur. When Swedes replaced the
king with a parliament and the aristocracy with a bureaucracy, they did not
replace deference with self-assertiveness. The democratic Swedish state was
in every sense a *state,* that is, a supreme institution that was expected to rule
and was accustomed to ruling. The deference once accorded an absolute
monarch was now given to a set of elected and appointed officials. The fact
that Sweden, beginning shortly after the advent of parliamentary democracy,
was ruled by the same political party, the Social Democrats, for nearly a half
century and that this party put in place policies that transformed Swedish
society made it easy for people to transfer their deferential habits to the new
state.

This interpretation of Swedish political culture is not unique to Kelman;
studies by Thomas Anton, Donald Hancock, and Sten Johansson have come

---

**There is very little systematic, quantitative research that analyzes cultural differences in
the beliefs of bureaucrats and almost none that discusses how those different beliefs affect
behavior. One interesting comparative study of international differences in work-related values
among several thousand members of a large, multinational corporation is Geert Hofstede, *Culture's Consequences* (Beverly Hills, Calif.: Sage Publications, 1980). There are certain parallels
between Hofstede's description of "power distance," "uncertainty avoidance," and "individualism" and my account of deference, formalism, and individualism.

to much the same conclusion. The consequences of this culture for Swedish politics are clear: Swedes accord government officials high status, do not participate (except by voting) in many political associations, and believe that experts and specialists are best qualified to make governmental decisions. Given these attitudes it is hardly surprising to find that industrial health and safety regulations were not hammered out in the legislature after long, bitter fights between contending interests but rather were written by a series of small committees consisting of experts drawn from government, business, and labor who met privately, resolved their differences by discussion, and did not have to confront any challenges in the legislature. Nor is it surprising to learn that in enforcing these regulations the Swedish ASV relied heavily on the exchange of expert advice among like-minded people in government, a particular firm, and its associated labor union. Experts—people with professional training—are an important influence on American bureaucracies, but they seem to be the dominant influence on Swedish ones.

American political culture hardly could be described as deferential. Americans value expertise but they do not defer to it; an expert who takes an unpopular position or acts contrary to the self-interest of an individual or a group will be treated as roughly as any other adversary. Americans admire their form of government but do not admire or accord high status to the officials who work for it.[†] Americans always entertain the suspicion that the government is doing something mischievous behind their backs and greet with outrage any indication that important decisions were made in a way that excluded any affected interest, no matter how marginal. Americans define their relationship with government in terms of rights and claims and are prepared to hire lawyers or complain to the newspapers at the slightest hint that a right has been violated or a claim ignored.

A deferential political culture is not unique to Sweden; elements of it can be found in all Scandinavian nations, in Germany, and in Great Britain. Steven L. Elkin, who had learned in this country that urban development decisions are the result of an intensely political process involving much conflict, aroused publics, and weak planners, was not prepared for what he found when he studied the same subject in London, where major new projects were undertaken with scarcely any public notice, much less participation. Large tracts were cleared and new developments begun on the unchallenged authority of town planners employed by the London County Council. The reason was clear: Such decisions, though they affected the lives, property, and incomes of thousands of people, were made in London by officials who were effectively insulated from community pressures. There seemed to be general agreement that the job of the government was to govern; there was little doubt as to who should make the decisions (appointed officials), how they should be made ("rationally"), or even what they should be. If American city

---

†More precisely, they tend to revere the presidency but not the president; they dislike Congress and "the bureaucracy" though usually they have a good opinion of their own representatives and report that their experience with a given bureaucrat was satisfactory.

planners had moved to London they would have thought they had died and gone to heaven.

An adversarial political culture is not unique to the United States, but it is in the United States that the political institutions—the separation of powers, judicial review, and federalism—allow it full expression and reinforce its central features. Everywhere, of course, institutions and the incentives they create interact with culture and the habits it fosters. American political culture and institutions are remarkably congruent, however, so much so that it is hard to imagine parliamentary institutions being transplanted here.

Formality Versus Informality

In France, and no doubt in many other countries as well, the members of an organization deal with one another formally; that is to say, impersonally, at arm's length, and with close attention to rank and titles. Though two workers may become friends they do not form cliques or voluntary associations that cut across hierarchical lines or carry over to life outside the workplace.

Crozier attributes this to the influence of French cultural traits that place great emphasis on the formal equality of individuals and impart a strong desire to see the exercise of power constrained by formal rules rather than by consultation or organizational checks and balances. At each level in a bureaucratic hierarchy authority is absolute; the limits on the exercise of that absolute authority are the written rules and prescribed procedures that the holder of authority must obey. The informal give-and-take of an American workplace is less common in France; by the same token, French citizens are far less likely than Americans to join or be active in voluntary associations. The employees of the two agencies he studied were prepared to obey the rules but not submit to the rulers; submission, cooperation, or informality would threaten the autonomy of each individual.

Crozier follows Alexis de Tocqueville in locating the source of these attitudes in the experiences of the French peasantry under the old monarchy. France was ruled from the center—that is, from Paris—by a succession of kings eager to consolidate their power and to extract from the people tax monies with which to finance the nation's endless wars. Taxes were collected in a way that gave to every taxpayer a powerful incentive to spy upon his neighbors and notify the tax collector whenever he saw any increase in their wealth. Local self-government existed in name only; nothing of consequence could happen unless at the direction or with the approval of the king's ministers in Paris. Where meaningful local government does not exist the people will have little incentive or opportunity for forming local associations to voice their grievances or press their claims; there is no one at hand who can hear the grievance or grant the claim. There were no assemblages in which peasants might learn to work with one another or become acquainted with officers and aristocrats. The French political system split French society into "small, isolated, self-regarding groups."

When the French Revolution swept away the monarchy and discredited the aristocracy it left behind a mass of people who sought to use their new powers to guarantee by law and constitution the formal equality of all citizens. A French Assembly replaced the king; then Napoleon replaced the Assembly; then a bewildering succession of emperors and new assemblies replaced the old ones. But throughout it all Paris never lost its grip over the countryside nor its insatiable appetite for tax monies with which to finance the army and the bureaucracy. Under these circumstances it is hardly surprising that the French people had little reason to trust one another, let alone their government. Power existed; that was a fact. Power was centralized; that had been decreed by history. Now power had to be checked, but the only check with which the French had any experience was that embodied in laws and rules. Crozier argues that this legacy of legalism and formalism has shaped the conduct of French government agencies right down to the present.

The United States followed a different path. There was scarcely any central government before, during, or for long after the Revolution; survival depended not on the benevolence of some lord but on the individual farmer's capacity for work and cooperation. Cooperation was encouraged not only by the need to subdue the wilderness but by the policy of self-government on which the Protestant churches were founded. This tradition of political and religious self-government was so strong that it nearly prevented the ratification of the Constitution, and even that document created a government of limited and circumscribed powers. The founders knew very well the views of the citizens and did not for a moment consider relying on legal rules and formal rights as the chief guarantee of individual liberty; in the words of James Madison these were but "parchment barriers" that without "auxiliary precautions" would not prevent the exercise of absolute power. The additional precautions were, of course, the separation of absolute power. The additional precautions were, of course, the separation of powers and the system of checks and balances. These devices insured that officials would negotiate with one another because without a negotiated policy there would be no policy at all. Consultation, participation, and informal arrangements all were part of the general scheme of government and, before that, of the daily life of Americans in their towns and villages. Small wonder that they should be part of the ongoing administration of the agencies spawned by that government.

As with deference, so also with formalism: What culture may have begun, institutions keep intact. The informality of Americans is a fact of their daily life. (Is there any other nation where everybody, including waiters and flight attendants, is so insistent on addressing people by their first names?) This informality animates our governing institutions but in turn is reinvigorated by them.

Groups Versus Individuals

The importance of consultation to American bureaucrats is great, but it pales into insignificance compared to its place in the lives of Japanese bureaucrats. This is puzzling because Japanese administration is far more hierarchical than its American counterpart. In this country, officials of different ranks meet frequently and address one another by their first names rather than by their titles. Such informality would be shocking to a Japanese; dealing with hierarchical unequals as if they were social equals simply is not done. The "formal status of every member, in relation to other members in the hierarchy, is defined in the most meticulous and particularistic manner."

Within a given hierarchical level, however, Japanese management takes the form of group decisionmaking in which individual differences are minimized and individual assertions of leadership or claims of credit are discouraged. The group emphasis is evident, as Gregory Noble notes, even in the physical arrangements. Where an American administrative unit usually will consist of a large office for the head and separate offices down the hall for each of his or her subordinates, a comparable Japanese unit will consist of a chief whose subordinates sit in the same room with him, their desks arranged in a semicircle. Organizational charts in the United States often contain the names of each official; in Japan, many of the boxes have no names in them. American police officers wear badges with individual identification numbers and sometimes personal nameplates as well; Japanese officers wear no badges, numbers, or nameplates. Americans leave their offices and go home when they have finished their individual tasks; Japanese linger in their offices until all members of the group have finished their duties and when they finally leave, it is often to go to a bar where they relax and drink together.

The importance of the group is heightened by the way new employees are socialized on the job. Individual job descriptions are vague or even nonexistent. Training emphasizes character-building and group loyalty as much or more than technical knowledge. When a group must make a decision, individual argumentation and personal clashes are avoided in favor of indirect suggestion and protracted exploration. The emphasis is on harmony and consensus, even at the cost of speed and decisiveness.

Lewis Austin explored these values in his probing interviews of Japanese and American elites. Though the sample was small, the differences were striking. Whereas the Americans attached the greatest importance to individualism, equality, and competition, the Japanese assigned greater weight to hierarchy, group solidarity, and harmony. The Americans described a good leader as someone who was decisive, knowledgeable, and willing to delegate; the Japanese described him as someone who was sincere, warm, and willing to listen.

When group-centered decisionmaking is linked with rigid hierarchical distinctions the problem of coordination can become acute. The loyalty to one's work group can impede communication with someone else's group; the need to achieve unanimity in one office before acting can get in the way of negotiating an agreement with another office. Chalmers Johnson, a perceptive student of Japanese politics, describes the problem this way: "[T]he most difficult coordinating task in the Japanese policymaking process is among ministries and agencies themselves; once an interministerial agreement has been reached, the chores of taking the proposal or bill through the party, cabinet, and Diet stages are relatively less onerous."

The reason why dealing with party politicians is easier than coordinating with fellow bureaucrats is that in Japan the bureaucracy *is* the government. The bureaucracy drafts the laws, determines the budget, has a near monopoly on the relevant information, enjoys great esteem, recruits the most talented graduates of the best universities, and has remained virtually intact (despite wars, military occupation, and constitutional change) for over a century. Only one party has governed Japan since 1955; its leaders have had much the same view as the bureaucracy. Moreover, Japan, like Sweden, is a nation governed by the principle that the state is supreme and expert knowledge is decisive. The bureaucracy embodies the traditional understanding of what governance is all about, and so it is the central institution of the regime rather than, as in the United States, a necessary but unwelcome contrivance that exists in order to implement the decisions of others.

The power of the bureaucracy may have structural or political explanations, but its manner of operating clearly has cultural roots. The importance of the group is not an administrative invention, much less one that can be exported; it is the institutional embodiment of a pervasive communalism in Japanese life, a communalism that can be found in the family, the schools, and the neighborhood.

Impersonal and Personal

Most nations of Latin America have a government bureaucracy at least as large and complex as anything found in Sweden or France; most also intervene in their economies—by regulation, licensing, and subsidy—at least as much as did England in its most mercantilist phase. But unlike the administrative systems of these European countries, most Latin American bureaucracies do not embody the tradition of the rule of law, the expert application of general rules to particular cases, or the impersonal conduct of public affairs.

Almost every scholar who has written about Latin American government stresses the patrimonial nature of administrative rule. The Spanish colonies in the New World were treated as the personal property of the monarch; the advent of independence replaced distant kings with local presidents, but many of the presidents maintained the feudal view that they were patrons and the people their clients. "Not only is politics concentrated in the office and person of the president, but it is by presidential favors and patronage that contracts are determined, different clientele are served, and wealth, privilege, and social position parceled out. The president is *the* national *patron,* replacing the local landowners and men on horseback of the past." Lower-level officials are selected on the basis of their loyalty to the chief or his political faction. Though nepotism is not as widespread as is sometimes alleged, recruitment on the basis of personal friendship and political sympathies is.

The administrative system is centralized as well as paternalistic. Though a few Latin nations such as Argentina, Brazil, Mexico, and Venezuela have in theory a federal regime with important powers left in local hands, in reality the central government either is wholly dominant or it reserves (and exercises) the right to intervene in local affairs whenever it chooses.

Patrimonial or personalistic rule embedded in a centralized state has at least three consequences. First, all power is lodged at the top so that even the most routine decisions and minor memoranda must be referred to the head of the agency. As a result, communication is slow and difficult, decisions are delayed and often ill informed. Second, when power is both concentrated and used at the pleasure of the leader, the incentives for corruption are very large. The corruption can take the form of lower-ranking officials accepting bribes from people who wish to short-circuit the official procedures or high-ranking officials using their great powers to line their own pockets. Third, to facilitate central control over a rapidly growing bureaucracy and to reduce the degree of corruption, Latin governments have relied on formal rules and increasingly complex laws and procedures. Every official is enmeshed in a web of regulations, many either ambiguous or contradictory. This legalism amounts to what one scholar called the "code fetish."

American bureaucrats weary from the effort of coping with a tangle of procedural constraints might feel their spirits lift (even if their burdens do not lighten) by contemplating the vastly greater procedural morass in which their Latin American counterparts find themselves, made all the more frustrating because it does not serve its goals. Multiplying rules neither enhances central control nor eliminates corruption. What it does instead is revealed by anecdotes such as this: "In Brazil a patient citizen waited ten years for a tax refund, only to see inflation reduce it to virtually nothing. When he offered to donate the pittance to the government, a bureaucrat begged him to keep the money in order to avoid another prolonged procedure." The Guatemalan bureaucracy consists of "executives seated behind mounds of petitions, memos, and requests for rulings; besieged by clients, friends, and subordinates begging assistance or opinions; and occasionally and hastily scanning a report that is probably largely fictitious and almost inevitably devoid of critical analysis."

To cope with the cumbersome, rule-bound, and overcentralized government bureaucracy, many Latin nations have created autonomous government corporations and independent agencies to which are given tasks that the central bureaucracy has shown itself unable to perform. Some produce steel or oil; others set prices and wages; still others distribute water, run schools, or administer social security. Some have become models of efficiency and integrity; others merely have increased the administrative confusion in which the government is immersed. But all contribute to the enormous power of the state. We expect the government to dominate the economy in Marxist Cuba or Nicaragua, but the dominance is nearly as great in Brazil, Colombia, or Mexico. For example, Howard Wiarda and Harvey Kline estimate that the Brazilian government either by itself or through public corporations generates 55 to 60 percent of that country's gross national product.

The United States has had its share of patrimonial rule in urban political machines, corrupt county courthouses, and overcentralized Washington bureaucracies. It has responded to these problems by attempting to use formal procedures and legal constraints to convert patrimonial rule into impersonal rule, but with a difference: In the United States, the state is limited (it produces only about 11 percent of the gross national product) and it is decentralized (state and local governments play large roles and exercise substantial powers).

Statist and Nonstatist Regimes

The limited scope and constrained powers of the American government are its most important features. By European standards it is not truly a "state"— that is, a sovereign body whose authority penetrates all aspects of the nation and brings each part of that nation within its reach. In this country sovereignty belongs in theory to the people and in practice to no one; to the extent that it exists at all it is shared by national and state governments. What someone once said of the British House of Commons—that it could do anything except change a man into a woman—can be said of no institution in this country.

State-centered regimes are executive-centered regimes, and executive-centered regimes are dominated by their bureaucracies. Despite their many important differences, the governments of France, Japan, Sweden, Brazil, and Mexico are alike in making the administrative apparatus the center of official action. The bureaucracy, with relatively few checks from either the legislature or the judiciary, drafts laws, issues regulations, allocates funds, and guides policy. Since no single institution can manage affairs unaided, the bureaucracy enlists or commands the help of private interests—corporations, labor unions, and professional groups. These join together comfortably or warily to propose policies and implement laws. There is an easy interchange of ideas and people between the public and private sectors, so easy an interchange that it is hard to draw any clear line between them. Senior government bureaucrats not only deal routinely and privately with their corporate counterparts, they join the corporate world as top executives upon retiring from government service (the Japanese call it "descending from heaven").

Political scientists have begun to describe such states as corporatist, meaning nations in which the major segments of society (especially industry and labor) are represented directly in the state as bodies entitled to participate in making laws and carrying them out. In the United States business and labor among all other segments of society form interest groups or lobbies to press claims on the government; in corporatist states these interests need not press anything, for they are, informally if not formally, part of the government. The formulation and implementation of industrial safety policies in Sweden with which this selection began is an example of corporatism in action.

Cultural and constitutional factors produce important differences in state-centered regimes and in how their bureaucracies function. In Japan and northern Europe the statist governments might be called rationalistic: They embody the values of deference, impersonal rule, and administrative discretion. In the state-centered regimes of Latin America the governments reflect the values of an adversary culture, personal rule, and procedural formalism.

In the first case the bureaucracy tends to operate on universalistic principles (the expert application of neutral rules), to recruit its members on the basis of achievement criteria (skills acquired by education and training), and to serve collective interests (the state, society, and the ruling party). In the second case the bureaucracy tends to operate on particularistic principles (give each person what he deserves), to recruit its members on the basis of ascriptive criteria (hire people with the "correct" social, familial, or political connections), and to serve personal interests (those of the president, the junta, or the minister).

Of course these sweeping distinctions oversimplify reality, ignore important exceptions (there is corruption even in Sweden!), and neglect trends that are moving once-different nations toward greater similarity. To do justice to these matters would require a book on comparative government. But simple as they may be, they highlight what is distinctive about the bureaucracy of the United States. It serves a weak state (though one that is growing stronger); it copes with an adversarial rather than a deferential culture; it has found its exercise of discretion falling under suspicion and so has tried to constrain that discretion by formal procedures and elaborate rules; it is staffed by individualistic rather than group-oriented workers; and it has taken steps to reduce personal and patrimonial rule in favor of impersonal and rationalistic standards.

Comparative Law

RUDOLPH SCHLESINGER, HANS BAADE, MIRJAN DAMASKA, AND PETER HERZOG

[In the trialogue that follows, Smooth and Edge are American common lawyers; Comparovich is an American comparative law scholar who has been retained by Smooth and Edge as a consultant to help them deal with issues arising in a civil law jurisdiction. Because the material and sources on comparative law are unfamiliar to most American students and lawyers, I have retained the footnotes for explanatory and bibliographic purposes.]

Public-Law Disputes*

The Civilians' Dichotomy Between Private-Law and Public-Law Litigation

SMOOTH: In many countries, we sell our products by vending machines installed in public places, or from specially equipped small trucks in the streets. Ordinarily, this requires an official license of some kind. If the issuance of such a license is refused, or if a license is revoked by the administrative authorities, we often get into litigation. In our system, of course, such cases would be handled by ordinary courts under writs of certiorari, mandamus, or prohibition, or some statutory version of these ancient writs. We noticed, however, that in continental countries such cases are not considered by the ordinary courts about which we have been talking, but by administrative tribunals.

COMPAROVICH: Apart from criminal matters, the jurisdiction of the ordinary courts in most civil law countries is essentially limited to "private law" disputes. A controversy involving the validity or propriety of an administrative act (such as the refusal or revocation of a business license), involves issues of public law and under their system will be determined by a separate hierarchy of administrative tribunals.[1]

---

*The materials in this section deal with *judicial* remedies available to a party who feels aggrieved by an act of governmental bureaucracy. The reader should keep in mind, however, that judicial remedies are not the only possible ones, and that in many countries there are less formal, nonjudicial channels through which a citizen can seek redress against official arbitrariness. A comparative study of these channels (which may exist in addition to, or in lieu of, judicial remedies) is of great interest. The classical work is W. Gellhorn, *Ombudsmen and Others: Citizens' Protectors in Nine Countries* (1966). For more recent contributions see, for example, G. E. Caiden (ed.), *International Handbook of the Ombudsman* (1983); J. C. Juergensmeyer and A. Burzynski, *Parliamentary and Extra-Administrative Forms of Protection of Citizens' Rights* (1979) (covers Poland, East Germany, West Germany, Sweden, U.S.A., U.S.S.R., and Yugoslavia); M. Wierzbowski, *Administrative Procedure in Eastern Europe*, 1 *Comparative Law Yearbook* 211 (1977–78); D. Rowat, *The Ombudsman Plan: Essays on the Worldwide Spread of an Idea* (1973); L. N. Brown and P. Lavirotte, *The Mediator: A French Ombudsman?* 90 *Law Quarterly Review* 211 (1974).

On American counterparts, see S. V. Anderson, *Ombudsman Paper: American Experience and Proposals* (1969); A. J. Wyner (ed.), *Executive Ombudsmen in the United States* (1973).

[1] For a comprehensive comparative treatment of the subject, see Max-Planck-Institut für ausländisches öffentliches Recht und Völkerrecht, *Judicial Protection against the Executive* (H. Mosler, ed., 1971), especially Vol. 3.

The civilian tradition of separate administrative courts is seldom followed in socialist countries of Eastern Europe. In some (U.S.S.R., Czechoslovakia, East Germany) there is virtually no judicial review of administrative acts. In others (Bulgaria, Hungary, Rumania, Yugoslavia) a more or less wide range of administrative acts is reviewable in ordinary courts, or in special sections of supreme courts. In 1980, Poland reintroduced a special administrative court of the traditional continental genre. This court has broad jurisdiction to set aside administrative acts that violate the law; but certain acts, such as the expulsion of an alien, or denial of a passport, are exempted from the court's jurisdiction. The impact of this reintroduced system of administrative

EDGE: It thus appears that under their system the respective subject-matter jurisdiction of ordinary courts and administrative courts depends on whether the dispute belongs to the realm of private law or that of public law. As a consequence, the distinction between private law and public law must be of much greater practical significance in civil law systems than under our law.

COMPAROVICH: You are basically right. But this difference between the civilians and us should not be exaggerated: we also assign practical importance to the distinction between private and "public rights" disputes, although for a more limited purpose. The legislative branch of the federal government has the power (not infrequently exercised) to have cases involving public rights—that is cases ordinarily involving a dispute between the government and a nongovernmental party—adjudicated by special legislative courts or by administrative agencies. The latters' decision may be subject to review by ordinary courts, but such review can be limited to issues of law, in much the same manner as the reviewing court's scope of review is limited in jury cases. On the other hand, Congress has no similar power to assign cases involving purely private rights to legislative courts or administrative agencies.[2]

But let me return to continental administrative tribunals. Their members usually have a high degree of expertise in administrative matters; but ordinarily they are independent of the executive and of the administrators over whose acts they sit in judgment.[3] They are judges in every sense of the word.

In Germany, where a high value is placed on specialized judicial expertise, there are several separate hierarchies of tribunals dealing exclusively with public law disputes: tax courts, social security courts, and "administrative" courts.[4] The latter deal with all public law disputes not involving tax or social security matters.

In France, jurisdiction over public law disputes is less fragmented. In particular, there is no separate hierarchy of tax courts, with the result that tax controversies normally come before the regular administrative tribunals.[5] The

---

adjudication seems to be considerable. For a survey of Eastern European systems (save Yugoslavia), see M. Weirzbowsky and S. C. McCaffrey, "Judicial Control of Administrative Authorities: A New Development in Eastern Europe," 18 *International Lawyer* 645 (1984).

[2] See, for example, Northern Pipeline Construction Co. v. Marathon Pipeline Co., 458 U.S. 50, 67, 69, 102 S. Ct. 2858, 2869, 2870 (1982). Nor can Congress, when dealing with actions at law that involve *private* disputes, take the fact-finding function away from the jury. See *Atlas Roofing Co., Inc. v. Occupational Safety and Health Review Commission,* 430 U.S. 442, 450, 97 S. Ct. 1261 (1977).

[3] See, for example, R. Drago, "Some Recent Reforms of the French Conseil d'État, 13 *International and Comparative Law Quarterly* 1282, 1287 (1964).

[4] Altogether, Germany has five separate judicial hierarchies: ordinary courts, labor courts, tax courts, social security courts, and administrative courts. Each of these hierarchies includes tribunals of first instance, intermediate appellate courts, and a court of last resort. In addition, there is the Federal Constitutional Court. . . . For a description of the entire system, see H. G. Rupp, "Judicial Review in the Federal Republic of Germany," 9 *American Journal of Comparative Law* 29 (1960).

[5] The statement in the text, supra, should be understood as referring to direct taxes. There are some peculiarities with respect to indirect taxes. For details see 1 Auby and Drago, *Traité de Contentieux Administratif,* §§569–71, 623–30 (3d ed., 1984).

advantages of judicial specialization, however, are not necessarily lost under the French system. The Conseil d'État, the administrative court of last resort,[6] normally sits in panels, called *sous-sections;* matters requiring specialized knowledge, such as tax cases, always are referred to particular panels.[7]

EDGE: Do civil-law systems generally provide for a hierarchy of administrative courts, distinct from the hierarchy of ordinary courts?

COMPAROVICH: A few of the developing countries, though otherwise influenced by French legal thinking, declined to adopt or retain the continental system of separate administrative courts. They made this policy decision on the ground that because of scarcity of trained judicial manpower it would be undesirable for them to maintain several hierarchies of tribunals. In the Republic of Senegal, for instance, "the courts of first instance have jurisdiction over administrative as well as civil matters, and the Supreme Court performs the functions of both the French Cour de Cassation and the Conseil d'État, although they are performed by separate sections of the Court."[8]

Subject to this observation concerning some of the developing nations, and subject also to certain reservations with respect to Latin America,[9] the answer to your question is in the affirmative: In the overwhelming majority of civil law countries the administrative courts form a judicial hierarchy separate and distinct from that of the ordinary courts. This, indeed, can be regarded as one of the earmarks distinguishing civil law from common law systems.[10]

---

[6] It should be noted that not all of the business of the Conseil d'État is judicial. Of its five Sections, only one, the *Section du Contentieux,* has judicial functions. What is said in the text about the Conseil d'État as the highest administrative court of France, refers entirely to the work of its *Section du Contentieux.* The work of the four nonjudicial Sections is discussed by L. N. Brown, "The Participation of the French Conseil d'État in Legislation," 48 *Tulsa Law Review* 796 (1974). This article is particularly instructive because its author was "privileged to witness in person the functioning of the Conseil d'État in its less publicized role of legal consultant to the French government and its Ministers."

[7] See S. Grevisse, France: Le Conseil d'État, in A. Tunc (ed.), "La Cour Judiciaire Suprême—Enquête Comparative," 30 *Revue Internationale de Droit Comparé* 1, at 217, 223 (1978).

[8] E. A. Farnsworth, "Law Reform in a Developing Country: A New Code of Obligations for Senegal," 8 *Journal of African Law* 6, 11 (1964).

[9] The attitude of Latin-American countries toward the jurisdictional separation of "private law" and "public law" disputes is somewhat wavering. These countries, it will be remembered . . . , belong to the civil law orbit with respect to their private law; but in the area of public law, the United States model has exercised considerable influence. It is not too surprising, therefore, to find that the Latin-American legal systems are almost evenly divided on the issue whether public law disputes should be determined by the ordinary courts, as in this country, or by administrative tribunals of the continental type. See H. Clagett, *Administration of Justice in Latin America* 61 ff. (1952). See also Note, "The Writ of Amparo: A Remedy To Protect Constitutional Rights in Argentina," 31 *Ohio State Law Journal* 831 (1970); P. P. Camargo, "The Claim of 'Amparo' in Mexico: Constitutional Protection of Human Rights," 6 *California Western Law Review* 201 (1970); C. E. Schwarz, "Rights and Remedies in the Federal District Courts of Mexico and the United States," 4 *Hastings Constitutional Law Quarterly* 67 (1977).

[10] For a suggestion that the civil law approach be adopted here, see C. H. Fulda, "A Proposed 'Administrative Court' for Ohio," 22 *Ohio State Law Journal* 734 (1961).

EDGE: I have familiarized myself with the main features of the French system of administrative law and administrative courts, which is described in a large number of English-language books and articles.[11] Is it safe to assume that in most other civil law countries the French model is followed?

COMPAROVICH: There is considerable diversity as to many points of jurisdiction, procedure and substantive law; almost every civil law system presents some unique features in its administrative law.[12] It seems, nevertheless, that most civilians share a basic approach to the subject and readily understand each others' systems. This is clearly demonstrated by the experience of the European Communities.[13]

The two systems of administrative law and administrative courts which have been most influential throughout the civil law world, are the French[14] and, to a lesser degree, the German.[15] Before we go into the differences between them, let me briefly point out some of the important aspects in which they are similar to each other.

1. Both in France and in Germany, virtually *every* administrative act is subject to review by the administrative courts, and may be *annulled* in case of unlawfulness or abuse of discretion.[16]

---

[11] For example, L. N. Brown and J. F. Garner, *French Administrative Law* (2d ed., 1973); C. E. Freedeman, *The Conseil d'État in Modern France* (1961); B. Schwartz, *French Administrative Law and the Common Law World* (1954); C. J Hamson, *Executive Discretion and Judicial Control* (1954); Letourneur and Drago, "The Rule of Law As Understood in France," 7 *American Journal of Comparative Law* 147 (1958); M. J. Remington, "The *Tribunaux Administratifs:* Protectors of the French Citizen," 51 *Tulsa law Review* 33 (1976).

A short but helpful introductory discussion can be found in H. J. Abraham, *The Judicial Process* 277–82 (5th ed., 1986). For a masterful brief description of the workings of the Conseil d'État, written by an insider but containing some comparative sideglances at similar institutions in other countries, see M. Lagrange, "The French Council of State," 43 *Tulsa Law Review* 46 (1968).

[12] See, for example, G. Treves, "Judicial Review in Italian Administrative Law," 26 *University of Law Review* 419 (1959); R. Parker, "Administrative Law Through Foreign Glasses: The Austrian Experience," 15 *Rutgers Law Review* 551 (1961).

For a comparative survey of the administrative courts of the six original member nations of the Common Market, see J.-M. Auby and M. Fromont, *Les recours contre les actes administratifs dans les pays de la Communauté économique européenne* (1971).

[13] See E. Stein, P. Hay, and M. Waelbroeck, *European Community Law and Institutions in Perspective: Text, Cases and Readings* 109–71 (2d ed., 1976 and Supp. 1985 at 33–44).

[14] The French system of administrative courts and administrative law is said to have served as model in The Netherlands, Belgium, Luxembourg, Spain, Turkey, Egypt, and a number of other countries. See Lagrange, supra n. 11, at 57.

[15] English-language literature on German administrative law is less voluminous than that dealing with the French system. There are, however, some useful accounts. See Bachof, "German Administrative Law With Special Reference to the Latest Developments in the System of Legal Protection," 2 *International & Comparative Law Quarterly* 368 (1953); W. Feld, "The German Administrative Courts," 36 *Tulsa Law Review* 495 (1962), where further references can be found.

For an interesting comparison of the French and German systems, see H. G. Crossland, "Rights of the Individual to Challenge Administrative Action Before Administrative Courts in France and Germany," 24 *International & Comparative Law Quarterly* 707 (1975).

[16] If the challenged act is a negative one (e.g., a refusal to issue a license or a passport), its annullation may in effect amount to an affirmative command, not unlike an order of mandamus.

2. In both countries, the administrative courts ordinarily have power to annul the administrator's action for errors of fact as well as law.

3. The procedure of the administrative courts is informal, inexpensive, and largely inquisitorial.[17] The court can direct the administrator to submit his entire *dossier* relating to the matter at hand. In addition, parties are given an opportunity to introduce, or suggest the taking of, new evidence which was not before the administrative agency. Arguments of counsel for both sides usually are presented in writing and orally in open court. The administrative courts in both countries have a high reputation for impartiality and efficiency, and their operation is regarded as essential to the maintenance of the rule of law.

4. Neither German nor French administrative law has been systematically codified. In both countries there are relevant statutues, especially on matters of procedure;[18] but the bulk of the rules determining the legality and propriety of administrative acts is judge-made.[19] When we speak of the civil law orbit as a world of codes, we must always remember that the statement requires qualification with respect to public law.

5. Both French and German lawyers are plagued by the difficulty of defining the line of demarcation between the jurisdiction of the ordinary courts and that of the administrative courts.

EDGE: How do they draw the line?

[17] The inexpensiveness of proceedings in continental administrative courts is emphasized by K. C. Davis, *Discretionary Justice: A Preliminary Inquiry* 156 (1971). According to Professor Davis, one of the reasons for the inexpensiveness of those proceedings is that the party suing the government is not only legally but also practically able to proceed without the help of a lawyer, because "clerks or lawyers attached to the court help parties to prepare their written complaints." Moreover, if the citizen-litigant has an attorney, and if his action is successful, German law permits him to recover counsel fees from the defeated government. See ibid. Compare 28 U.S.C.A. §2412 (1978 and Supps.).

[18] In Germany, the procedure of administrative courts is governed by an elaborate and systematic federal statute, the *Verwaltungsgerichtsordnung* of 1960, as amended. In its organization and terminology, the *Verwaltungsgerichtsordnung* is patterned after the Code of Civil Procedure. This makes the procedure before the administrative courts comparable, and in some respects similar, to ordinary court procedure. But the greater simplicity of administrative court procedure has been preserved.

While the *Verwaltungsgerichtsordnung* regulates the procedure of administrative courts, the practice of the administrative agencies themselves is governed by a more recent statute, the *Verwaltungsverfahrensgesetz* of 1976.

[19] This is especially true of France, where a single central institution, the Conseil d'État, during the last 150 years has *created* a modern body of administrative law, in much the same way in which the common law courts created the law of England.

In Germany, the situation is somewhat different. Administrative law is partly state law. Moreover, it was only after World War II that a federal administrative court of last resort was created. Until then, no appeal to a federal court could be taken from the decisions of the various states' highest administrative tribunals. Thus no single court was able to acquire a position of nationwide importance and prestige comparable to that of the French Conseil d'État. Therefore (and again, comparable developments in earlier periods of history easily suggest themselves), academic influence on German administrative law has been strong, much stronger than in France.

COMPAROVICH: Concerning this point, the French system and the German system are very different from each other.

In the first place, they use different procedures for resolving a conflict of jurisdiction.[20] The French submit such a conflict to a special "Tribunal of Conflicts," composed of judges drawn from the Cour de Cassation and the Conseil d'État.[21] Under the present German system, both the ordinary and the administrative courts have power to determine their own jurisdiction. If an ordinary or administrative court, in a decision no longer subject to appeal, has affirmed or denied its own jurisdiction, such decision is res judicata in all subsequent proceedings that may be brought concerning the same matter in the same or any other court (even if the other court belongs to a different judicial hierarchy).[22] A court which considers itself incompetent on the ground that the case should have been brought before a court belonging to a different judicial hierarchy, may upon plaintiff's motion transfer the case to the latter court; the decision to transfer, unless it is reversed on appeal, is binding on the transferee court.[23]

EDGE: It seems to me that by their method the Germans aim to *avoid* head-on conflicts between ordinary and administrative courts, while the French provide machinery to *resolve* such conflicts.[24]

---

[20] Such a conflict may be "positive," if both the ordinary and the administrative courts *claim* jurisdiction; or it may be "negative," in a case in which both *decline* jurisdiction.

[21] For an interesting and typical example of a case decided by the Tribunal of Conflicts see Note, 67 *Law Quarterly Review* 44 (1951).

[22] See Gerichtsverfassungagesetz § 17; Verwaltungsgerichtsordnung §§ 40, 41. See also infra n. 84.

Similar problems of conflict of jurisdiction can arise in our law, for example, between a Workmen's Compensation Board and an ordinary court. Our solution appears to be similar to the German one: Each set of tribunals has the power to determine its own jurisdiction, and such determination, once it becomes final, is binding on the parties, with the result that the judgment, even if erroneous on the jurisdictional point, cannot be collaterally attacked on the ground of lack of subject-matter jurisdiction. See, for example, Murray v. New York, 43 N.Y.2d 400, 407, 401 N.Y.S.2d 773, 776, 372 N.E.2d 560, 563 (1977); Scott v. Industrial Accident Comm., 46 Cal.2d 76, 293 P.2d 18 (1956). Cf. Doney v. Tambouratgis, 23 Cal.3d 91, 151 Cal. Rptr. 347, 587 P.2d 1160 (1979).

[23] There is controversy concerning the extent to which the transferee court is bound. According to most of the commentators, the transferee court *must* take jurisdiction. For example, if an administrative court has transferred to an ordinary court, then the latter court, in the opinion of the commentators, *must* proceed to determine the case on the merits. Many courts have held, however, that in such a situation the ordinary court is bound only to the extent that it may not dismiss or transfer the case on the ground that the administrative court has jurisdiction; under this view, it would still be open to the transferee court (the ordinary court in our example) to hold that the case is not justiciable at all, or that a tribunal belonging to a third hierarchy, for example, a tax court, has jurisdiction, and to dismiss or again transfer accordingly. See Baumbach-Lauterbach-Albers-Hartmann, Zivilprozessordnung, GVG § 17, Anno. 3B (45th ed., 1987) and authorities there cited. Such complexities (and this is not the only one) are part of the price which the German system must pay for its multiplicity of judicial hierarchies and for the advantage of operating without a separate Tribunal of Conflicts superimposed upon them.

[24] It should be noted, however, that the German statute (Gerichtsverfassungsgesetz § 17a) authorizes the states to create special tribunals similar to the French Tribunal of Conflicts. It seems that at present such a tribunal exists in Bavaria, but that otherwise this system is not widely

COMPAROVICH: If you like pithy generalizations, you might put it that way.[25] From a policy point of view, good arguments can be made for and against either method. Certain it is, however, that any legal system that creates several separate judicial hierarchies, must provide some method by which jurisdictional conflicts can be avoided or resolved.

EDGE: Do the French and the Germans agree on the criteria by which they seek to distinguish between private and public disputes?

COMPAROVICH: No, on this subject, again, there is a great deal of disagreement between them. Relatively little difficulty arises under either sytem in cases in which the plaintiff seeks to have an administrative act annulled; ordinarily there is little doubt that this is a matter of public law, regardless of whether the administrative act to be reviewed is an act of the police[26] or of any other public authority.[27] Really tough problems of demarcation, however, are encountered

used in Germany. Whether the powers of the Bavarian Tribunal of Conflicts can be invoked in a case in which a court, of whatever hierarchy, already has made a final (i.e., no longer appealable) jursidictional determination, is a question which, again, has caused much doubt and controversy.

[25] Mr. Edge's generalization, although it captures the traditional spirit of the two methods, is no longer quite accurate. Pursuant to a Decree of 1960, a ruling of the French Tribunal of Conflicts can now be obtained even though no actual ("positive" or "negative") conflict between an ordinary and an administrative court has as yet arisen. The decree authorizes the Tribunal of Conflicts to issue a preliminary ruling on where the action *should be* brought; and this, it seems, is now the most common procedure. See Crossland, supra n. 15, at 715.

[26] The statement in the text must be qualified with respect to police activities in the course of formal criminal proceedings, which are subject to the jurisdiction and supervision of ordinary rather than administrative courts. In West Germany, a distinction is drawn between informal police inquiries and those that have turned—following the intervention of the Public Prosecutor's Office—into formal criminal proceedings. With respect to the former, administrative courts have jurisdiction. In a series of decisions these courts have held, for example, that a person under informal police investigation can, by a proceeding before an administrative court, obtain information from police files, or may even compel the police to destroy the relevant records and discontinue further inquiries. See, for example, the decision of the VGH München of 9–27–1983, Neue Juristische Wochenschrift 1984, p. 2235. The line of demarcation between formal and informal criminal proceedings can be difficult to draw, leading to uncertainties as to whether administrative or ordinary courts have jurisdiction. For an interesting article, critical of the above-cited decision, see A. Schoreit, *Verwaltungsstreit um Kriminalakten, Neue Juristische Wochenschrift 1985*, p. 169.

In France, a distinction is drawn between the *police judiciaire* (which deals with criminal cases and whose acts are, at least to some extent, subject to the control of the prosecutor and the criminal courts), and the *police administrative* (which deals with noncriminal matters and whose acts can be reviewed by the administrative courts). See L. H. Levinson, Enforcement of Administrative Decisions in the United States and in France, 23 *Emory Law Journal* 11, at 13–15, 90 (1974); C. L. Blakesley, Conditional Liberation (Parole) in France, 39 *Louisiana Law Review* 1, at 16–17 (1978).

But there are troublesome border areas. When the public authorities act in such a border area, the affected citizen may well be in doubt as to whether he should turn to the ordinary or the administrative courts for review. A good example is incarceration for "vagrancy." See the fascinating and highly instructive Vagrancy Cases decided by the European Court of Human Rights in 1971 and extensively discussed by S. A. Cohn, International Adjudication of Human Rights and the European Court of Human Rights: A Survey of Its Procedural and Some of Its Substantive Holdings, 7 *Georgia Journal of International Comparative Law* 315, at 377–408 (1977).

[27] Difficult questions arise in connection with the activities of nationalized enterprises and of other public corporations. For a comparative treatment of the important question of judicial

in the innumerable cases—of contracts, of torts, and of eminent domain—in which the redress sought by the individual is a money judgment against the State or one of its subdivisions. Are such disputes private or public? On this point, the French and the Germans part company.

SMOOTH: We have run into this problem repeatedly. Both in France and in Germany our trucks have been damaged in one-vehicle accidents caused by bumps and potholes that were the result of negligent maintenance of public roads. We found that we had to sue the French government in the administrative courts, while in Germany such an action must be brought before the Civil Chamber of the ordinary court of first instance.[28] What is the reason for this difference?

COMPAROVICH: The French tried to draw a distinction along functional lines. If the damage is caused by some malfunctioning of the public service (including human failure of the State's agents, high or low), the State is liable, and must be sued in the administrative courts. If the official or employee who caused the damage, is personally at fault, he is liable as an individual, and his liability can be enforced in the ordinary courts.

EDGE: I suppose such personal fault will be found where an official has maliciously caused defendant's injury or loss.

COMPAROVICH: Yes, but the notion of personal fault, upon which the jurisdiction of the ordinary court as well as the official's personal liability is bot-

control (by ordinary or administrative courts) over the activities of such corporations, see W. Friedmann (ed.), *The Public Corporation,* passim (1954).

[28] Frequently these jurisdictional problems are treated incidentally in books and articles primarily dealing with the substantive aspects of tort liability for the acts of public officials and employees. The most extensive comparative treatment to date is the 1964 Colloquium on "Liability of the State for Illegal Conduct of Its Organs" published by the Max-Planck-Institut für Ausländisches Offentliches Recht und Völkerrecht (1967). For comparative discussions in the English language see, for example, H. Street, *Governmental Liability* (1953); Z. Szirmai (ed.), *Governmental Tort Liability in the Soviet Union, Bulgaria, Czechoslovakia, Hungary, Poland, Roumania and Yugoslavia* (1970), reviewed by H. R. Hink in 21 *American Journal of Comparative Law* 616 (1973); S. B. Jacoby, "Federal Tort Claims Act and French Law of Governmental Liability," 7 *Vanderbilt Law Review* 246 (1954); K. Kautzor-Schroeder, "Public Tort Liability Under the Treaty Constituting the European Coal and Steel Community Compared With the Federal Tort Claims Act," 4 *Villanova Law Review* 198 (1958–59); R. Braband, "Liability in Tort of the Government and Its Employees: A Comparative Analysis With Emphasis on German Law," 33 *New York University Law Review* 18 (1958); H. R. Hink, "Service-Connected Versus Personal Fault in the French Law of Government Tort Liability," 18 *Rutgers Law Review* 17 (1963); id., "The German Law of Governmental Tort Liability," 18 *Rutgers Law Review* 1069 (1964).

Under German law (as in the United States under the Federal Tort Claims Act), the substantive aspects of the State's tort liability are essentially governed by general tort principles, that is, by private law. See infra n. 52. In France, on the other hand, it was held that the provisions of the Civil Code are not applicable to governmental liability, which is governed by principles peculiar to public law. This divergence of the relevant sources also leads to substantive differences between French and German law with respect to governmental tort liability. For an example, see infra n. 34. Cf. K. Lipstein, "The Law of the European Economic Community" 323 (1974); id., "Some Practical Comparative Law: The Interpretation of Multi-Lingual Treaties with Special Regard to the EEC Treaties," 48 *Tulsa Law Review* 907, 913–14 (1974).

tomed, is not limited to cases of malice. A policeman who in arresting a person commits hideous and unnecessary acts of brutality, can be personally sued in the ordinary courts, regardless of whether he was motivated by personal hatred of his victim or by an excess of zeal in the performance of his duties. Even a nonintentional but grossly negligent act can constitute personal fault.[29] Some of the cases so holding involved careless use of firearms by soldiers or policemen.[30]

SMOOTH: Are the situations in which the State is liable, and those in which the official is individually liable for his intentional or negligent wrongdoing, mutually exclusive?

COMPAROVICH: They were under the older doctrine, which prevailed until the beginning of this century. But since then, and with evergrowing liberality, the French courts have recognized that there are countless situations in which the State's servant is sufficiently blameworthy to be subjected to personal liability, and in which, at the same time, the damage is so closely connected with the public service that the State, also, should be held liable.

SMOOTH: This is similar to our doctrine of respondeat superior. In cases to which that doctrine applies, you can sue the servant *and* the master.[31]

COMPAROVICH: Correct; but with respect to governmental liability, the French have gone further. "Cumulative liability" of the official and of the government often exists in French law even though in terms of our phraseology the wrongdoing official has not acted within the scope of his employment.[32] As examples one can cite a number of cases in which the plaintiff's injury or loss was caused solely[33] by what we would call a "frolic" of the official; in these cases the government as well as the individual official was held liable under French law, on the ground that the wrongdoing functionary had acted on the occasion of an official function or had used an instrumentality entrusted to him by the government. In one such case a high official, while inspecting a police station, had engaged in target practice and negligently wounded a person who happened to pass by. Although it was clear that the shooting exercise had been purely for fun and had nothing whatever to do with the wrongdoer's official duties, this was held to be a case of cumulative liability, on the ground that the official had acted on the occasion of his inspection visit and had used one of the service guns kept at the police station.[34]

[29] See A. de Laubadère, *Traité de droit administratif,* §§ 1193–94 (8th ed., 1980).

[30] Ibid. For further examples see Hink, supra n. 28, at 27.

[31] This is the general rule. For a statutory exception see infra at n. 41.

[32] See de Laubadère, op. cit. supra n. 29, at §§ 1202–08.

[33] In cases in which the damage is caused (a) by the wrongdoing official's personal fault *and* (b) by some malfunctioning of the public service that is separable from the conduct of the primary wrongdoer (e.g., failure of higher officials properly to supervise the malefactor), it was even less difficult for the French courts to justify the imposition of cumulative liability. See ibid.

[34] See M. Waline, *Droit Administratif,* Sec. 1458 (9th ed. 1963), where the reader also will find further references to relevant cases.

In Germany, where governmental tort liability is controlled by private law (see supra n. 28), a more restricted rule has been adopted. Under relevant principles of private law, which in this

SMOOTH: I take it that in cases of this kind, in which both the State and the individual public servant are liable, both can be sued in the same court?

COMPAROVICH: No, even in these cases the public authority must be sued in the administrative courts, while an action against the individual is within the jursidiction of the ordinary courts.[35] As a matter of practice, plaintiffs prefer to sue in the administrative courts, out of fear that the individual tortfeasor may be judgment-proof, although the administrative courts have the reputation of being conservative in determining the amount of damages to be awarded.

SMOOTH: Having satisfied the judgment of the administrative court, can the public authority recover indemnity or contribution from the negligent functionary?

COMPAROVICH: Yes. Overruling older decisions to the contrary, the highest administrative court held in 1951 that the public authority *in its own right* is entitled to such recovery.[36] In addition, it seems to be the practice of the administrative courts, in the action brought by the victim against the public authority, to condition an award in favor of the victim upon the latter's *assigning* to the public authority his tort cause of action against the official who was guilty of "personal fault." The public authority usually does not collect the assigned claim. The purpose of the assignment is to prevent the victim from suing the official rather than to have the public treasury reimbursed. This practice has been criticized on the ground that in effect it relieves the guilty individual of civil responsibility.[37]

respect are somewhat similar to our notion of "scope of employment," the German courts have declined to hold the State liable in situations where, as in the French case mentioned in the text, the accident was caused by a frolic involving misuse of firearms kept for official purposes. See RGZ 105, 230 (1922). The fact that the frolicking functionary acted on the occasion of an official activity, and that he used a service gun or other State-owned instrumentality, in the German view is insufficient to support governmental liability in such cases. See Palandt, *Bürgerliches Gesetzbuch*, § 839, Anno. 2Ac(bb) (43rd ed., 1984), and cases there cited. See also infra at nn. 51–53.

For futher substantive details of the French and German doctrines of governmental tort liability, the reader is referred to the books and articles cited supra n. 28. The aim of the present discussion is merely to throw some light on the jurisdictional demarcation lines between the business of the ordinary and of the administrative courts.

[35] The general jursidictional principles discussed in the text, supra, no longer apply to those actions against the State or its subdivisions in which the plaintiff seeks recovery for personal injury or property damage caused by a "vehicle." The question of jurisdiction over such actions now is governed by a special statute of relatively recent vintage. See infra, at nn. 39–47.

[36] See Waline, supra n. 35, Sec. 1617; Laubadère, supra n. 29, Sec. 1213.

Compare United States v. Gilman, 347 U.S. 507, 74 S. Ct. 695 (1954). It is interesting to observe that in dealing with this important question the French Conseil d'État and the U.S. Supreme Court, both acting at about the same time, effected radical changes—in opposite directions.

[37] See Waline, supra n. 35, Sec. 1615. It should be noted, however, that today (i.e., since 1951, see supra at n. 36) this criticism is justified only if the State enforces neither the assigned claim nor the reimbursement claim which it has in its own right. The reported cases (see infra nn. 38, 44, and 45) show that at least in some instances the State's own reimbursement claim in fact has been enforced.

SMOOTH: If the public authority chooses to sue its negligent functionary, must the suit be brought in the administrative or the ordinary court?

COMPAROVICH: Cute question. I can see you're getting into the spirit of these refinements. It seems that the public authority, if it sues as the assignee or subrogee of the victim (which is rarely done), stands in the shoes of the victim, and hence must sue in the ordinary court. Where, however, the public authority in its own right seeks contribution or indemnity, the administrative courts have jurisdiction, according to a much-criticized decision of the Tribunal of Conflicts.[38]

SMOOTH: These rules of French law must cause grave difficulties in automobile accident cases when one of the vehicles involved is owned by the State or one of its subdivisions. In such a case, it would follow from the rules just discussed that the plaintiff is unable to join the driver and the owner of the government vehicle as defendants in the same action. Moreover, in cases of collisions between private cars and government vehicles, when multiple parties assert crisscrossing claims based on tort, indemnity, contribution, and subrogation, the resulting jurisdictional problems must be staggering.

COMPAROVICH: They used to be staggering indeed; but in 1957 the legislator came to the rescue.[39] The statute of December 31, 1957, addressing itself to this specific problem, provides that—regardless of the involvement of the government or some other public authority—all actions seeking recovery for injury or damage caused by "a vehicle" shall be governed by principles of private law and must be brought in the ordinary courts.[40] The statute further contains a provision, comparable to a section of our Federal Tort Claims Act,[41] which in effect deprives the victim of any cause of action against the driver of the vehicle, thus rendering the public authority exclusively liable in cases of this kind.

EDGE: Does this create an exception to the general French rule, explained by you a minute ago,[42] that the public authority, having satisfied the injured person's claim, can obtain reimbursement from the derelict functionary?

---

[38] Id., Sec. 1617.

[39] For the text of the 1957 statute, and an instructive discussion of its provisions, see 1 Auby and Drago, supra n. 5, §§ 563–65.

[40] The statute applies regardless of whether the "vehicle" travels on land, in the water, or in the air. Nor does it make any difference whether the vehicle serves military or civilian purposes. If the vehicle is used by a public servant for purposes of the public service, the statute applies even though the public servant rather than the State or its subdivision owns the vehicle. Tribunal of Conflicts, November 20, 1961, D.1962, 265.

[41] 28 U.S.C.A. § 2679(b), as amended in 1961 and 1966. For a discussion of this statute, see Anno., 16 A.L.R.3d 1394 (1967). See also the removal provision in 28 U.S.C.A. § 2679(d), discussed in Anno., 41 A.L.R.Fed. 288 (1979).

Prior to the enactment of these statutory provisions, the driver of a U.S. government vehicle could be sued in a state court, and sometimes only in a state court, while the government was suable only in a federal court. This created difficulties somewhat similar to those experienced by the French before 1957. See, for example, *Falk v. United States*, 264 F.2d 238 (6 Cir. 1959).

[42] See supra at n. 36.

COMPAROVICH: No, the statute renders the public authority's liability exclusive only insofar as the victim is concerned. With respect to the guilty functionary's duty to reimburse the public authority, the vehicular accident cases are governed by the general rule. Thus, if it so chooses, the public authority can obtain total or partial[43] reimbursement from the negligent driver.[44]

EDGE: Am I correct in assuming that in this situation, where a vehicular accident covered by the 1957 statute is involved, the public authority's reimbursement claim against its agent (even though the public authority asserts the claim in its own right) has to be pursued in the ordinary courts?

COMPAROVICH: No. In a 1965 decision—of questionable soundness—the Tribunal of Conflicts has held that even though the injured party had recovered from the State by way of an action in the ordinary court under the 1957 statute, the State must proceed in the administrative court in order to obtain reimbursement from the negligent driver.[45]

EDGE: The legislator's obvious aim—to subject all of the legal consequences of vehicular accidents to the jurisdiction of the ordinary courts—thus is partly thwarted.

COMPAROVICH: I tend to agree with you on this point. Moreover, there are other rough edges which indicate that even in the limited subject-area of vehicular accidents the statute did not overcome all of the doubts and super-refinements stemming from the jurisdictional dichotomy.

Though a vehicle is damaged, or a person riding in a vehicle is injured, the 1957 statute by its terms does not apply unless the damage or injury is "caused *by* a vehicle" (emphasis added). Thus the one-vehicle accidents caused by holes in the road, to which Mr. Smooth referred a while ago, are not covered by the statute; the action against the public authority responsible for road maintenance must still be brought in the administrative court.

There remain tricky borderline problems, exemplified by the following two cases:

Case No. 1: Two private cars collided on a narrow bridge. Access to the bridge was controlled by a traffic light at one end, and by a traffic-directing officer at the other. The car coming from the direction of the traffic light entered the bridge because the light was green. The driver of the other car entered because the officer, without paying sufficient attention to the light at the other end, had waved him on. Clearly, for any resulting litigation among the owners and drivers of the two cars, the ordinary courts had jurisdiction. But what of an action against the governmental unit responsible for the act of the officer? Before 1957, such an action doubtless had to be brought in the administrative court. Did the 1957 statute effect a change in this regard?

---

[43] In determining the amount of such reimbursement, the court takes into account the extent to which the victim's damage has been caused (a) by the driver's "personal fault," and (b) by some malfunctioning of the public service not attributable to him personally.

[44] 1 Auby and Drago, supra n. 5, § 564.

[45] Tribunal of Conflicts, November 22, 1965, Préfet de la Seine-Maritime, D.S.1966, 195.

Case No. 2: An official of the State's Bureau of Mines was engaged in checking the brakes of a truck belonging to a private mining company. He was not himself in the truck, but directed the driver to engage in certain go-and-stop maneuvers. One of these maneuvers, a sudden stop ordered by the official, caused an accident. Here again, an action of the victim against the owner or driver of the truck (an action of doubtful chances) would have to be brought in the ordinary courts. But what court has jurisdiction over the victim's suit against the State?

SMOOTH: In my opinion the two cases are indistinguishable from each other. In both situations the public official, while not actually driving the vehicle, had assumed control over its movements. Thus in both cases the vehicles were in effect operated by an official, and the 1957 statute should be applied.

COMPAROVICH: Your reasoning is of Gallic finesse; but the Tribunal of Conflicts has cut it even more finely. It held that in Case No. 2 the official was "associated with the operation" of the truck, with the result that the 1957 statute applied and that the State could be sued in the ordinary court.[46] But Case No. 1 was held to be distinguishable on the ground that the officer directing traffic did not really control the movements of the car, and that by his single, momentary act of letting the car pass, the officer did not become "associated with the operation" of the vehicle. Therefore, in this case the statute was not applicable, and the public authority had to be sued in the administrative court.[47]

SMOOTH: I realize that the jurisdictional problems arising in cases of torts committed by public officials are rarely simple. Even in our legal system, we encounter some difficulties in this respect, because suits against the State may have to be brought in a special court (Court of Claims), while actions against private persons, though arising from the same occurrence, must be instituted in the ordinary courts.[48] Perhaps the subject is inherently complex; but the

[46] Tribunal of Conflicts, March 5, 1962, Dame Boule et Cie. d'Assurances "La France" c. Etat, J.C.P. 62, II, 12593.

[47] Tribunal of Conflicts, June 28, 1965, Del Carlo c. Laurent, Req.N.1866, J.C.A. Fasc. 710.

[48] The difficulties to which Mr. Smooth refers are illustrated by cases such as Horoch v. State of New York, 286 App. Div. 303, 143 N.Y.S.2d 327 (3d Dep't, 1955); Smith-Cairns Motor Sales Co. v. State, 45 Misc.2d 770, 258 N.Y.S.2d 51 (Ct.Cl. 1965). See also J. J. McNamara, "The Court of Claims: Its Development and Present Role in the Unified Court System," 40 *St. John's Law Review* 1, 24–25 (1965).

In cases where jointly or concurrently with a private tortfeasor the federal government is liable, it is sometimes—but not always—possible to join or implead all parties in a U.S. District Court. An example is *Ayala v. United States,* where the government was sued for injuries sustained in an explosion of government-owned boxcars, and plaintiffs attempted to join the manufacturer of the boxcars as defendant. Some of the plaintiffs were diverse, some nondiverse. The latters' claims against the manufacturer were dismissed, the court holding that there was no "pendent party" jurisdiction. 550 F.2d 1196 (9th Cir. 1977), cert. dism. 435 U.S. 982, 98 S. Ct. 1635 (1978). But there seems to be some District Court authority *contra.* See J. Cound, J. H. Friedenthal, A. R. Miller & J. E. Sexton, *Civil Procedure—Cases and Materials* 268 (4th ed., 1985).

French approach, it seems to me, compounds the difficulty. Is the German solution any simpler?[49]

COMPAROVICH: Perhaps a little. Basically, the German solution is not radically different from the French. The Germans, also, must draw a line between those acts which are purely personal wrongs of the employee (e.g., the driver of a mail truck intentionally runs down a bicyclist whom he suspects of illicit relations with his wife), and those acts for which the State should be liable;[50] but according to the German view, liability of the State, where it exists, completely supplants the liability of the individual official or employee.[51]

SMOOTH: Does the victim of the tort bring his action in the ordinary or the administrative court?

COMPAROVICH: An action against the individual naturally must be brought in the civil courts. Contrary to the French rule, moreover, positive provisions of the German Constitution and of the German Civil Code confer upon the ordinary courts jurisdiction over tort actions against the State or its subdivisions.[52]

[49] In speaking of "German" law, the text refers to the law of the West German Federal Republic. In East Germany, as in other socialist countries, judicial remedies against wrongdoing functionaries and against the State itself tend to be more restricted. See §§ 11, 330 and 331 of the 1975 Civil Code of the German Democratic Republic; Szirmai, supra n. 28; W. Gray, "Soviet Tort Law: The New Principles Annotated," 1964 *University of Illinois Law Forum* 180, 193–96.

[50] See infra n. 52. . . .

[51] This is a general rule, not limited to vehicular accident cases. The existence of governmental liability constitutes a complete defense in an action by the victim against the individual official or employee. . . .

The government, having satisfied a claim for damages, may recover indemnity from the official or employee who caused the damage intentionally or by gross negligence; see infra n. 52. As in France, however, it cannot be taken for granted that in practice such indemnity will always be collected.

For a comparative discussion, see G. A. Bermann, "Integrating Governmental and Officer Tort Liability," 77 *Columbia Law Review* 1175 (1977).

[52] See Art. 34 of the Basic Law of the German Federal Republic: "If any person, in exercising the duties of a public office entrusted to him, violates his official obligation towards a third party, liability shall in principle rest with the state or his employing authority. If such person has acted willfully or with gross negligence, the [state's] right of recourse [against him] shall be reversed. In respect to the claim for damages and in respect to the right of recourse, the jurisdiction of the ordinary courts must not be excluded."

Sect. 839 of the Civil Code (which went into effect January 1, 1900) originally provided that the public servant should be personally liable in such a case. In subsequent statutes, in Art. 131 of the Weimar Constitution, and again in Art. 34 of the Basic Law of 1949, the State assumed this liability of its servants; but the essentially private nature of the liability was not changed (see supra n. 28), and the ordinary courts retain their jurisdiction. A 1981 federal statute created a new type of *direct* state liability. See the Staatshaftungsgesetz of June 26, 1981 (B.G.Bl. I 553). But this statute was declared unconstitutional by the Federal Constitutional Court, which, in its decision, strongly emphasized the private law character of the existing system of state liability. See Decision of Oct. 19, 1982, BVerfGE 61, 149, NJW 1983, p. 25.

In this connection, it must be remembered that German administrative law is largely state law, and that until recently there was no federal administrative court of broad jurisdiction. Private law, on the other hand, was essentially unified and federalized by the enactment of the Civil Code; the civil court hierarchy has had a single federal court of last resort at its apex since 1879.

SMOOTH: I prefer the German system to the French. The latter—in addition to the other difficulties to which we have alluded already—must produce very doubtful cases whenever a public servant commits a tort of a kind (such as assault and battery) which has only a slight or problematic connection with his official duties.[53] The plaintiff may have to choose at his peril, not only between possible defendants, but also between two different tribunals—unless as a matter of precaution he brings two actions in two different courts.

COMPAROVICH: Yes, an enormous amount of lawyers' time and clients' money is spent to resolve such doubts. Dichotomies in the law always lead to such wasted effort. Our dichotomy between law and equity illustrates the same point.

SMOOTH: The German system seems to avoid some of these difficulties.

COMPAROVICH: That is true. The French, on the other hand, will argue that the German system requires two actions, one in the administrative and the other in the ordinary court, if the aggrieved individual desires to annul an administrative act *and* to recover damages in tort.[54] In France, he can get both kinds of relief in one proceeding before the administrative court.

EDGE: Our business involves many contracts with foreign governments.[55] Are such contracts to be enforced in the ordinary or the administrative courts?

Therefore, to allocate governmental tort liability to the sphere of "private" law, has had the effect (a) of federalizing and unifying the applicable substantive rules, and (b) of entrusting their development to a single federal court of last resort.

In the United States, where private law is not nationally unified, the adoption of ordinary (private) tort law principles in the Federal Tort Claims Act has had the opposite effect: to preserve the motley variety of state laws even with respect to claims against the federal government.

[53] Under the German system, too, these borderline cases may produce doubts as to whether the State or the wrong-doing official is liable; but since both the State and the individual tortfeasor can be sued in the same court, the plaintiff can join both as defendants, even though his action will be successful only against one or the other. The point is somewhat controversial; but according to the better view such alternative joinder of parties is permissible. See Rosenberg-Schwab, Zivilprozessrecht, § 65 IV 3 b (13th ed., 1981).

[54] . . . An analogous problem often troubles American courts: Can the petitioner in a certiorari or mandamus proceeding, or in a statutory proceeding of similar nature, obtain *damages* as well as an order directing the doing or undoing of an administrative act? The cases, reflecting a quaint common law rule as well as a variety of pertinent statutes, are numerous and disharmonious. See Anno., 73 A.L.R.2d 903 (1960). The excellent analysis of the problem in Weinstein-Korn-Miller, New York Civil Practice ¶7806.01 (1967 and Supp.), though limited to a discussion of the unfortunate New York statute, carries more general implications. It shows that jurisdictional and procedural difficulties, comparable to those encountered in some of the continental countries, also exist in our law, but that often these difficulties are overshadowed by the niggardliness and obscurity of our substantive rules regarding governmental liability for official wrongdoing.

[55] For comparative treatments of British, United States, and French law dealing with government contracts see Street, *Governmental Liability* 81 ff. (1953, repr. 1975), and Mitchell, *The Contracts of Public Authorities* (1954), reviewed by Pasley, 41 *Cornell Law Quarterly* 342 (1956). A good comparative survey of administrative (or public law) contracts in various countries of Western Europe can be found in C. C. Turpin, Public Contract, Vol. VII (Contracts in General), *International Encyclopedia of Comparative Law,* Chap. 4, at pp. 27–31 (1982).

COMPAROVICH: Both in France and in Germany, it is recognized that contracts between a private party and a public authority are sometimes enforceable in the administrative courts, and sometimes in the ordinary courts, depending on the public or private nature of the particular contract.[56]

EDGE: What criteria are used to attribute a contract to the public or the private sphere?

COMPAROVICH: Here, again, the French and the Germans are not in agreement. In general, one can say that the French extend the jurisdiction of the administrative courts over a much wider range of contracts than the Germans do. In searching for a workable basis of distinction, the French avowedly use a flexible and somewhat casuistic method, while the Germans have tried to develop a single abstract criterion.

The German theory is predicated on the essential nature of the legal relationship between the parties.[57] If that relationship is one which could equally well exist between private individuals or corporations, then any dispute resulting from it is "private." Applying this principle to contractual relationships, the German courts have found that some contracts involve the enforcement or distribution of public burdens, or relate to some other subject with which private parties, by themselves, could not deal under any circumstances,[58] contracts of this kind are "public." Most contracts between the government and a private party, however, involve transactions which in the abstract (and the German criterion emphasizes the abstract nature and the typical features of the relationship more than the particular facts of the case at hand) could equally well be concluded among private parties. Contracts for the sale or purchase of goods, or involving sales or leases of real property, are the classical exmaples; such contracts are private and hence enforceable in the ordinary courts, even though one of the contracting parties is the government.

Although its precise formulation as well as its application to borderline situations remains somewhat controversial,[59] this German theory seems to furnish a reasonably clear line of demarcation.[60]

---

[56] The French distinguish between "administrative contracts" and *contrats de droit commun*. The Germans use terms such as "public law contract" and "private law contract."

[57] For references, see Baumbach-Lauterbach-Albers-Hartmann, *Zivilprozessordnung*, GVG § 13, Annos. 3 and 4 (45th ed., 1987).

[58] For example, the owner of a city plot, who desires to build an apartment house but pursuant to the provisions of a statute can get a building permit only on condition of providing a certain number of parking spaces, enters into a contract with the city; under the terms of the contract, the city will issue the permit unconditionally and will itself provide the parking spaces, while the owner will pay the city a specified sum of money. This was held to be a "public law contract," enforceable only in the admnistrative courts. BGHZ 32, 214 (1960).

[59] There is some doubt, for example, concerning contracts which publicly owned suppliers of water, gas, or electricity conclude with their customers. See E. Eyermann and L. Fröhler, Verwaltungsgerichtsordnung, § 40, Annos. 50–53 (8th ed., 1980).

[60] The theory seems to work even in complex situations, for example, where a statute, as part of a subsidy program, authorizes the government to make loans to private persons for the construction of low-rent housing or for similar social purposes. When an application for such a loan has been denied, it is clear that judicial review of the denial must be sought in the administra-

EDGE: In virtually all cases of what we would call government contract disputes, the effect of the German doctrine must be to bring such disputes before the ordinary courts.

COMPAROVICH: That is correct.

EDGE: You told us that the French, in distinguishing between administrative and private contracts, use an approach more flexible than the Germans. Is the French distinction similar to that which our courts draw between governmental and proprietary functions of municipal corporations?[61]

COMPAROVICH: During the nineteenth century, the French used a similar formula. Since the beginning of this century, however, they have abandoned the attempt to encapsule the guiding criteria in a formula, and have adopted a pronouncedly pragmatic approach.[62] According to the present French view, the nature of a contract as administrative or private may be determined by statute,[63] by decisional law,[64] or, to a certain extent, by the terms of the contract.

EDGE: Do the parties have the power to stipulate in their contract that the administrative courts (or the civil courts) shall have jurisdiction?

COMPAROVICH: No. This stipulation, as such, would have no effect because the parties do not have the power to change the pertinent jurisdictional rules. But the entire contract will be scrutinized for indications of so-called "*clauses exorbitantes*," that is, terms which are unusual in, and perhaps inconsistent with the nature of, a contract between private parties. In this connection, an express stipulation providing for jurisdiction of the administrative court may assume some significance. Much more frequently, however, a "*clause exorbi-*

tive court. But once a contractor has obtained such a loan, the relationship between him and the government is the ordinary one of debtor and creditor, with the result that the government's action for repayment of the loan must be brought in the ordinary court. In spite of doubts expressed by some authors, this rule has been generally recognized in judicial practice. See the cases cited by Eyermann Fröhler, supra n. 59, § 40, Anno. 46.

[61] Cf. Langrod, Administrative Contracts—A Comparative Study, 4 *American Journal of Comparative Law* 325, 328 (1955).

A comparable distinction (between acts *jure gestionis* and acts *jure imperii*) is drawn by the courts of many countries in applying the international law doctrine of sovereign immunity. See, for example, H. G. Maier, "Sovereign Immunity and Act of State: Correlative or Conflicting Policies?" 35 *University of Cincinnati Law Review* 556, 561 ff. (1966). See also 28 U.S.C.A. §§ 1602 to 1605, added in 1976.

[62] 1 De Laubadère, Moderne, and Devolvé, *Traité des Contrats Administratifs* 126–28 (2d ed., 1984).

[63] A statute, for instance, provides that contracts for public works are administrative. But a seller who merely furnishes paving stones to a public authority, without participating in the work of paving, can sue in the ordinary courts. So can the financier whose loan enables a municipality to carry out such a project. Id., at 265–66.

[64] The rules have grown in casuistic fashion, and the distinctions are as subtle as any to be found in the common law. Sales of movables to a municipal or other local authority are "private" unless the contract contains an "exorbitant" element (see n. 65 infra). In the case of the sale of movables to the central government, however, the opposite presumption applies: such a contract is administrative unless its terms or certain circumstances external to the contract show that the government intended to subject itself to the principles of private law. Id., at 148, 270–71.

tante" will be found in contract terms which give the public authority far-reaching powers of supervision, direction, and unilateral cancellation.[65] The presence of a *"clause exorbitante"* will as a rule mark the contract as an administrative one,[66] while its absence (in cases not governed by a special statute) is at least a strong indication of the private nature of the agreement.[67]

SMOOTH: At this moment, one of our subsidiaries is about to sue the French government in order to obtain payment for two million bottles of Dulci-Cola delivered to military canteens. It seems that soft drinks have met with a measure of customer resistance; the French government now claims the merchandise was unfit for human consumption. According to what you said, I take it that we have to bring the action in a civil court?

COMPAROVICH: Was there an elaborate, formal contract?

SMOOTH: No. Our subsidiary received an informal, routine order.

COMPAROVICH: In that event, the action has to be brought in the civil court. It was so held in a very similar case.[68]

[65] The criteria of what constitutes a *"clause exorbitante"* are not wholly clear and in part controversial. Id., at 213–29. However, the practical significance of the clause seems to be declining. Currently, the distinction between public and private contracts turns less on "exorbitant" elements introduced into the contract by the parties, than on elements external to the contract that render contractual relationships "exorbitant" (e.g., the impact of governmental regulations). Id., at 131. On the evolving distinction between the *"régime exorbitant"* and the *"clause exorbitante"* within the larger notion of *"éléments exorbitants,"* see id., at 159, 183, 229–35.

[66] This rule, which is well-established as to actions against the domestic government, has an interesting side effect in cases where a foreign government is sued in a French court. The point is illustrated by the decision of February 7, 1962, rendered by the Court d'Appel of Paris in the case of Perignon c. Etats Unis d'Amérique, 89 Journal du Droit International 1016 (1962) (noted ibid., in French and English, by J. B. Sialelli), aff'd on December 8, 1964, by the Cour de Cassation, 92 id. 416 (1965) (again noted by J. B. Sialelli). In that case, the U.S. government was sued by a French contractor who had done construction work on buildings (near Paris) intended to house U.S. government personnel dealing with foreign aid. Defendant's claim of immunity was sustained on the ground that its contract with the plaintiff contained the dispute settlement clause ordinarily used in U.S. government contracts as well as other terms which in the French view constituted *"clauses exorbitantes."* Read together with earlier cases (referred to ibid.), the decision seems to indicate that in the absence of any *"clause exorbitantes"* the court would have been more inclined to treat the U.S. government as having entered into a purely private transaction and as having waived its immunity in this way. The doctrine of *"clause exorbitantes"* thus radiates into the field of international law.

[67] For examples and discussion, see Auby and Drago, supra n. 5, at 466–69, and 590–93; De Laubadère, supra n. 23, § 576.

There is a further and somewhat controversial refinement. In spite of the absence of a *"clause exorbitante,"* the contract may be treated as an administrative one where the State has delegated the performance of inherently governmental functions or duties to the private party (e.g., to provide food for refugees confined by the government in a "repatriation center"). See the cases cited by De Laubadère, *Traité élémentaire de droit administratif,* Sec. 527 (4th ed. 1967).

[68] See 1 De Laubadère, *Traité Théorique et Pratique des Contrats Administratifs* 30–31 (1956). Even though the contract may have been made with the central government, the persumption mentioned supra n. 64, would be overcome by the absence of a *"clause exorbitante,"* or other exorbitant elements, and by the business-like manner in which the transaction was concluded and carried out.

SMOOTH: It is getting late, and we should not keep our distinguished visitor much longer. Before we adjourn, however, I would like to raise a more general question which has bothered me throughout our discussion of public law disputes.

Where you have two separate judicial hierarchies, as in France, (or five, as in Germany), each with its own court of last resort, it must happen—perhaps not too infrequently—that on some point of law the several courts of last resort take conflicting positions. When that occurs, how is the citizen or his legal advisor to know what "the law" is?

COMPAROVICH: This is an important and difficult question. Before I attempt to answer it, let me point out, by way of a caveat, that when the Cour de Cassation and the Conseil d'État deal with what seems to be the same problem, their respective solutions sometimes can differ from each other without actually being in conflict. For example, in connection with administrative as well as private contracts the question may arise whether a radical and unforeseeable change of circumstances, which has occurred after the formation but before the complete performance of the contract and has rendered such performance unduly burdensome for one party, gives such party a right to demand cancellation or modification of the contract. Over the last half-century, the Conseil d'État has developed an elaborate body of rules (the so-called doctrine of *imprévision*) pursuant to which the administrative court has broad powers[69] to adjust the parties' rights and duties under an administrative contract in such a case.[70] The Cour de Cassation, in dealing with private contracts, has declined to follow this doctrine; by and large, it has adhered to the view that, unless performance has become literally impossible, the contract should be enforced no matter how painfully the parties' expectations have been frustrated by virtue of a change of circumstances.[71] Thus we find that on this important point of contract law the position of the Cour de Cassation differs drastically from that of the Conseil d'État. Yet it would be misleading, in my opinion, to call this a "conflict." It merely means that there is one rule for private contracts, and a very different one for administrative contracts. It is not difficult to present policy arguments supporting this difference in the treatment of the two types of contracts; and regardless of what one may think of the strength of these arguments, it is clear that we deal here with an attempted distinction rather than an outright conflict.

---

[69] In observing the resourcefulness with which the Conseil d'État has fashioned remedies in cases of this kind, a lawyer brought up in the common law is strongly reminded of the powers and the practice of courts of equity.

[70] See De Laubadère, supra n. 62, §§ 637–39.

[71] See Amos and Walton's *Introduction to French Law* 165 (3d ed. by F. H. Lawson, A. E. Anton and L. N. Brown, 1967); 2 K. Zweigert & H. Kötz, *An Introduction to Comparative Law* 195–98 (1977); De Laubadère, supra n. 29, at 392. With respect to private contracts, the French position on this point is more conservative than that of most other legal systems in the civil law as well as the common law orbit. See P. Hay, "Frustration and its Solution in German Law," 10 *American Journal of Comparative Law* 345, especially at 346–56 (1961). . . .

SMOOTH: There must be cases, however, in which the several courts of last resort take truly conflicting positions.

COMPAROVICH: Yes, cases of such interhierarchy conflicts do occur in civil law countries.[72]

Thereotically such a conflict could arise, first of all, with respect to a question of constitutional law. It appears, however, that in the great majority of those civil law countries which provide for judicial review of the constitutionality of statutes, such review is entrusted to a separate Constitutional Court, which stands outside of, and in a sense above, the top courts of the several judicial hierarchies. If the power of constitutional review were not thus concentrated in a single court, it might easily happen in a civil law country that the same statute is upheld by the highest administrative court but declared invalid by the ordinary court of last resort (or vice versa). The necessity of obviating this kind of interhierarchy conflict was one of the most potent reasons for the creation—by Austria, Germany, Italy, and other civil law countries—of a single separate supercourt dealing only with constitutional matters.[73] By adopting that system, they have avoided such conflicts with respect to the most fundamental issues—those of constitutional law.[74]

Outside of the constitutional area, true conflicts can arise when the top courts of several judicial hierarchies take inconsistent positions on the same issue of substantive law. A French case, which at the time caused some flurry in the daily press, may serve as an *Example:*[75]

[72] For a general discussion of the interaction of administrative decisions and decisions of the ordinary courts in France, see L. H. Levinson, "Enforcement of Administrative Decisions in the United States and in France," 23 *Emory Law Journal* 11, at 47–79, 102 (1974).

[73] Even though the Constitutional Court always has the last word on questions of constitutional law, it is not entirely impossible that a conflict might arise between the Constitutional Court and one of the other top courts. Suppose, for instance, that the Constitutional Court, in upholding the constitutionality of a statute, gives it a particular construction. In Italy, it seems, the Court of Cassation has not considered itself bound by such construction. See J. H. Merryman and V. Vigoriti, "When Courts Collide: Constitution and Cassation in Italy," 15 *American Journal of Comparative Law* 665 (1967); G. Bognetti, "The Political Role of the Italian Constitutional Court," 49 *Notre Dame Lawyer* 981, 995–99 (1974). Conflicts of this type seem to be rare, however; they can be avoided by a properly drawn constitutional or statutory provision. Cf. German Bundesverfassungsgerichtsgesetz § 31.

[74] In Greece, the system of judicial review is similar to ours in the sense that every trial or appellate court that is called upon to apply a given statute, has the power to examine the constitutionality of that statute and in a proper case to declare it unconstitutional. At the same time, as a typical civil law country, Greece has several judicial hierarchies, each of which has its own court of last resort. It is possible, therefore, that the several courts of last resort differ on the question whether a particular statute is unconstitutional. In that event, the question is submitted to a Special Highest Court, consisting of eleven judges drawn from the courts of last resort of several hierarchies, and of two law professors. The decision of this special court is binding on all courts. See F. Spiliotopoulos, "Judicial Review of Legislative Acts in Greece," 56 *Temple Law Quarterly* 463, 496–501 (1983).

[75] *Le Monde,* Dece. 6, 1962, p. 14. Additional examples can be found in G. A. Bermann, "French Treaties and French Courts: Two Problems in Supremacy," 28 *International and Comparative Law Quarterly* 458, 477 (1979).

The mayor of Lyon had issued an ordinance which reserved certain parking spaces, near the city's Produce Market, for the trucks of vendors of vegetables. An automobilist not belonging to the privileged class, who had parked in one of the reserved spaces and was criminally prosecuted and fined for this violation of the ordinance, carried his case all the way to the Cour de Cassation, where his conviction was reversed on the ground that there was no proper statutory authority for the mayor's ordinance.[76] In another case, the Conseil d'État refused to nullify the mayor's ordinance (the very same ordinance involved in the case before the Cour de Cassation), holding that it was valid and duly supported by an authorizing statute.

Here we observe a direct clash of conflicting decisions.

EDGE: Can this conflict be resolved by the Tribunal of Conflicts?

COMPAROVICH: No. That Tribunal resolves only jurisdictional conflicts. It does not deal with cases where the Cour de Cassation and the Conseil d'État, each clearly acting within its jurisdiction, reach inconsistent results on a point of substantive law.

EDGE: What machinery does the French system provide for the resolution of conflicts of the latter kind?

COMPAROVICH: None whatsoever, unless the legislator acts to resolve the disputed question.[77]

EDGE: As a practical matter, then, it would seem that the ordinary court's view will prevail in the end. It is true that after the mayor has won his case in the highest administrative court, nobody can force him to withdraw the ordinance, or to discontinue his attempts to enforce it. Yet everybody who has heard of the—probably much-publicized—decision of the Cour de Cassation, will merrily park in the reserved spaces; and to prosecute violators will be a hopeless undertaking, since the lower criminal courts know that no conviction would survive an ultimate appeal to the Cour de Cassation.

COMPAROVICH: On the other hand, if the mayor orders the police to erect a physical barrier preventing access to the reserved spaces, and to lift the barrier only to admit vegetable trucks, the public authority may have the last laugh. Any legal action challenging the use of the barrier would have to be brought in the administrative court, with predictable results.

---

[76] There is no doubt under French law that a criminal court in which the defendant is charged with violation of an *ordinance,* has the power and duty to examine the validity of such ordinance (even though the court could not review the constitutionality of a *statute*). See L. H. Levinson, "Presidential Self-Regulation through Rulemaking: Comparative Comments on Structuring the ChiefExecutive'sConstitutionalPowers,"10*VanderbiltJournalofTransnationalLaw*1,at17–18(1977).

[77] In a pre-1957 automobile accident case the victim of a collision between a private and a government vehicle was denied relief in the civil courts on the ground that the accident had been caused by the governmental car; the administrative courts also denied relief, holding that the driver of the private vehicle was to blame for the accident. In this instance the legislator intervened, conferring power on the Tribunal of Conflicts to resolve a conflict of this kind (even though the conflict is a substantive rather than a jurisdictional one). But this statute has lost much of its significance due to the enactment of the 1957 statute mentioned supra at n. 39. See P. Herzog, *Civil Procedure in France* 117–18 (1967).

EDGE: How does the German legal system deal with this problem?

COMPAROVICH: In an entirely different way. The parliament of the German Federal Republic, implementing a constitutional mandate,[78] has enacted a statute setting up machinery for the resolution of conflicts between courts of last resort.[79] The statute is based on the idea that conflicts among the top courts of the several judicial hierarchies should be treated in a similar manner as conflicts between several panels of a single court of last resort. Both types of conflicts are to be resolved by a superpanel of judges; the only difference is that in the case of an interhierarchy conflict the superpanel is composed of judges representing all five hierarchies, that is, the presidents of the five top courts plus four additional judges (two from each of the hierarchies involved in the particular dispute).

EDGE: I take it that if a case like that of the Lyon parking imbroglio occurred in Germany, and the highest administrative court found itself unable to agree with the position taken by the ordinary (criminal) court of last resort, it would have to certify the controversial question of law to the superpanel.

COMPAROVICH: Precisely.[80] It should be noted, moreover, that in Germany the same machinery can be utilized to resolve interhierarchy conflicts on questions of jurisdiction.[81]

SMOOTH: In bringing this discussion to a close, I should like to thank you, Professor, for having enhanced our understanding of foreign and—seen from where I sit—rather strange institutions and approaches. In the area of public law disputes, in particular, there appears to be a real chasm between civil law and common law.

COMPAROVICH: If you view them in a broader perspecitve, you may find that the differences reflected in this "chasm," although they are remarkable and of great significance in legal practice, relate only to matters of technique. The

[78] Basic Law, Art. 95, par. 3, as amended in 1968.

[79] Law of June 19, 1968, BGBl I 661. The provisions of the statute are reproduced and discussed in Baumbach-Latuerbach-Albers-Hartmann, *Zivilprozessordnung*, Appendix following § 140 GVG (4th ed. 1987).

[80] It has been held that certification to the superpanel is necessary even though the prior decision of the other court (from which the certifying court desires to deviate) was rendered prior to the enactment of the 1968 statute creating the superpanel procedure. See Baumbach, supra n. 79.

[81] The reader will recall (see supra, at nn. 22–24) that in Germany the resolution of interhierarchy jurisdictional conflicts as a rule is not entrusted to a Tribunal of Conflicts, and that generally the jurisdictional question can be determined, in binding fashion, by the court—of whatever hierarchy—in which the action has been brought. But the binding effect of the jurisdictional determination thus made is limited to the case at hand. For example, if in case A the administrative court of last resort affirms the jurisdiction of its hierarchy, that is the end of the matter insofar as case A is concerned; but if subsequently case B, involving identical facts and issues, is instituted by another party in an ordinary court, the decision rendered in case A is not binding on the latter court. Thus the jurisdictional question may again be litigated up to the highest court—this time the ordinary court of last resort. If that court, in dealing with case B, wishes to deviate from the position taken in case A by the highest administrative court, it must submit the question to the superpanel.

fundamental social problem to be resolved is everywhere the same: to strike a balance between the interests of individuals and of voluntary associations, on the one hand, and those of the community, on the other; and to do so in an era of accelerating industrialization and urbanization, necessarily accompanied by the constant growth and proliferation of the tentacles by which various public authorities reach into our lives. The responses to this problem that have been fashioned by civil law and common law systems, do not seem to be basically dissimilar. In both types of systems it has been recognized that strengthened judicial protection of the individual against an increasingly powerful and ubiquitous government plays a central role in the required balancing of interests. To the extent that there remains a measure of disagreement on this fundamental point, the dissenters are to be found in some of the socialist systems rather than in the civil law or common law orbit.[82]

NOTES

1. A comparative source not cited in either excerpt is Stephen Legomsky, *Immigration and the Judiciary: Law and Politics in Britain and America* (1987), which emphasizes the nature of judicial review of immigration controls in the two countries. Legomsky finds that judges in both systems exercised much greater restraint and deference to administrators in immigration cases than in public law cases generally. While noting some important differences in their treatment of certain immigration issues, Legomsky is most struck by the remarkable parallels between the United Kingdom and the United States in the specifics of judicial deference in this area, including the reasoning techniques and even the rhetoric employed by the courts.

Another comparison between the United States and the United Kingdom is directed at the two countries' styles of social regulation. Reviewing the empirical studies and explanations of the differences, Keith Hawkins, an English legal sociologist, finds that American regulation tends to be more legalistic, rule-bound, politicized, and public than its more discretionary, conciliatory, low-visibility British counterpart. Hawkins, "Rule and Discretion in Comparative Perspective: The Case of Social Regulation," 50 *Ohio State Law Journal* 663 (1989).

2. For a comparison between American and German administrative law, see Dieter Lorenz, "The Constitutional Supervision of the Administrative Agencies in the Federal Republic of Germany," 53 *Southern California Law Review* 543 (1980), and Lee Albert's comment on that article, id. at 583.

3. A comparison of the politics and policies of toxic chemical regulation in the United States, Britain, France, and the Federal Republic of Germany includes a discussion of the role of courts in reviewing agency decisionmaking. The authors find that judicial review in the other three countries is very uncommon as a result of significant prior consultation of the agency with industry (Britain, France, and Germany), voluntary safety requirements (Britain), and low reliance on ambient workplace standards (Britain and France), all of which give industry little reason to take agencies to court. They also find that "[t]hough there are significant similarities in the law of judicial review across the four countries, the U.S. courts, often acting in partner-

---

[82] See the broad-based comparative study by V. Bolgar, "The Public Interest," 12 *Journal of Public Law* 13 (1963). For further references see supra n. 1.

ship with the legislature, have consistently interpreted the relevant legal principles in ways that enhance their power to supervise the administrative agencies." Ronald Brickman, Sheila Jasanoff, and Thomas Ilgen, *Controlling Chemicals: The Politics of Regulation in Europe and the United States* (Ithaca: Cornell University Press, 1985), chapter 5.

4. The role and strength of political parties vary in different systems, and these variables surely affect the structure, functions, and behavior of administrative agencies in different countries, as well as the nature of the legal controls to which they are subject. Scholarly writing on parties tends to treat their influences on administration as only indirect, channeled through the legislature and the chief executive. For a study comparing senior bureaucrats and legislative politicians in the United States and six western European countries, including their relations with party leaders, see Joel D. Aberbach, Robert D. Putnam, and Bert A. Rockman, *Bureaucrats and Politicians in Western Democracies* (1981), especially chapter 7. For a speculative discussion of the differences between parliamentary systems and separation-of-powers systems, see Terry M. Moe, "Political Institutions: The Neglected Side of the Story," 6 *Journal of Law, Economics & Organization* 213, at 238–48 (1990). For recent discussions of the American party system, see Warren E. Miller, "Party Identification, Realignment, and Party Voting: Back to Basics, " 85 *American Political Science Review* 565 (1991); and sources cited in Michael A. Fitts, "The Vices of Virtue: A Political Party Perspective on Civic Virtue Reforms of the Legislative Process," 136 *University of Pennsylvania Law Review* 1567 (1988).

The Future of Administrative Law

In this final chapter, the magisterial *tour d'horizon* by Martin Shapiro and the critical commentary by E. Donald Elliott bring together some of the major themes of contemporary administrative law, while attempting to discern its future course. They perceive certain continuities, especially the overriding concern with administrative discretion, as well as some new directions, such as a focus on informal agency action and nonjudicial controls. They also foresee further fragmentation of the field.

Administrative Discretion: The Next Stage

MARTIN SHAPIRO

The history of American administrative law consists in large part of a game of procedural catch-up. Courts and legislatures attempted to control agencies' autonomy only after agencies came to wield substantive authority. This tardiness stemmed initially from the fundamental antipathy of Anglo-American jurisprudence to administrative law: Allegiance to the "rule of law" demanded that government officers be subject not to special rules invented for their benefit but to the same common law rules that governed private persons. This attitude finally crumbled under the weight of the New Deal's administrative activity. A decade after the New Deal had endowed the agencies with vast substantive power, Congress provided procedural rules for the exercise of that power in the Administrative Procedure Act of 1946 (APA).

Because American administistative law represents such a tardy reaction, it has never pretended to be a complete body of law. Most procedural requirements are found in the variously worded and incomplete procedural provisions tacked onto the organic acts that establish particular agencies or programs. The APA is residual law for courts to use when there are gaps in those organic procedural provisions.

Because the APA was meant to control the greatly expanded adjudicatory activities of regulatory agencies, it contains relatively detailed rules for administrative adjudication, but says little about other forms of administrative action. In the three decades after passage of the APA, administrative agencies shifted increasingly from quasi-judicial adjudication to quasi-legislative rulemaking as their primary way of making policy. The APA contains only a few cryptic words about rulemaking: It requires only that rules be made after "notice" and "submission of written data, views, or arguments," be published in the Federal Register, and be accompanied by a "concise general statement of their basis and purpose." During the 1960s and 1970s, the courts, led by the Court of Appeals for the D.C. Circuit, caught up to the substantive reality of greatly increased rulemaking by writing a detailed, judge-made code of administrative procedure for rulemaking. Suitably winnowed, that case law seems likely to be incorporated into the revised APA that Congress will probably enact within the next few years.

Now that the courts have caught up with adjudication and then with rulemaking, what next? The APA essentially divides administrative action into three parts: quasi-judicial adjudication; quasi-legislative rulemaking; and a residual category, which I shall call "informal action." This article discusses

Reprinted by permission of the Yale Law Journal Company and Fred B. Rothman & Company from *The Yale Law Journal,* Vol. 91, pp. 1487–1522.

the nature of administrative actions in this residuum. First, it explores the "discovery" of administrative discretion and the courts' initial articulation of standards of review for discretionary action. It then locates the issue of judical control of discretionary agency action in the larger context of American political development. Finally, it examines the various forms that discretion can take and speculates on the courts' possible responses to these different types of agency behavior.

The Idea of Agency Discretion

Giving the Residuum a Name: Discovering Administrative Discretion

The APA sets up a strange gravitational attraction between the nonadjudicative, nonrulemaking residuum and discretion. It provides no procedural rules for the residuum and no specific standards for judicial review of informal actions. If courts are to review the residuum at all, they cannot do so for its adherence to the procedures of Section 553 or Sections 556 and 557. Instead, they must look to the APA's general catchall standard of review. Courts may review any administrative action to determine whether it was "arbitrary, capricious, [or] an abuse of discretion." "Arbitrary and capricious" is the standard that courts normally use to review rulemaking under Section 553; unless they wish to equate the standard by which they review action under Section 553 and the standard by which they review the residuum, they must review such informal action for abuses of discretion. Now, if the courts are looking for *abuses* of discretion, it must follow that what they are looking at is *exercises* of discretion. Thus, although it is not logically or even empirically true that all informal agency actions are exercises of discretion, courts will typically label informal actions discretionary because the standard by which the actions are reviewed uses that label. . . .

All of this is not to say that [*Citizens to Protect*] *Overton Park* [*v. Volpe*] by substituting "clear error" for "abuse of discretion," compels courts to review informal nonrulemaking action more actively than they review informal rulemaking under the traditional "arbitrary and capricious" standard. Nevertheless, the use of "clear error" language appears to be a movement toward the less deferential end of the spectrum. As my use of the metaphors of spectrum, neighborhood, and family resemblance is meant to suggest, it would be quite unrealistic to create a formal system of degrees of deference out of these open-textured legal materials. They do, however, seem to yield the result that informal nonrulemaking activity is to be reviewed under a standard at least as strict as that for informal rulemaking, and perhaps an even stricter one.

Although the Supreme Court itself has not appeared very fond of the "clear error" language in *Overton Park,* the D.C. Circuit has been very fond of it indeed. If we combine "clear error" with "so long as there is law to apply," we have constructed the springboard from which the next great leap forward of judicial review of administrative decisionmaking may occur. The previously invented standards—"substantial evidence" and "arbitrary and capricious"—have done triple duty as evidentiary standards, procedural standards, and substantive review standards. There is every reason to believe that "clear error" has the same potential to spread the new judicial review of informal action into all three realms.

Thus, the historical tendency of administrative law's move from adjudication to rulemaking to informal action means we should now focus on the informal actions of government that used to be called discretionary. All students of administrative law know, however, that elements of what was, and often still is, called discretion are also liberally scattered through administrative adjudication and rulemaking; there are often a number of alternatives, no single one of which is dictated by the law and the facts. Such situations arise in the procedural aspects of adjudication (e.g., agency decisions to limit cross-examination) as well as throughout the rulemaking process. Our concern with control of agency discretion therefore encompasses a broader area than merely informal agency action.

Administrative Discretion in Its Wider Political Context

So far this discussion has proceeded as if administrative law, once triggered by the New Deal, developed autonomously. Outside forces were also important, however, and they are particularly important in explaining the current situation. At the same time that administrative law is apparently moving toward greater restraints on agency discretion, a number of commentators have been been calling for greatly expanded agency discretion. Bruce Ackerman and William Hassler have suggested a return to the New Deal's broad delegations to those agencies whose enormous discretion is justified by their technical expertise; Richard Stewart has proposed a shift away from "command" processes, like adjudication and rulemaking, toward bargaining and negotiation. Stewart acknowledges that such a shift would increase agency discretion, and calls for judicial review. Both proposals seem opposed to the prediction that administrative law is now poised to catch up with and limit discretion as it earlier caught up with and limited adjudication and rulemaking.

To some extent, however, these crosscurrents prove my point. The calls for increased discretion recognize that discretion, particularly in rulemaking, has been so restricted by legal rules, emanating from both Congress and the courts, that it is time to take a new tack.

The Tension Between Democracy and Technocracy

The Ackerman-Hassler and Stewart proposals reflect a cycle of American politics that alternately assists and impedes the movement toward greater legal restraints on agency action. From the founding of the republic, Americans have embraced two opposing modes of public administration, the democratic and the technocratic. The former, which we might term the Jacksonian tradition, calls for government by the common people themselves, or at least by administrators directly representative of and responsive to the people.

The opposing, Federalist tradition, first advocated by Hamilton, stresses the need for efficient government and thus the need for an administration staffed not by an ever-changing stream of Know-Nothings, but by experts. The Progressive movement inherited this tradition, and its civil service "reform" rooted out the spoils system. . . .

The Current Situation

By 1980, however, technocratic themes and a certain disillusion with democratic administration began to reemerge. The contempt for technocracy of the 1960s and 1970s had been bolstered by the widely held belief that we produced so much and were so efficient that we could willingly accept considerable technical inefficiency as the price of bringing technocrats to heel.

By 1980, productivity was once again a major American concern, and we were casting eyes at the Japanese as models of technological efficiency. In such a setting, renewed allegiance to technocracy was not simply a nostalgic "return" to the New Deal but also the next episode in a technocratic tradition stretching back to the founding of the republic. If productive efficiency is again our watchword, then discretion must be vested in those who know how to do things.

Both the book by Ackerman and Hassler and the article by Stewart explicitly acknowledge that their purpose is to accommodate environmental goals to the need for economic efficiency. Indeed, Stewart argues that environmental improvement can come only through technical antipollution innovations piggybacked on improved productivity. These authors work against a background of increasing calls for rationality in the regulatory process. Regulatory impact statements and cost-benefit analyses are much in the air. Even judges have made a subtle shift from demanding that the agencies listen and respond to all outside comments to demanding that they learn all the facts and consider all the alternatives. That is precisely the shift from democracy, in its group-politics version, to technocracy. The agency need not act democratically, only rationally, correctly, and efficiently. In short, it can return to expert decisionmaking.

Ackerman and Hassler, reacting to the economic and technical irrationality of a Congress dominated by group politics, and Stewart, fearful of potential inefficiencies caused by government commands to industry, call for a return to technocratic discretion. Ackerman and Hassler apparently vest discretion solely in agency technocrats, although one version of their program would give some discretion to industry acting under market constraints. Stewart vests discretion not in the agency but in a bargaining or negotiation process in which agency and industry technocrats join. The goal is environmental and safety policies that are not only rationally cost-effective in and of themselves, but fully compatible with maximum economic growth.

Against this background, we should consider the wide range of discretionary situations, and speculate on how, and the degree to which, such discretion can be brought under legal rules. In the remainder of this article, I set out a tentative typology of administrative discretion and explore the courts' potential responses to these various forms of agency activity.

A Typology of Administrative Discretion

There are a variety of forms that discretionary agency action might take. One species—which we might term "traditional" discretion—occurs in a variety of normal agency activities. The other—the "new" discretion—has appeared as part of the current trend toward technocracy.

Traditional Discretion

Agency activity has long been marbled through with discretionary behavior. The courts' response to these various discretionary activities is likely to depend crucially on the particular situation involved.

Distributive Decisions

Some agency decisions involve the allocation of scarce resources when no legal rights or entitlements have been vested in particular individuals. . . .

That there is not enough to go around may justify random or first-come, first-served distribution; it does not justify arbitrary distribution. I have little doubt that courts will increasingly either demand articulated standards, adherence to precedent, and the provision of reasons for decision or insist upon a truly random allocation. This trend toward limited administrative discretion will, however, be slowed by the factors discussed in the following two sections.

High-Volume, Low-Level Decisions

Some agencies must make a great many decisions, none of which has a very significant effect. Due to the volume of cases, the costs of even minimum procedural guarantees are high. For this reason a great deal of "negative" discretion goes unchecked. Allowing agencies substantial negative discretion is most defensible when an agency is asked to allocate a great many insignificant goods to which no particular individuals have a legal entitlement. It is not so much that discretion is desirable as that the costs of controlling it seem to outweigh the meager benefits yielded by such controls. . . .

Subtle and Complex Assessments of Human Characteristics

Some agency decisions involve such subtle and complex human factors that the agency may argue that it needs unbounded discretion. . . .

. . . In some areas, such as child placement, agencies can claim expertise and demonstrate that their decisionmakers are given the time and information necessary to make highly particularized judgments. Courts hearing such cases often reject most constraints on agency discretion.

In the absence of such capacity, discretionary activity is likely to be subjected to substantial control even when we acknowledge that discretion ideally should be exercised. Courts are likely to make the agency produce standardized profiles of the most worthy applicants so that decisionmakers, in no position to make subtle, correct judgments, can make unsubtle, standardized judgments which can be reviewed for arbitrariness. When judgments are highly particularized and rest on the totality of a large number of factors, however, standards may in fact be useless because they can do no more than list a large number of factors to be considered; nearly any decision can somehow be justified in terms of that list. . . .

. . . In light of these difficulties, courts are likely to eschew substantive review of decisions involving complicated assessments of human characteristics, and instead impose procedural safeguards to ensure that sufficient attention is paid to each case by really expert, particularized decisionmakers and to guarantee that internal review proceedings, in which one expert checks on another, are available.

Agency Waivers

Where the uniform application of a rule generates a small number of random, unforeseeable inequities—either absolute hardships or instances of one person being treated more harshly than most others—we may want to give agencies the discretion to grant waivers. The less random and more anticipated the inequities, and the higher their numbers, the more we should be alert for pseudo-waivers—that is, for the employment of discretionary waivers when a new rule should be made. . . .

Faced with such agency behavior, courts must ask whether what they are asked to review resembles a highly particularized plea for equity or policymaking by exception. If they face policymaking by exception, courts are likely to insist on something like the rules for hybrid rulemaking that encourage third-party intervention. . . .

Thematic Statutes

I noted earlier that although the historical dynamic is moving administrative law beyond adjudication and rulemaking, huge areas of discretion remain uncharted. One of the most intransigent consists of circumstances in which agencies are faced with "thematic" statutory commands to take into account a number of goals or factors but are given no assignment of relative weights to those factors. One can imagine statutes of this sort arranged along a spectrum. At one end lie those statutes which, in announcing a number of purposes, state priorities or weights as clearly and exactly as they can. At the other end lie "lottery" statutes in which contending forces in the legislature, unable to agree on weights, placed all their contending preferences in the statute as the only available alternative to having no statute at all. . . .

In such situations, courts have two choices, both involving substantive judicial review. They may strike their own balance, declaring it the legislature's true intent. In the *Benzene [Industrial Union Dept., AFL-CIO v. American Petroleum Institute]* case, for example, the Supreme Court read a requirement of "significant risk" into the statute; the agency had to show such a risk before it could choose among a wide array of mixes of health and efficiency. Alternatively, courts can read the lottery out of the statute, claiming that the statute has a threshold single purpose. In *Overton Park*, for example, the Court denied that the statute set park preservation, neighborhood preservation, cost-saving, and highway construction as goals with unspecified weightings and instead found that park preservation trumped the other statutory purposes. In short, precisely because lottery statutes give agencies unlimited discretion, courts are likely to deny that they are lottery statutes. . . .

The courts are going to find it particularly difficult to engage in substantive review in high-technology areas. . . . Courts cannot take a hard look at materials they cannot understand nor be partners to technocrats in a realm in which only technocrats speak the language. I think we should anticipate that in the future courts will be less, not more, successful at bringing thematic discretion under judicial control than they have been in the past.

Decisions Under Conditions of High Uncertainty

Sometimes both government action and inaction entail unknowable risks and unknowable benefits. Without any real indication of the effects of various alternatives, these situations appear to be ones of pure discretion: The decisionmaker faces a high and equal level of uncertainty about the outcome of each alternative and cannot avoid uncertainty by doing nothing. The agency can be controlled by various procedural and reason-giving requirements, but these controls do not ultimately limit its discretion. Indeed, there seems to be neither a way of judicially limiting the agency's discretion nor reason to do so. . . .

Because administrative discretion cannot be subjected to judicial restraints in situations of high uncertainty, courts are likely to read congressional limitations into the statute. In the *Benzene* case, for example, the Court read into the statute a requirement that the agency establish "significant risk" for low-level benzene exposure before it set a maximum exposure level. In effect, this is a directive to the agency to resolve the high uncertainty about low-level exposures in favor of cost-saving and against worker safety until uncertainties can be substantially reduced. Thus Congress, through the mouth of the Court, strips the agency of its discretion.

The "significant risk" holding is a specific example of a more general mode of limiting agency discretion. Where uncertainties are very high, whoever bears the burden of proof loses. Congress and the courts may limit discretion by creating burdens of proof and assigning them to whatever purpose, interest, or value they least favor. . . . In any event, as the influence of technocratic administration increases, it will become more difficult for judges to fault an agency's choice among alternatives that all involve complex and nearly equal technical uncertainties. . . .

Perhaps acknowledgement and explanation is about the only limitation courts can honestly place on discretion created by uncertainty, since any more substantively based standard of review merely shifts the locus of discretion to the courts. . . .

The Discretion to Initiate Action

Administrative discretion to initiate or not to initiate action might be broken into two categories; the initiation of adjudications and the initiation of rulemaking proceedings or other policymaking actions. The former is closely analogous to prosecutorial discretion. Administrative agencies typically enjoy enforcement resources far too limited to bring adjudicatory actions against all worthy candidates. As Richard Stewart and Cass Sunstein have pointed out, this situation is about as pure an instance of "no law to apply" as one can find because, except in those rare instances where Congress has absolutely mandated enforcement, there is no law telling the agency which malefactors it should single out for investment of its scarce prosecutorial resources. Initiation of rulemaking might, in contrast, be analogized to congressional discretion to make or not to make laws since rulemaking is an exercise of Congress' delegated lawmaking power. Here, too, there is no law to apply in the sense that, aside from constitutional limitations, there is no control over which bills Congress should or should not consider.

The analogy is, however, incomplete. It is true that in many instances the delegation is so broad that the agency is placed very much in the position of a legislative body with a limited capacity to make laws and an almost infinite range of lawmaking options. In other instances, however, Congress states relatively clear and limited goals, and rulemaking partakes more of enforcement than of lawmaking. Particularly over time, as the agency builds up its body of rules, the next item on the agency's agenda may become clearer and clearer and the agency's rulemaking resources may appear more and more adequate to meet its remaining agenda. . . .

Particularly in those situations where the agency comes closest to meeting the conditions of the analogy of prosecutorial discretion on the one hand or of congressional lawmaking discretion on the other, private rights of initiation have probably not yet devleoped far enough to constitute a definitive check on discretion. Only under bizarrcly favorable procedural circumstances are courts likely to order an agency to promulgate regulations. Stewart and Sunstein themselves are anxious to hedge private rights to control agency prosecutorial discretion, lest private priorities be allowed to disrupt public ones. . . .

The New Discretion

A number of commentators on administrative law have recently suggested the creation of new modes of agency action. These new modes will create new reservoirs of agency discretion. In this section, I discuss how such discretion would operate and how courts may respond to it.

Negotiation, Mediation, and Arbitration

Richard Stewart proposes that agencies shift away from adjudication and rulemaking and toward more cooperative forms of regulation. In Stewart's scheme, discretion can be a tool for overcoming resistance from entrenched interests in both policymaking and implementation. Sometimes, regulated groups may resist implementation so strongly that they drive implementation costs up above levels acceptable to the regulator. In such cases, agencies can be granted discretion to seek less than full compliance with a law in order to reduce the costs of implementation. Such discretion must be relatively unbounded because it is precisely the resister's inability to predict the agency's response that will move it to reach a negotiated settlement.

Judicial review poses serious difficulties for the exercise of this form of discretion. First, the availability of review itself changes the terms of the game in ways that may be adverse to the government negotiator. Second, judges may strike down an agency's choice of less than full enforcement in return for cooperation. Where an agency tempers its demands in order to avoid amendments to its organic act, the bargain may appear particularly unsavory.

If courts see a range of individual degrees of compliance conformations that fall within the demands of the statute, they may uphold negotiation of each individual conformation. The very fact of a negotiated settlement, however, is likely to persuade judges (and the public) that there was a "correct" legal rule or level of implementation and that the agency's negotiations gave it away.

These considerations are further complicated when the agency is not itself directly involved in negotiations but instead is ratifying a mediated settlement among interested parties. Discretion not to approve a mediated settlement is likely to be subject at most to the requirement that reasons be given. Exercising discretion to ratify a private agreement might be subject to a higher level of judicial review because it involves public enforcement of a private agreement.

As rulemaking has become more a form of multiparty adjudication, more time-consuming, and more subject to judicial scrutiny, there has been increasing interest in negotiating processes that would arrive at an agreed-upon policy. Commentators like Stewart who espouse negotiation have argued for less searching review of negotiated outcomes. But how can we be sure that the agency has not given the store away in the course of negotiations? In many instances, of course, enough opposing interest groups will have participated to ensure that the agency has not given away any version of the store. In other instances, however, the structure of the bargaining conflict may leave the "public interest" unrepresented by anyone but the agency. If we do not employ judicial review to check administrative discretion in negotiation, how can it be checked? Of course, Congress could theoretically formulate a new standard for reviewing informal, negotiated agency action. I doubt, however, that judges would pay much attention to such an expression of congressional mood standing by itself.

One idea worth exploring is the employment of "reserve prices." Experts on bargain theory argue that one should always enter a bargaining session having in mind an outcome below which one will not be prepared to complete a bargain. In some cases, a precise dollar reserve price can be set. In others, a reserve price can only be stated so vaguely that it is meaningless as a guide to action. It is possible, however, to envision some areas of government action— for instance, the setting of water purity standards—in which the statute could set a reserve price or minimum standard below which the agency could not accept a negotiated outcome, but authorize the agency to set a much higher standard in a notice-and-comment proceeding. Under such a scheme, courts might be persuaded to curtail their review of negotiations because the agency's negotiating discretion would already have been constrained by statute.

It is difficult to say whether or not the bargaining process would be sharply skewed against the agency because its reserve price was known by the other players while theirs would be unrevealed. Still, there are strategies an agency could use to improve its position. It might announce before bargaining that it had itself set a goal of exceeding the statutory reserve price in at least half of its negotiations, and that it would reject a bargain and go to rulemaking in any particular proceeding if it felt that settling for its reserve price in that proceeding would place its goal in danger. In other words, the agency could set a secret reserve price *above* the statutory reserve price but not below it. Thus, Congress could constrain agency discretion and the agency could maintain much of its bargaining power. Particular where Congress is interested in "technology forcing," the combination of a reserve price and a "bargaining-upward" arrangement may be preferable to setting unrealistic statutory standards from which Congress itself must subsequently bargain a retreat.

If we move from bargaining before rulemaking to enforcement bargaining, we come close to prosecutorial discretion in conventional law enforcement. Agencies such as OSHA or EPA have limited enforcement resources and may prefer to offer something in the way of reduced enforcement as an inducement to voluntary compliance. Of course, agencies may sometimes prefer to bargain for another reason altogether. They may fear that full enforcement will mobilize sufficient political opposition to lead to congressional amendment of the statutes authorizing their programs. Most of us, and surely most courts, would find an explicit use of such "prudential" discretion illegitimate. In practice, such discretion may not be clearly identifiable. Regulated entities can engage in a wide range of resistance to agency activity. They can seek full administrative and judicial review; they can lobby Congress to curtail the agency's power. If an agency is entitled to take into account the degree to which its scarce resources will be consumed in responding to other modes of resistance, why shouldn't it take account of the drain on its resources that a defensive lobbying campaign in Congress will cause?

There has, however, been a tendency, both in statutory and case law, to treat regulatory benefits as entitlements. Under OSHA, for example, courts might discover an entitlement to a safe workplace. This approach tends to treat enforcement discretion as illegitimate and to demand full enforcement of the entitlement. Thus, there is a growing interest in private actions to challenge agency failure to initiate full enforcement proceedings or to challenge agency settlements, despite the traditional view that the agency's behavior resembles prosecutorial discretion.

This tendency is fueled by the ease with which "political" considerations may be confounded with considerations of scarcity. A particular administration may underenforce a particular program not because its resources are scarce but because it does not believe in the program, or at least not in the version of the program created by the rulemaking of the previous administration. If a new administration proceeds to offer reduced enforcement for voluntary compliance, it will claim scarcity. Its opponents will charge that it is modifying the program, not through aboveboard resort to new rulemaking proceedings or requests to Congress to amend the statute, but through devious underenforcement disguised as enforcement bargaining.

Faced with these problems, courts have a number of alternatives. They may create additional enforcement entitlements. They may decide whether the agency's true goal is deliberate underenforcement designed to undermine statutory mandates or bargaining to get more bang for the enforcement buck. Or they may adhere to the traditional analogy of prosecutorial discretion. The first alternative runs the risk that whoever gets to the courthouse door first will obtain full enforcement of his particular entitlement no matter what the cost and thus reduce excessively the enforcement resources available for the protection of others' entitlements. The second option raises the root problem of discretion—probing the true intentions of the administrator. A requirement that reasons be given is likely to lead to routine formulas about insufficient enforcement budgets. The third alternative leaves us no way to determine whether the agency is trading reduced enforcement for voluntary compliance in order to stretch its enforcement dollars or is not fully enforcing the statute for some less acceptable reason.

Some agencies absolutely eschew enforcement discretion; OSHA, for instance, tells its inspectors to write up every violation. Where some enforcement bargaining appears desirable, it is probably easiest to limit discretion by specifying in organic statutes the agenda and terms of bargaining; otherwise, any lapse in agency enforcement can be excused as an implicit bargain. The simplest form of bargaining may be a joint agency-enterprise agreement to a time-staged compliance plan explicitly displaying the tradeoffs between voluntary compliance and enforcement delay. In addition, the statute or the courts may require an agency to establish rules setting the general terms of tradeoffs between voluntary compliance and enforcement delay. The "blue book" with which assistant prosecutors work in some large district attorney's offices exemplifies such constrained discretion. In effect, it sets reserve prices below which a prosecutor cannot go. If offered less, he or she must seek full—or, in the agency context, immediate—enforcement.

Such deliberately structured bargaining in which the quid pro quo and the reserve price are matters of public record could provide adequate restraints on discretion wthout judicial review. Where bargaining is not so carefully structured, however, it is difficult to see what courts can do except either start down the *Overton Park* path of motive review or abandon review altogether by invoking the analogy to prosecutorial discretion.

Overton Park, however, very much reflects the glorification of the judge as lay representative, and the disdain for both group politics and technocracy, of the 1960s and 1970s. Moreover, in *Overton Park* the Department of Transportation did not even attempt a technical screen for its simple, essentially political decision to accept the solution propounded by a local government. In a more technocratic era in which agencies have learned to justify each bargain in terms of the long-term pursuit of maximum technical and economic rationality, courts may be reluctant to spot clear errors. Courts may be quite content if an agency can give reasons for the bargains it has struck.

Agency Delegation

In a recent book, Bruce Ackerman and William Hassler advocate a variant of the New Deal's image of discretion vested in expert agencies that pursue broadly stated goals set by Congress. They suggest that Congress set only time-staged performance goals, such as increasing life expectancy by one percent at the end of ten years. The goal having been set, and the lawmaking power having been delegated, the agency would then have almost unlimited discretion in deciding how to meet that target. Such technocratic proposals reflect society's swing back toward a greater respect for expertise. This shift toward technocracy, however, is bound to be greatly tempered both by the long-term trend toward bringing the executive branch's action under legal control and by the increasing disenchantment with regulation by command. Thus, it may be to Stephen Breyer's proposals for regulatory reform that we must turn to give content to Ackerman's vision.

Breyer proposes that in many areas of regulation the agency should set performance standards or output goals, and leave the instrumental decisions to those regulated. Combining the proposal of Ackerman and Hassler with that of Breyer may produce a system of "subspecification." For example, Congress may command an agency to save 10,000 lives by 1995. The agency may decide that the best way to save those lives by 1995 is to reduce lead pollution levels to X. It might then assign to each source, each corporation, or each industry an annual pollution-reduction target. If the agency were to assign targets to collective entities, then each entity would presumably subspecify targets to each individual source. These individual sources would then choose their own technology and/or transfer of pollution rights.

Viewed independently from Breyer, Ackerman and Hassler appear to be vesting great discretion in the agencies. Combining the proposal of Ackerman with that of Breyer, one might view the result as virtually eliminating both policymaking and implementation discretion from the agency; the combined proposal vests policy discretion almost entirely in the legislature and implementation discretion almost entirely in the market. Ultimately, individual firms would make market-dictated choices among the alternative means available to meet their assigned goals. Implementation discretion rests at whatever level makes actual instrumental choices. This assignment of discretion may raise *Schechter*-type [*Schechter Poultry Corp. v. United States*] problems, since it is unclear why the decentralization of discretion to private groups is a good thing. Under the Ackerman-Hassler-Breyer scheme, cartels rather like the old NIRA cartels may be making crucial allocation decisions, and we have no more reason to trust those cartels now than we did then. Of course, Congress or the agencies might eliminate the cartel problem by creating specialized sections within the agency to assign particularized performance goals to individual firms. Such highly particularized performance goals, however, might be as rigid and inefficient as command and process standards. The alternative is the market, with the specialized agency sections assigning marketable entitlements or duties to particular enterprises which in turn would buy and sell so as to reach efficient solutions.

Even if most implementation discretion can ultimately be returned to the market, a great deal of discretion will probably remain in the process of setting subgoals, time-staging them, and allocating them among various enterprises. It is almost impossible to believe that such suballocation could be done by agency fiat or by an internal agency "planning" process that was purely discretionary. Suballocation would almost certainly require notice-and-comment rulemaking. Solving the problems caused by discretion may be possible only at the cost of suffering many of the rigidities and delays that now accompany rulemaking. This rulemaking process would closely resemble the legislative process. Ultimately, Ackerman and Hassler's call for agency discretion comes down to the belief, stated clearly in their book, that the federal bureaucracy is a far better legislator than is Congress. The Ackerman-Hassler-Breyer proposal, then, is not so much a call for increasing administrative discretion as for shifting as much discretion as possible to individual enterprises under market constraints and shifting lawmaking from statute-making in Congress to rulemaking in the executive branch. If my prognostication about suballocations being done by rulemaking is correct, then determining the value of the Ackerman-Hassler proposal does not mean comparing the merits of rules to those of discretion, but comparing the merits of statute-making to those of rulemaking as an alternative legislative process. If Congress sets only long-term, general goals, however, we are faced once again with the problems raised in my discussion of thematic statutes. Faced with a truly blank ends-forcing statute, courts will not easily be able to read in thresholds or priorities. If courts therefore assign high burdens of proof to agencies and require them to prove that their particular allocations will meet congressional targets, the rulemaking process will be further rigidified.

To the extent that the agencies proceeding under ends-forcing statutes can avoid rulemaking, a good deal of suballocation will probably be done through negotiation. The problems with that form of discretion, are, as we have seen, rather troubling.

Conclusion

. . . By the 1970s, courts were trying to cure the increasingly noted pathology of group politics by insisting that the rulemaker listen to and respond to every argument put forward by every group. By about 1980, courts required instead that an agency consider all the facts and arguments that ought to be considered. The former is an insistence that government submit itself to democratic processes; the latter is an insistence that it act rationally or scientifically. To make the latter demand, however, at a time when technicians are regaining their ascendancy over laymen, is for judges a peculiarly self-destructive stance. . . .

Given these crosscutting tendencies, we need not expect a great judicial crusade against discretion. Indeed, in certain areas, such as the discretion entailed in regulatory negotiation and the setting and subspecification of general, long-term performance goals involved in the Ackerman-Hassler-Breyer proposal, we might expect gains rather than losses for discretion. In areas that combine high technology with high uncertainty, judges may demand that agencies demonstrate that they have acted as synoptically as possible, that agencies clearly identify areas of high uncertainty, and that agencies give reasons for electing the strategies they do elect in the face of that uncertainty. Such judicial demands will increase decisionmaking costs, but ultimately will do little to curtail discretion. The more activist judges will try to go further. We may expect some instances in which judges employ statutory interpretation to undercut agency decisions with which they disagree in their assignments of burdens of proof under conditions of high uncertainty or in line-drawing situations. . . .

The area of waivers or exemptions is one I would put forth as a prospective battleground. There seems likely to be increased interest in private causes of action to challenge waivers, and increased judicial sensitivity to pseudo-waivers leading to more demands that waivers be accomplished through "hybrid" proceedings which involve relatively easy access to third parties and approach full-scale adversary proceedings.

Thematic or lottery statutes, and the problems of the discretion that accompany them, also seem unlikely to disappear. When a statute is genuinely underdetermined, there is no legitimate mode of judicial control over agency discretion. It is difficult to predict the extent to which judges will "cure" their institutional incapacity by resorting to statutory interpretations that set firm priorities in the face of congressional waffling. If I am correct in predicting that the next few decades will witness some judicial retreat as technocracy flourishes, we might not expect a major growth in judicial control of this area of discretion.

It is in the first three areas of discretion I discussed—distributive decisions; high-volume, low-level decisions; and subtle and complex assessments of human characteristics—that I expect the thrust toward judicial review of informal agency action to conflict most strongly with the contrary pressures set up by the resurgence of technocracy. In part through the demand for substantive and procedural rules, and in part through requirements that reasons be given, we are likely to see more restraint on the kind of adminstrative discretion that flourishes when large numbers of small distributive decisions requiring individualized judgments must be made. Courts will place increasing pressure on agencies that defend discretion on the grounds of subtlety, complexity, and particularity of administrative decisions to prove that these agencies have the organization, personnel, and information necessary to make such decisions. Our new electronic capacity for storing, retrieving, and analyzing millions of small decisions has already resulted in demands that agencies follow their own past decisions. Stricter requirements to follow precedent, rather than proliferation of procedural rules, may become the major vehicle for limiting these areas of discretion over the next few years. If so, requirements that reasons be given will also proliferate, both because reasoned opinions are necessary for determining what counts as a precedent, and because the only way to allow an agency flexibility under such a regime is to allow it to deviate from precedent when it can give reasons for doing so.

It is distressingly clear to teachers of administrative law that they may have no subject matter to teach, that administrative procedure may be determined largely by the agency or even the policy involved. I certainly do not expect a uniform judicial or congressional reaction to the many forms of discretion. Particularly in high-technology areas, courts are likely to pause in any forward march toward limiting discretion. In other areas, especially where large numbers of small claims involving common human problems are central, the tendency for rules to catch up with administration is likely to continue.

The Dis-Integration of Administrative Law: A Comment on Shapiro

E. DONALD ELLIOTT

Martin Shapiro's proposal to break "administrative discretion" down into smaller units is a step in the right direction, but it does not go far enough. In this comment, I propose an alternative view which I call the "dis-integration" of administrative law. I do not mean that administrative law no longer exists, but rather that it is gradually becoming a more diffuse and less powerful force in the law. . . . Shapiro predicts both greater emphasis on "technocracy" *and* an expansion of court control over administrative discretion. I believe that these two tendencies conflict and that increased reliance on forms of expertise other than judge-made law will accelerate the "dis-integration" of administrative law. . . .

I do not share Shapiro's assumption that the traditional administrative law categories—"substantial evidence," "capricious and arbitrary," or "clear error of judgment"—determine the stringency of judicial review. Judicial review varies from one substantive area to another, with the nature of the subject matter and the realities of the agency involved influencing the "mood" of judicial review far more than the characterization of the administrative action as adjudication, rulemaking, or informal action. Courts should be hesitant to second-guess agency risk assessments, not because a matter involves rulemaking rather than adjudication, but because the agency's exercise of discretion is based in part on technical evidence which judges rarely understand. . . .

In the final analysis, administrative discretion and judicial review pull in opposite directions. One can imagine greater emphasis on administrative discretion and government by "technocracy" in the future *or* more court-imposed rules to control administrative discretion, but not both simultaneously. . . .

This brings me to the heart of my disagreement with Shapiro. Underlying Shapiro's proposed categories is the premise that it is appropriate to analyze administrative discretion from the Olympian perspective of an administrative law that transcends particular agencies and statutes to embrace broad categories of administrative action. I am not convinced. I doubt whether the traditional conception of administrative law as "embrac[ing] all governmental machinery for carrying out government programs" remains viable (if it ever was). Certainly there is *a* labor law, *a* law of social security and entitlement programs, *an* environmental law, and so on. What I question is whether it makes sense to think in terms of a unitary, overarching "trans-substantive" administrative law that controls all governmental decisionmaking.

Reprinted by permission of The Yale Law Journal Company and Fred B. Rothman & Company from *The Yale Law Journal*, Vol. 92, pp. 1523–36.

Thus, Shapiro and I fundamentally disagree about where administrative law is headed. Shapiro's predictions for the "next stage" of administrative law are based on a combination of outside forces and the internal dynamics of administrative law. I agree that law grows in response to both. I am even willing, for purposes of argument, to accept Shapiro's assumption that we are moving into a new technocratic episode in the "technocratic-democratic cycle." Where he and I differ is in our understandings of the nature of the "internal dynamic" influencing the course of administrative law. Shapiro argues that the historical trend in administrative law is toward greater judicial control over administrative discretion—first, over adjudication; then, over rulemaking; and, in the future, over informal, discretionary action. I believe that Shapiro is incorrect on two counts, one historical and the other institutional.

As an historical matter, the high-water mark of judicial control of administrative action was reached about a decade ago with cases such as *Citizens to Preserve Overton Park, Inc. v. Volpe* and *Association of Data Processing Service Organizations v. Camp*. These and other decisions of the middle 1970s marked the culmination of the most recent wave of agency creation. The "struggle for judicial supremacy" was complete once *Overton Park* brought informal, discretionary administrative actions within the reviewing power of the courts and once the Court of Appeals for the D.C. Circuit developed techniques for reviewing informal rulemaking.

But the culmination of one style necessarily sets the stage for the next. Since the mid-1970s, the direction of legal development has changed. During the last decade, the Supreme Court has sharply restricted the tools available to reviewing courts. The pendulum had clearly begun to swing away from judicial control and toward greater administrative autonomy in *Train v. Natural Resources Defense Council,* in which the Supreme Court held that a reviewing court cannot overrule an agency for what the court believes is a mistake in interpreting the law; the court must defer to the agency's interpretation of statutory langauge and legislative history if it is "reasonable." *Vermont Yankee Nuclear Power Corporation v. Natural Resources Defense Council* further limited the judiciary's power by holding that a court reviewing an administrative action may not impose additional procedural requirements if an agency, in the exercise of its discretion, has chosen not to require such procedures. The judiciary can no longer pioneer the development of administrative procedures—as the D.C. Circuit did in the area of informal rulemaking in the early 1970s. Most recently, the Court has indicated that it may be impermissible for federal judges to develop common law in areas touched by comprehensive regulatory statutes: Administrative discretion to make law is exclusive. Thus, the Supreme Court's direction in the dozen years since *Overton Park* does not support Shapiro's belief in the continuing expansion of judicial control over administrative discretion.

If only recent Supreme Court decisions were at issue, the "dis-integration" of a unitary, court-made administrative law might be written off as a passing phase, or one of Professor Rabin's oscillations which is likely to reverse itself as soon as a few new justices are appointed. There are, however, deeper institutional factors that suggest that the trend will continue.

As the administrative state matures, the locus of power naturally shifts from court-imposed administrative law, which was central during the formative era, toward particularized statutory goals and policies that are elaborated primarily by agencies. To some degree, the shift can be attributed to the sheer volume of administrative activity. Once an area of law reaches a certain "critical mass," the need to borrow concepts from outside decreases and the field moves toward greater self-sufficiency. The process resembles the way in which a natural species is formed as animals lose the ability to crossbreed.

A second process is at work as well. As the mass of statutes, regulations, and cases continues to grow beyond a certain point (defined perhaps by the amount of information that an Arthur Corbin or Kenneth Culp Davis can absorb in a lifetime), lawyers and judges can no longer relate all the parts to one another. They begin to think in terms of separable fields of law. Gradually, an NLRB [National Labor Relations Board] case just *seems* less binding on an environmental case. Through a kind of centrifugal force, a number of administrative subdomains gradually form, each having its own internal integrity. The boundaries are neither firm nor exact, but they are there. Ironically, the success and expansion of the administrative state leads naturally to the "dis-integration" of administrative law.

I am not saying that administrative law no longer exists in a literal sense; standing, notice-and-comment rulemaking, and all the rest are, of course, still with us. Rather, my point is that the center of gravity has shifted away from the broad, overarching generalizations of the administrative law of the 1960s toward more particularistic statutory and policy objectives. This phenomenon is not unique to administrative law; it occurs in many bodies of law as they mature.

Let me give a practical example. I see the Supreme Court's recent pronouncements in the *Benzene* [*Industrial Union Dept., AFL-CIO v. American Petroleum Institute*] case concerning the evidentiary burden that an agency must satisfy before regulating a carcinogen as having, at most, precedential value in the environmental area. I question whether this case, or most other current opinions involving administrative decisions, can be generalized across so broad a category as all "informal rulemaking" cases.

This trend toward the fragmentation of administrative law can also be seen at the statutory level. Instead of revising the 1946 Administrative Procedure Act (APA), Congress has written separate procedural codes into the organic acts establishing agencies and programs—creating a series of "mini-APAs." I suspect that Congress will continue to create procedures tailored to the particular needs of various substantive areas through separate statutes, rather than rewriting the APA so as to impose a unitary procedural code, as Shapiro predicts. Of course, "dis-integration" does not mean that the D.C. Circuit is about to go out of business. On the contrary, its task is becoming more challenging. In place of a single, unitary administrative law, we now have a series of administrative laws, which are developing more or less independently of one another.

As a result of these changes in the nature of administrative law, I doubt that we can make useful generalizations about "moods" of judicial review applicable across broad procedural categories of cases, as Shapiro suggests. What actually constitutes "substantial evidence" in environmental cases no longer bears much resemblance to what goes by the same name in labor or rate-making cases.

I must also confess some skepticism about the method that Shapiro uses in the early part of his article to analyze different standards of judicial review. To the extent that productive generalizations can be made about categories as broad as all informal administrative action, one has to go beyond an analysis of the wording of legal "standards" to study actual behavior, as Shapiro and other political scientists have demonstrated over the years. I doubt that empirical study would bear out Shapiro's suggestion that informal agency action is now reviewed at the most demanding point on the spectrum. Nor is this problem avoided by recasting the subject of the generalization as only a "mood," not a rule. . . .

Let me add a word or two about the implications for lawyers of the changes I have been describing. The causes of the "dis-integration" of administrative law. . . . are related to the rise of the activist state and public law. As long as we viewed administrative law as involving *disputes* between an individual and a governmental official—a tradition that goes back to *Marbury v. Madison*, and *Ex parte Young*—it made sense to fashion an administrative law along the lines of judicial procedure, with emphasis on formal regularity, fairness to individuals, and confining decisionmakers within the scope of their authority. As a "public law" or "activist state" conception emphasizing *implementation of policy* to improve the general welfare gradually supplants the dispute resolution model, the courts' claim to fashion a unitary "administrative law" is undermined. Courts and lawyers may be experts at assessing fairness to individuals; they hold no special brief in fashioning policies to serve the general welfare.

Now, and increasingly in the future, there will be less law and more administration in administrative law. The ambitions—perhaps it would be more accurate to say the *conceits*—proposed for judicial review by an earlier generation of administrative lawyers have failed. Louis Jaffe, for example, believed that courts could impose a system of administrative law that would bring integrity and coherence to administrative decisions and proclaim the premise that agencies would be "brought into harmony with the totality of the law. . . ." Jaffe's image of courts as supervising, coordinating, and integrating administrative action is irreconcilable with Landis's vision of politics and expertise, not common law logic, guiding the exercise of administrative discretion. Landis, not Jaffe, has been borne out by history. Today, the function of coordinating and integrating exercises of administrative discretion is no longer being performed primarily by courts. Instead, new forms have emerged to fill the vacuum. The Office of Management and Budget now wields power far greater than any court's over the substance of agency decisions. Until recently, the legislative veto gave broad supervisory power to committees of Congress. In addition, a number of specialized review institutions, such as the Science Advisory Board within the Environmental Protection Agency, have evolved to constrain administrative discretion, the role that Jaffe and his contemporaries claimed as the natural inheritance of courts and the system of administrative law.

What I have been calling the "dis-integration" of administrative law does not mean that administrative lawyers must abdicate in an activist state. It does mean, however, that judicial review and the procedural arsenal of the administrative lawyer of the 1960s are no longer sufficient. In place of judicial review under the Administrative Procedure Act, the primary tools of an administrative lawyer's trade are increasingly a variety of new techniques for controlling administrative discretion: actions for damages and injunctive relief; the Freedom of Information Act; advisory committees; legislative oversight; and cost-benefit analyses by the Office of Management and Budget.

Now and in the future, administrative lawyers must master new skills so that they can represent their clients before legislative committee staffs, in technical policy formulation, in the substance of administrative rulemakings, and in interagency review within the Executive Branch. This requires not only that lawyers acquire a speaking knowledge of economics and politics, but also that they overcome their "technophobia" and become comfortable making arguments in the technical languages of administrative discretion as well as the legalistic language of administrative law.

Assuming that the legal profession meets these challenges, lawyers will continue to play a major role in designing the programs of the activist state. But we should not overlook the equally important role for lawyers in mediating between a technocratic, activist state and individuals' claims of right. In our enthusiasm for the activist state, we should never lose sight of the traditional aspiration of lawyers to protect individual rights. This mission does not evaporate in an activist state; it only becomes less popular.

NOTES AND QUESTIONS

1. In his article, which was published in 1983, Shapiro predicted that the APA would probably be amended in the next few years to codify much of the case law imposing additional "rationality" requirements on rulemaking, a prediction that Elliott doubted. In this, Elliott has proved the keener prognosticator. The Administrative Procedure Act has been amended only once since then—in 1990 to provide for negotiated rulemaking, which proceeds from a consensus model rather than a rationality model. See Chapter VI, section 5. Do the judicial review provisions of the negotiated rulemaking statute bear out Shapiro's analysis?

2. Shapiro envisions that market-oriented regulation, which confers broad discretion on private firms concerning how to comply with standards, could raise "*Schechter-*

type problems." Do you agree? Shortly after his article was published, the Supreme Court issued its *Chevron U.S.A., Inc. v. Natural Resources Defense Council, Inc.* decision (discussed by Farina in Chapter VI, section 2) upholding the Environmental Protection Agency's "bubble" regulations, which permitted precisely this kind of private discretion. Ronald Cass's article in Chapter VI, section 6 also bears on this question.

3. In a footnote not reproduced here, Elliott disputes Shapiro's reading of *Citizens to Protect Overton Park v. Volpe* with respect to the standard of review question as well as Shapiro's interpretation of the differences among the "clear error of judgment" and "capricious and arbitrary" formulations in their effects on judicial review of discretionary decisions.

4. Elliott cannot imagine a simultaneous increase in *both* technocratic discretion and judicial rules to control it. Why not? Is he suggesting that they are inconsistent logically? Politically? In terms of their effectiveness?